K KIRBY



# Electronic Circuits

GIVEN TO MARILYN TAYLOR

*Electronic Circuits* is particularly useful for students in the following branches of engineering: Electrical, Electronics, Electronics & Communication/Instrumentation, Telecommunication, Bio-Medical Engineering, Computer Engineering and Information Technology.

# Electronic Circuits

Dr. H.N. Shivashankar
Principal, R.N.S. Institute of Technology, Bangalore

B. Basavaraj
Department of Electronics/Physics,
S.J.R. College, Bangalore

JAICO PUBLISHING HOUSE
Mumbai • Delhi • Bangalore • Kolkata
Hyderabad • Chennai • Ahmedabad • Bhopal

Published by Jaico Publishing House
121 Mahatma Gandhi Road
Mumbai - 400 023
jaicopub@vsnl.com
www.jaicobooks.com



ELECTRONIC CIRCUITS
ISBN 81-7992-256-1

First Jaico Impression : 2004
Second Jaico Impression : 2004



Printed by
Sanman & Co.
113, Shivshakti Ind. Estate, Marol Naka,
Andheri (E), Mumbai - 400 059.

# PREFACE

This book entitled **"Electronic Circuits"** is designed to cater to the students of engineering (CSE, ISE, EE, EC, IT, TC, BM and ML). All chapters of the book are written in a simple language with an emphasis on clarity of concepts. Each chapter is self-contained and comprehensive. A large number of solved problems have been included in each chapter. Important points, a list of formulae, questions and exercises are included at the end of each chapter. There are numerous illustrations to guide the students through concepts and techniques.

We are grateful to Dr. M.R. Holla, Director, RNSIT Former Principal, RVCE, Bangalore, for his constant encouragement to this venture.

We express our gratitude to S.V. Uma, M.E., RNSIT for her assistance in correcting the proof and thank Mrs. Renuka B. Raj who assisted and encouraged in the preparation of the manuscript.

We thank "Friends Data Impressions" for their typesetting assistance. We would like to thank Mr. Akash Shah, Publisher; Mr. Hemanth Sharma, Regional Manager, South India; J. Umakanth, Executive; Mr. Radhakrishnan, Manager, Educational Division; and Mr. Shambu Bobbi, Representative, Educational Division; Jaico Publishing House.

Suggestions for further improvement of this book will be gratefully acknowledged.

Bangalore
19th October 2003

**Dr. H.N. Shivashankar**
**Prof. B. Basavaraj**

# CONTENTS

# 1 DIODE CIRCUITS

## Chapter Outline



## INTRODUCTION

When *p*- and *n*-type silicon (or germanium) are joined, a *p-n* junction diode is formed. This two-element device has a unique characteristic; the ability to pass current readily in only one direction. Figure 1.1 shows the schematic symbol of a rectifier diode. The *p* side is called the anode, and the *n* side the cathode. The diode symbol looks like an arrow that points from the *p* side to the *n* side, from the anode to the cathode. Because of this, the diode arrow is a reminder that conventional current flows easily from the *p* side to the *n* side.

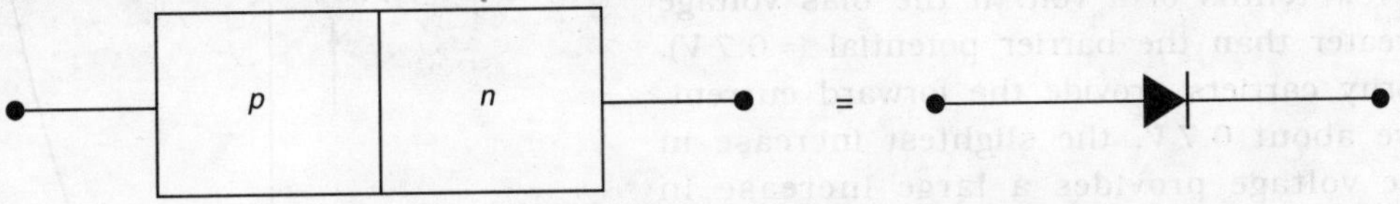


**Fig. 1.1 Schematic symbol of a diode**

The *n*-side of the *p-n* junction has more free electrons and the *p*-side has more holes. The net charge of *n*-region or *p*-region is zero and hence both are electrically neutral. When the *p-n* junction is formed, diffusion will cause random motion of the free electrons and holes. There will be a net flow of the free electrons from the *n*-type region into the *p*-type region and holes from the *p*-type region into the *n*-type region. Each time an electron crosses the junction a positive ion is created in the *n*-type and a negative ion in the *p*-type. When sufficient ions are created, there is a net positive charge on the *n*-side and negative charge on the *p*-side. These charges produce an electric field and is directed such that it opposes further flow of electrons from *n*-region to *p*-region. This field acts as a barrier and is called the **potential barrier.** It is about 0.7 *V* for silicon and 0.3 *V* for germanium. Thus a thin layer on each side of the junction has no free electrons or holes. Since this thin layer is depleted of free charges it is called **depletion layer**. The distance between the two ends of the depletion layer is called width of the barrier. Its width is about 1 μ*m*. Depending on the nature of the semiconductor and its doping the width of depletion layer may change.

The potential barrier opposes this crossing of the majority carriers from one side to the other. But the minority carriers (electrons in the *p*-type and holes in the *n*-type) can move through this junction. The barrier potential is so self-adjusted that the flow of minority carriers is exactly nullified by the flow of majority carriers across the junction.

The electric field created at this junction is very large because of the small thickness of the depletion layer. For a potential barrier of 0.7 *V* and depletion layer width $10^{-6}$ *m* the electric field is $E = V/d = 0.7/10^{-6} = 0.7 \times 10^6$ *V*/*m*. For further flow of electrons through the junction external voltage must be applied. The barrier potential changes with temperature. The barrier potential decreases by 2.0 *mV* for each degree Celsius rise. Mathematically, the decrease in barrier voltage is

$$\Delta V_B = -0.002 \times \Delta t \qquad (1)$$

where $\Delta t$ is the increase in temperature in °*C*.

Consider a *p*-*n* junction connected to an external *dc* voltage source. It is then called **biased *p*-*n* junction**. A biased *p*-*n* junction finds a variety of applications in semiconductor devices.

A *p*-*n* junction can be biased in two ways:

1. If the positive terminal of the *dc* battery is connected to the *p*-side and negative terminal to the *n*-side, the *p*-*n* junction is said to be **forward biased.**
2. If the positive terminal of the *dc* battery is connected to the *n*-side and negative terminal to the *p*-side, the *p*-*n* junction is said to be **reverse biased.**

Figure 1.2 shows the graph of *V*-*I* characteristics of a silicon diode. In a forward-biased condition, the current is small for the first few tenths of a volt. If the bias voltage is greater than the barrier potential (≈ 0.7 *V*), majority carriers provide the forward current. Above about 0.7 *V*, the slightest increase in diode voltage provides a large increase in current. The voltage where the current starts to increase rapidly is called the **knee voltage** of the diode. The *p*-*n* junction offers a very low resistance called the forward resistance.

In a reverse-biased diode, minority carriers provide the small reverse current called **reverse saturation current.** The depletion region of a *p*-*n* junction device widens. The junction offers a very high resistance called the reverse resistance.

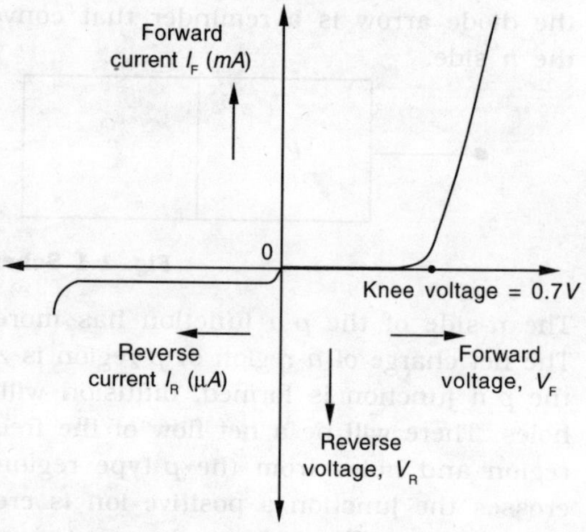


**Fig. 1.2 Current-voltage characteristics of a *p*-*n* junction diode**

The forward current $I_F$ is related to the voltage $V$ by the equation,

$$I_F = I_o\left(e^{V/\eta V_T} - 1\right) \longrightarrow (2)$$

where $I_o$ = a constant current (reverse saturation current)

$V_T$ = volt-equivalent of temperature $= \dfrac{T}{11,600} = 26\ mV$

$\eta$ = 1 for Ge and $\eta$ = 2 for silicon for 300 $K$

The reverse current $I_R$ is related to the voltage $V$ by the expression,

$$I = I_o\left(e^{-V/\eta V_T} - 1\right) \longrightarrow (3)$$

**Cut-in Voltage:** The voltage below which the diode forward current $I_F$ is very small (say, less than 1 percent of maximum rated value) is called the **cut-in, offset, break-point** or **threshold voltage.** Beyond cut-in voltage the current rises very rapidly. For Si diodes, the cut-in voltage at room temperature is 0.6 $V$ while for Ge diodes, it is 0.2 $V$.

## JUNCTION CAPACITANCE

An unbiased diode consists of a $p$-type region containing holes as majority carriers and an $n$-type region containing electrons as majority carriers, separated by a depletion layer, containing virtually no free carriers. The depletion layer acts as a dielectric medium between $p$- and $n$-regions. This is analogous to a capacitor consisting of two conductors separated by an insulator or dielectric.

The total negative charge on one side of the junction equals the total positive charge on the other side. Therefore, there exists a certain amount of capacitance of the junction. Hence, the capacitance formed in a junction area is called **depletion layer capacitance** or simply **junction capacitance.** This arises from the widening or narrowing of the depletion layer as a function of the junction voltage.

Capacitive effects are exhibited by $p$-$n$ junction when they are either forward biased or reverse biased.

### Space Charge or Transition Capacitance ($C_T$)

Space charge capacitance exists when a $p$-$n$ junction is reverse biased. When a $p$-$n$ junction is reverse biased, the depletion width increases. The depletion capacitance per unit area of the junction is the ratio of the change of the charge per unit area to the change of applied junction potential. Thus

$$C_T = \frac{dQ}{dV}$$

The value of transition capacitance is given by the relation,

$$C_T = \frac{k}{(V_B - V_R)^n} \longrightarrow (4)$$

where $k$ = a constant depending on semiconductor material.

$V_B$ = barrier voltage.

$V_R$ = applied reverse voltage.

$n = \frac{1}{2}$ for alloy junction and $n = \frac{1}{3}$ for diffused junction.

## DIFFUSION CAPACITANCE

The capacitance which exists in a forward biased *p-n* junction is called a **diffusion** or **storage capacitance.** It is also defined as the rate of change of injected charge with voltage.

By definition,

$$C_D = \frac{dQ}{dV}$$

$$C_D = \tau \frac{dI}{dV} = \tau g = \frac{\tau}{r} \longrightarrow (5)$$

where $g = \frac{dI}{dV}$ is the diode incremental conductance.

Since $g = \frac{I}{\eta V_T} = \frac{1}{r}$, $\boxed{C_D = \frac{\tau I}{\eta V_T}}$ $\longrightarrow (6)$

where $r = \frac{dI}{dV}$ called the diode incremental resistance and $\eta = 1$ for Ge and $\eta = 2$ for Si and $V_T = \frac{T}{11{,}600} = 26\ mV$.

It is seen that the diffusion capacitance is proportional to the current *I*.

If $C_{D_p}$ be the diffusion capacitance due to holes, i.e., caused by holes diffusing across the potential barrier, then

$$C_{D_p} = \tau_{pn}\, g_p = \frac{dQ_p}{dV} \longrightarrow (7)$$

where $\tau_{pn}$ is the lifetime of holes in the *n*-region.

Similarly, diffusion capacitance due to electrons is given by:

$$C_{D_n} = \tau_{np}\, g_n = \frac{dQ_n}{dV} \longrightarrow (8)$$

where $\tau_{np}$ is the lifetime of free electrons in the *p*-type region.

The total diffusion capacitance is the sum of its two components.

$$C_D = C_{D_p} + C_{D_n} = \tau_{pn}\, g_p + \tau_{np}\, g_n \longrightarrow (9)$$

For forward biased junction, the diffusion capacitance is usually very much greater than the transition capacitance. Thus, the transition capacitance $C_T$ can be neglected when the diode is forward biased. On the other hand, when the junction is reverse biased, the converse is true and the diffusion capacitance can often be neglected.

The diffusion capacitance of a forward biased junction is usually large value; for instance a typical value is $20\ \mu F$. It is about million times larger than the transition capacitance. However, this is shunted by the very low junction resistance. Thus, at low-frequencies, $C_D$ often has very little effect.

Despite the large value of $C_D$, the time constant $\tau = rC_D$ may not be excessive because the dynamic forward resistance $r = \frac{1}{g}$ is small.

$$\therefore \qquad \tau = rC_D \qquad \longrightarrow (10)$$

Hence the diode time constant is equal to mean lifetime of minority carriers. The value of $\tau$ lies in the range of nanoseconds to hundreds of microseconds.

### Diffusion Capacitance for an Arbitrary Input

Suppose that a forward bias is applied. The potential barrier is reduced and many holes diffuse across it into the $n$-type region. The hole density $p_n$ now takes on the form shown in Fig. 1.3. Adjacent to the junction, the density is much greater. However, as the distance from the junction increases, the hole density decreases exponentially toward the equilibrium. In Fig. 1.3, the curved marked (1) is the steady-state value of hole density for an applied voltage $V$. If the voltage at time $t$ is increased by $dV$ in the interval $dt$, then $p'_n$ changes. It is indicated by the curve marked (2) at time $t + dt$. The increase in charge $dQ'$ in the time $dt$ is proportional to the heavily shaded area in Fig. 1.3. If the applied voltage is maintained constant at $V + dV$, the stored charge will continue to increase.

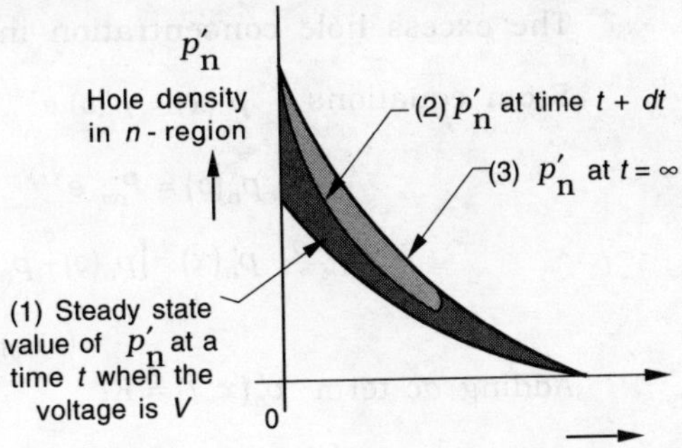


**Fig. 1.3 The transient buildup of stored excess charge**

At $t = \infty$, the hole density $p_n$ follows the curve marked (3). The steady-state injected charge $dQ$, due to the increase in voltage by $dV$, is proportional to the total shaded area in Fig. 1.3. The charge $dQ - dQ'$ is represented by the lightly shaded area and hence $dQ > dQ'$.

Since $dQ'$ is the charge injected across the junction in time $dt$, the current is given by:

$$i = \frac{dQ'}{dt} = C'_D \frac{dV}{dt} \qquad (\text{because } C = QV) \longrightarrow (11)$$

where $C'_D = \frac{dQ'}{dV}$, called the **small-signal diffusion capacitance.**

Notice that the diode current is not given by the steady-state charge $Q$ or static capacitance $C_D$.

That is,

$$i \neq \frac{dQ}{dt} \neq C_D \frac{dV}{dt} \longrightarrow (12)$$

Since $dQ' < dQ$, then $C'_D < C_D$.

From the above discussion we conclude that the dynamic diffusion capacitance $C'_D$ depends upon how the input voltage varies with time. To determine $C'_D$ the equation of continuity must be solved for the given voltage waveform. This equation controls how hole density $p_n$ varies both as a function of $x$ and $t$ and from $p_n$ $(x, t)$ we can obtain current. If the input varies in an arbitrary way, it may not be possible to define the diffusion capacitance in a unique manner.

## Diffusion Capacitance for a Sinusoidal Input

If the excitation to a diode is sinusoidal, the diffusion capacitance $C_D$ may be found from the reactive component of current.

Assume that *p*-region of a diode is doped much more heavily than the *n*-region, so that we need only to calculate the hole current. Consider that the external voltage is

$$V = V_1 + V_m\, e^{j\omega t} \quad \longrightarrow (1)$$

where $V_1$ represents a bias voltage and $V_m$ is the peak value of sinusoidal voltage. If $g$ and $C_D$ represent dynamic conductance and diffusion capacitance of the diode, then the current must be of the form:

$$I = I_1\, g\, V_m\, e^{j\omega t} + j\omega\, C_D\, V_m\, e^{j\omega t} \quad \longrightarrow (2)$$

The excess hole concentration in the *n*-type region consists of a *dc* term.

From equations $p'(x) = p'(0)\, e^{-x/L_p} = p(x) - p_o$ and

$$p_n(0) = P_{no}\, e^{V/V_T}, \text{ we have}$$

$$p_n'(x) = \left[p_n(0) - p_{no}\right] e^{-x/L_p} = p_{no}\left(e^{V_1/V_T} - 1\right) e^{-x/L_p}$$

Adding *ac* term $p_n'(x, t) = Ke^{-(1+j\omega\tau_p)^{\frac{1}{2}} x / L_p}\, e^{j\omega t}$ to the above equation, we get

$$p_n'(x) = p_{no}\left(e^{V_1/V_T} - 1\right) e^{-x/L_p} + Ke^{-(1+j\omega\tau_p)^{\frac{1}{2}} x / L_p}\, e^{j\omega t} \quad \longrightarrow (3)$$

To calculate $K$, put $x = 0$ and also use the law of the conjunction with $V$ given by Eq. (1). Then,

$$p_n(0) - p_{no} = p_{no}\left(e^{V_1/V_T} - 1\right) + K\, e^{j\omega t}$$

$$= p_{no}\, e^{(V_1 + V_m e^{j\omega t})/V_T} - p_{no}$$

Assuming $V_m / V_T << 1$,

$$\text{we have } p_n(0) - p_{no} = p_{no}\, e^{V_1/V_T}\left[1 + \frac{V_m\, e^{j\omega}}{V_T}\right] p_{no} \quad \longrightarrow (4)$$

$$K = p_{no}\, \frac{V_m}{V_T}\, e^{V_1/V_T} \quad \longrightarrow (5)$$

The diffusion current crossing the junction,

$$I_{pn}(0) = -AqD_p\left(\frac{dp_n'}{dx}\right)\Bigg|_{x = 0}$$

Using the value of $K$ in Eq. (5), we get

$$I_{pn}(0) = \frac{AqD_p\, p_{no}}{L_p}\left(e^{V_1/V_T} - 1\right) + \frac{AqD_p\, p_{no}\, V_m\, e^{V_1/V_T}}{V_T} \times \frac{\left(1 + j\omega\tau_p\right)^{1/2}}{L_p}\, e^{j\omega t} \quad \longrightarrow (6)$$

## Special Cases:

1. At very low frequencies, i.e., $\omega\tau_p << 1$,

$$I_{pn}(o) = I_1 + \frac{AqD_p\, p_{no}}{L_p} \frac{V_m}{V_T} e^{V_1/V_T} \left(1 + \frac{j\omega\tau_p}{2}\right) e^{j\omega t} \longrightarrow (7)$$

where $I_1 = \left(\frac{Aq\,D_p\,p_{no}}{L_p}\right)\left(e^{V_1/V_T} - 1\right)$ is the bias current.

Comparing Eq. (7) with Eq. (2), we get

$$g = \frac{Aq\,D_p\,p_{no}\,e^{V_1/V_T}}{L_p\,V_T} = g_e = \text{low-frequency conductance} \longrightarrow (8)$$

and $$C_D = \frac{g\tau_p}{2} \longrightarrow (9)$$

2. At very high frequencies, i.e., $\omega\tau_p >> 1$,

$$\left(1 + j\omega\tau_p\right)^{1/2} \cong \left(j\omega\tau_p\right)^{1/2} = \left(\omega\tau_p\right)^{1/2} e^{j\pi/4} = \left(\frac{\omega\tau_p}{2}\right)^{1/2} (1+j)$$

and from Eq. (6),

$$I_{pn}(o) = I_1 + g_o \left(\frac{\omega\tau_p}{2}\right)^{1/2} (1+j)\,V_m\,e^{j\omega t} \longrightarrow (10)$$

Comparing the above equation with Eq. (2), we get

$$g = g_o \left(\frac{\omega\tau_p}{2}\right)^{1/2} \longrightarrow (11)$$

and $$C_D = g_o \left(\frac{\tau_p}{2\omega}\right)^{1/2} \longrightarrow (12)$$

Hence it is clear that both the diffusion capacitance and the conductance are functions of frequency.

**Note:**

The transition and diffusion capacitance must be considered when high frequency signals are applied to the diode. In general, we can analyse the effect of the capacitances by assuming that a capacitance

$$C = C_T + C_D$$

is connected in parallel with the diode. This capacitance therefore reduces the effectiveness of the diode as a rectifier at high frequencies.

Diodes are often used in digital circuits, where the signal varies between two values. Then, the diode will be switched from forward bias to reverse bias, or vice versa. Even if the applied signal switches instantaneously, the diode response will not be instantaneous.

The capacitance in shunt with the diode will limit the rate at which the diode can switch. It is conventional to say that the forward bias switches the diode on while reverse bias switches it off.

## THE DIODE AS A CIRCUIT ELEMENT

The *p-n* junction diode is considered as a circuit element. The basic diode circuit, shown in Fig. 1.4, consists of a diode in series with a load resistance $R_L$ and an input signal source $v_i$. This circuit is now analysed to find the instantaneous current *i* and the instantaneous diode voltage $v$, when the instantaneous input voltage is $v_i$.

**The Load Line:** From Kirchhoff's voltage law,

$$v = v_i - iR_L \qquad \longrightarrow (1)$$

The above equation is not sufficient to determine the two unknown values $v$ and $i$. However, a second relation between these two variables is given by the static forward characteristic of the diode. The simultaneous solution of Eq. (1) and the diode characteristic is indicated in Fig. 1.5(a). The straight line that is represented by Eq. (1) is called the **load line.** The load line passes through the points $i = 0$, $v = v_i$ and $i = v_i/R_L$, $v = 0$. The intercept with the voltage (X-axis) axis is $v_i$ and with the current (Y-axis) axis is $v_i/R_L$.

$$\text{Slope of the load line} = -\frac{1}{R_L}$$

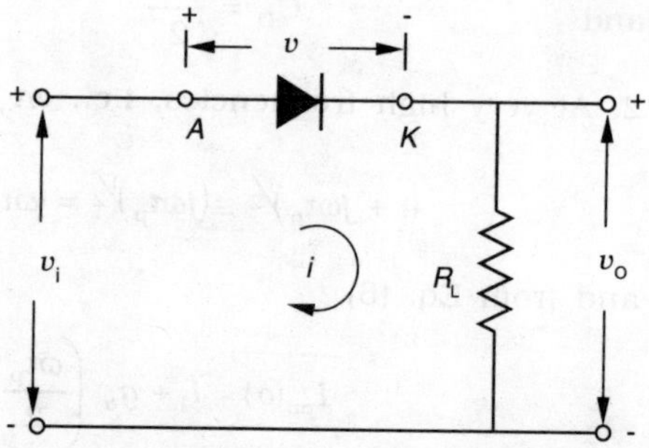


**Fig. 1.4 The basic diode circuit**

The point of intersection A of the load line and the static characteristic curve gives the current $i_A$ that will flow under these conditions. This construction determines the current in the diode when the instantaneous input voltage is $v_i$.

Since $i = v_i/R_L$ is too large to appear on the printed V-I curve supplied by the manufacturer, a slight complication may arise in drawing the load line. Hence it is better to choose an arbitrary value of the current $I'$ which is on the Y-axis of the printed characteristic. Then the load line is drawn through the point P, where $i = I'$, $v = v_i - I'R_L$ and $i = 0$, $v = v_i$.

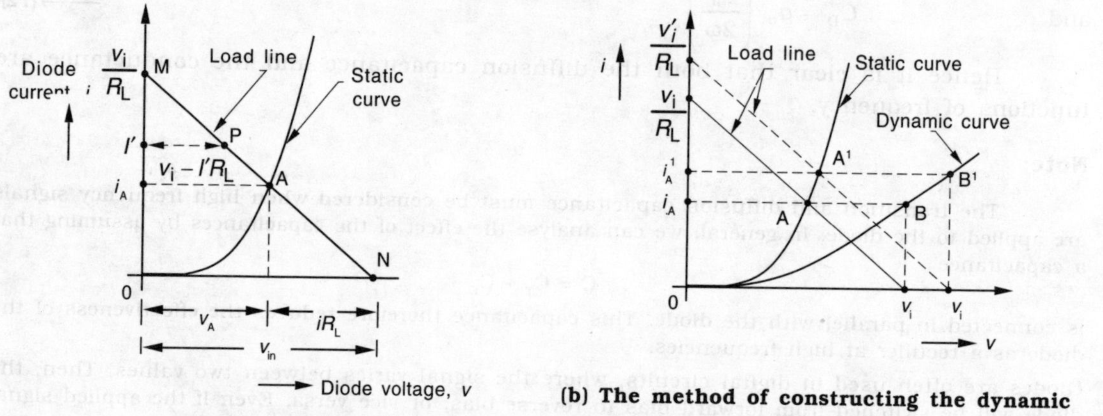


**(a) Load line concept**

**(b) The method of constructing the dynamic curve from the static curve and the load line**

**Fig. 1.5**

## The Dynamic Characteristic

Consider that the input voltage is variable. The above procedure is then repeated for each voltage value. A plot of current vs. input variable voltage, called the **dynamic characteristic** [(Fig. 1.5(b)], may be obtained as follows: The current $i_A$ is plotted vertically above $v_i$ at point B. As $v_i$ changes, the slope of the load line does not change since $R_L$ is fixed. Thus, when the applied voltage is $v_i'$, the corresponding current is $i_A$. This current $i_A$ is plotted vertically above $v_i'$ at $B'$. The resulting curve OB B' that is obtained as $v_i$ varies is the dynamic characteristic.

**The Transfer Characteristic**

The curve which relates the output voltage $v_o$ to the input voltage $v_i$ of a circuit is called the **transfer characteristic**. Since in Fig. 1.4, $v_o = iR_L$, then for this particular circuit the transfer curve has the same shape as the dynamic characteristic.

Note that, regardless of the shape of the static *V-I* characteristic or the waveform of the input signal, the resultant output waveshape can always be found graphically (at low frequencies) from the transfer curve. This is illustrated in Fig. 1.6. The input signal waveform (not necessarily triangular) is drawn with its time axis vertically downward, so that the voltage axis is horizontal. Let the input voltage $v_{iA}$ be indicated by the point A at an instant $t'$. The corresponding output voltage $v_{oA}$ is obtained by drawing a vertical line through A. This value of $v_o$ is then plotted at an instant $t'$. Similarly, points *b,c,d,* ..... of the output waveform correspond to points *B,C,D,* ..... of the input voltage waveform are obtained. Note that $v_0 = 0$ for $v_i < V_\gamma$, so that the diode acts as a **clipper** i.e., a portion of the input signal does not appear at the output. There is also distortion introduced into the output in the neighborhood of $v_i = V_\gamma$ because of the nonlinearity of the transfer curve in this region.

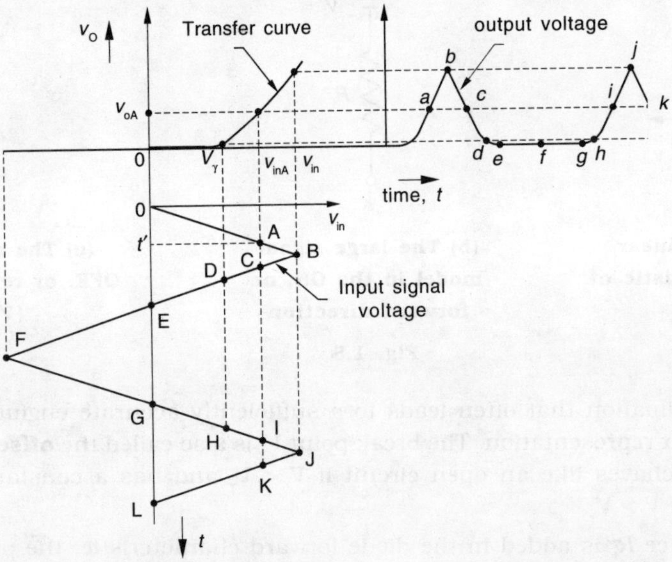


**Fig. 1.6 The method of obtaining the output voltage waveform from the transfer characteristic for a given input signal voltage waveform**

## THE LOAD-LINE CONCEPT

The external circuit at the output of almost all devices consists of a *dc* (constant) supply voltage *V* in series with a load resistance $R_L$ as shown in Fig. 1.7. Applying *KVL* to the output circuit, we have

$$v = V - iR_L \longrightarrow (1)$$

The above equation gives a straight line relationship between output current *i* and output voltage $v$. The load line passes through the point $i = 0$, $v = V$ and has a slope equal to $-1/R_L$ independent of the device characteristic. A *p-n* junction diode possesses a single volt-ampere characteristic at a given temperature.

However, most other devices such as transistor, photodiode, FET, etc., possess a family of curves. But the output circuit is identical with that shown in Fig. 1.7, and the graphical analysis begins with the construction of the load line.

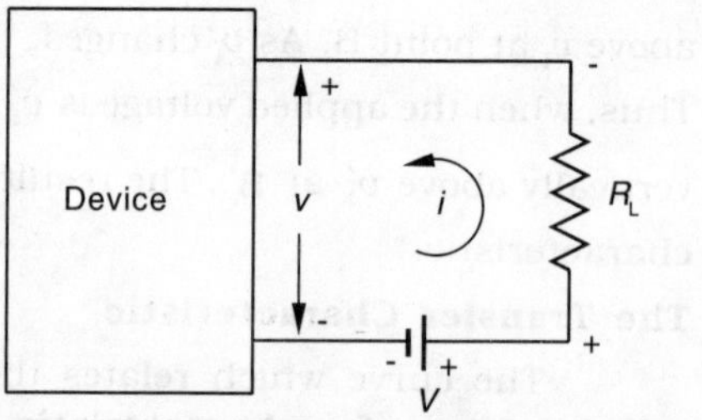


**Fig. 1.7 The output circuit of a device consisting of a supply voltage V in series with load resistance $R_L$**

## THE PIECEWISE LINEAR DIODE MODEL

The models that ignore the direct components of voltage and current are called **small-signal models** because they are used to study the effects of small changes in voltages or currents in the circuit. Large-signal models can be used for much larger changes in voltages and currents. Note that the small signal models will be very accurate if the signal swings are small.

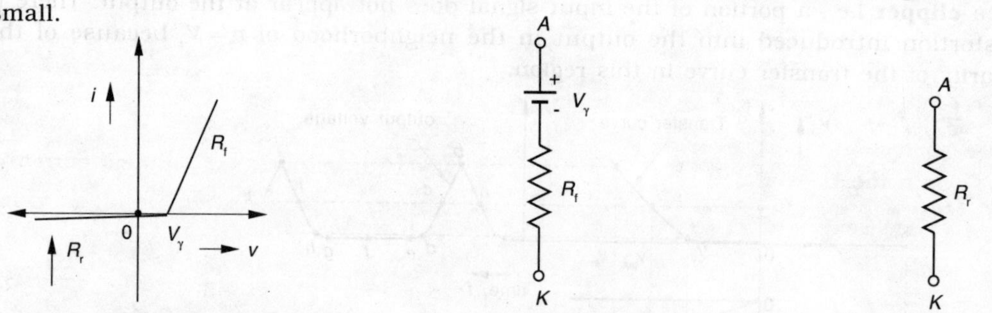


**(a) The piecewise linear volt-ampere characteristic of a *p-n* diode**

**(b) The large signal model in the ON, or forward direction**

**(c) The model in the OFF, or reverse direction ($V < V\gamma$).**

**Fig. 1.8**

A large-signal approximation that often leads to a sufficiently accurate engineering solution is the **piecewise linear** representation. The break point $V_\gamma$ is also called the **offset** or **threshold voltage**. The diode behaves like an open circuit if $V < V_\gamma$ and has a constant incremental resistance $r = dV/dI$ if $V > V_\gamma$.

If the reverse resistance $R_r$ is added in the diode forward characteristic, the piecewise linear and continuous volt-ampere, characteristic of Fig. 1.8(a) is obtained. The diode is a **binary** device, it means that it can exist in only one of two possible states, i.e. the diode is either ON or OFF at a given time. If the voltage applied across the diode exceeds the cutin potential $V_\gamma$, the diode is forward-biased and is said to be in the ON state. The large-signal model for the ON state is shown in Fig. 1.8(b) as a battery $V_\gamma$ in series with the low forward resistance $R_f$. For a reverse bias i.e., $v < V_\gamma$ the diode is said to be in its OFF state. The large-signal model for the OFF state is shown in Fig. 1.8(c) as a large reverse resistance $R_r$ ($\cong \infty$).

**A Simple Application.** In the basic diode circuit of Fig. 1.1, let the input signal be sinusoidal, so that $v = V_m \sin \omega t$. Assume that the piecewise linear model of Fig. 1.8 (with $R_r = \infty$) is valid.

The equivalent circuit of a diode in the ON state is shown in Fig. 1.9(a). The current in the forward direction ($v_i > V_\gamma$) is given by

$$i = \frac{V_m \sin \omega t - V_\gamma}{R_L + R_f} \longrightarrow (1)$$

for $v = V_m \sin \omega t \geq V_\gamma$ and $i = 0$ for $v_i < V_\gamma$.

The input voltage and the rectified current waveforms are shown in Fig. 1.9(b). The cutin angle $\phi$ is given by

$$\phi = \arcsin\left(\frac{V_\gamma}{V_m}\right) \longrightarrow (2)$$

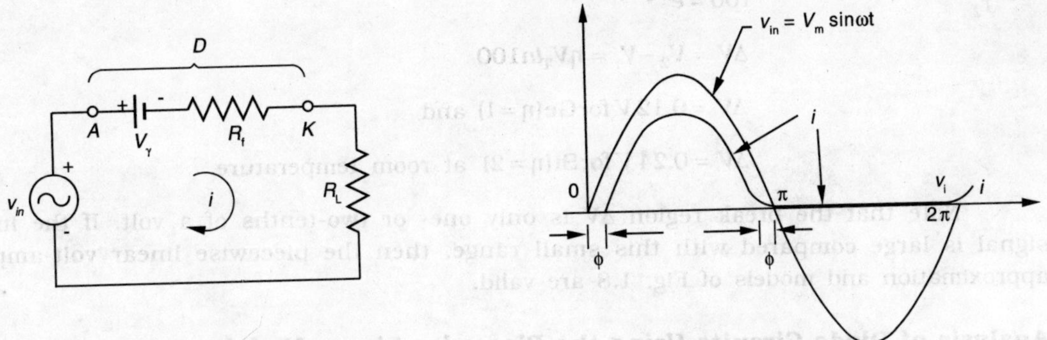


(a) **The equivalent circuit of a diode (in the ON state)**

(b) **Input waveform $v_i$ and rectified current $i$**

**Fig. 1.9**

If, for example, $V_m = 2V_\gamma$, then $\phi = 30°$. Hence a cutin angle of 30° is obtained for very small peak sinusoidal voltages; 1.2*V* for Si and 0.4*V* for Ge. On the other hand, if $V_m \geq 10V$, then $\phi \leq 3.5°$ (1.2°) for Si (Ge) and the cutin angle may be neglected; the diode conducts for a full half cycle.

The circuit of Fig. 1.9(a) may be used to charge a battery from an *ac* supply line. The battery $V_B$ is placed in series with the diode *D* and $R_L$ is adjusted to supply the desired *dc* (average) charging current. The instantaneous current is given by Eq. (1), with $V_B$ added to $V_\gamma$.

## The Break Region

The piecewise linear approximation shown in Fig. 1.8(a) shows an abrupt discontinuity in slope at $V_\gamma$. Actually, the transition of the diode from the OFF state to the ON state is not abrupt. Therefore the waveform transmitted through a clipper or a rectifier does not show an abrupt change of attenuation at a break point, but instead there exists a break region, that is, a region over which the slope of the diode characteristic changes gradually from a very small to a very large value.

The break point is defined at the voltage $V_\gamma$, where the diode resistance changes discontinuously from the very large value $R_r$ to the very small value $R_f$. Hence, we can arbitrarily define the break region as the voltage change over which the diode resistance is multiplied by some large factor, say 100.

The incremental resistance is given by

$$r = \frac{dV}{dI} = \frac{1}{g}$$

Since $g = \frac{I_o e^{V/\eta V_T}}{\eta V_T}$, $\quad r = \eta \frac{V_T}{I_o} e^{-V/\eta V_T}$

If $V_1$ and $V_2$ are the potentials at $r_1$ and $r_2$, then

$$\frac{r_1}{r_2} = e^{(V_2 - V_1)/\eta V_T}$$

For $\frac{r_1}{r_2} = 100$, $\quad 100 = e^{(V_2 - V_1)/\eta V_T}$

$$\Delta V = V_2 - V_1 = \eta V_T ln 100$$

$\Delta V = 0.12\, V$ for Ge ($\eta = 1$) and

$\Delta V = 0.24\, V$ for Si ($\eta = 2$) at room temperature.

Note that the break region $\Delta V$ is only one- or two-tenths of a volt. If the input signal is large compared with this small range, then the piecewise linear volt-ampere approximation and models of Fig. 1.8 are valid.

### Analysis of Diode Circuits Using the Piecewise Linear Model

Consider a circuit consisting several diodes, resistors, supply voltages and sources of excitation. A general method of analysis of such a circuit is in assuming the state of each diode. For the ON state, replace the diode by a battery $V_\gamma$ in series with a forward resistance $R_f$. For the OFF state, replace the diode by the reverse resistance $R_r$. After the diodes have been replaced by these piecewise linear models, the entire circuit is linear and the currents and voltages everywhere can be calculated using Kirchhoff's voltage and current laws. The assumption that a diode is ON can then be verified by observing the sign of the current through it. If the current is in the forward direction, the diode is indeed ON and the initial guess is justified. However, if the current is in the reverse direction, the assumption that the diode is ON has been proved incorrect. Under this circumstance the analysis of a circuit must begin again with the diode assumed to be OFF.

We can test the assumption that a diode is OFF by finding the voltage across it using the above trial and error method. If this voltage is either in the reverse direction or in the forward direction but with a voltage less than $V_\gamma$, the diode is indeed OFF. However, if the diode voltage is in the forward direction and exceeds $V_\gamma$, the diode must be ON and the original assumption is incorrect. In this case the analysis must begin again by assuming the ON state for this diode.

## CLIPPING (LIMITING) CIRCUITS

**A circuit which shapes waveforms by removing (or clipping) a particular portion of the input signal above or below a certain level is known as a clipping circuit, limiter, amplitude selector or simply slicer.**

Figure 1.10(a) shows a diode clipping circuit which transmits that part of the

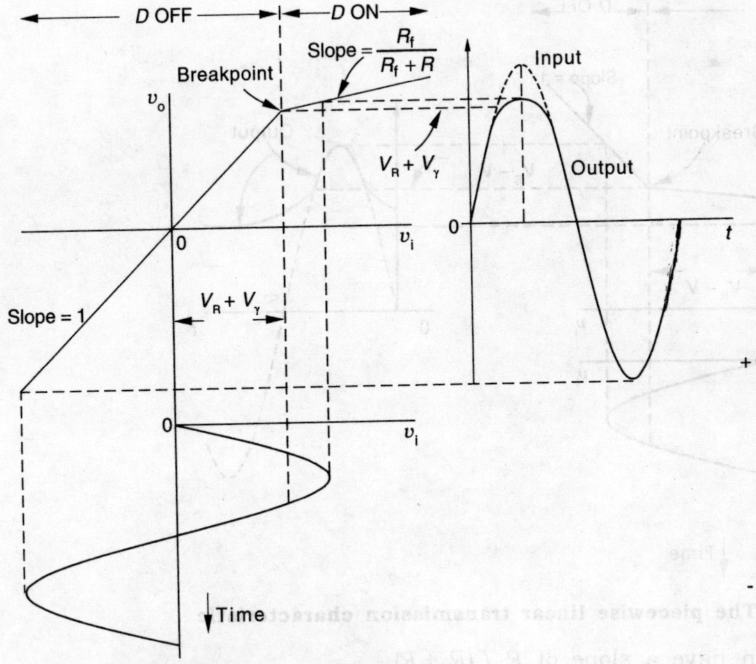


Fig. 1.10(b) The piecewise linear transmission characteristic

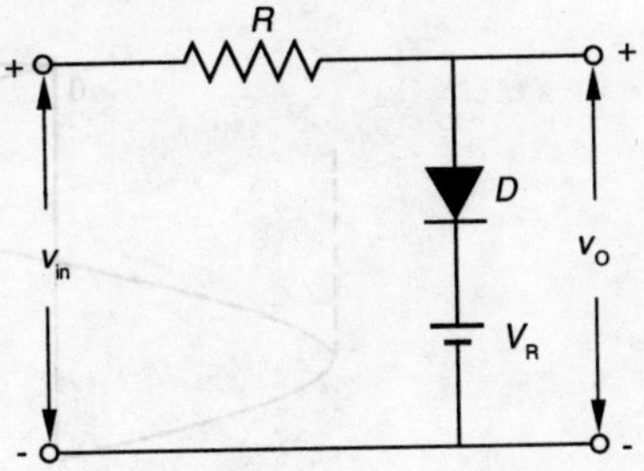


Fig. 1.10 (a) Clipping circuit

waveform more negative than $V_R + V_\gamma$. $V_R$ is the limiting reference voltage. If diode voltage $v < V_\gamma$ and $v_{in} < V_R + V_\gamma$, the diode is OFF. There is no current flowing through $R$ and $v_o = v_{in}$. This argument justifies the linear portion (with slope unity) of the transmission characteristic extending from arbitrary negative values to $v_{in} = V_R + V_\gamma$. If $v_{in}$ is greater than $V_R + V_\gamma$, the diode conducts and it then behaves as a battery $V_\gamma$ in series with a resistance $R_f$, so that increments $\Delta v_i$ in the input are attenuated and appear at the output as increments $\Delta v_o = \Delta v_{in} R_f/(R + R_f)$. This verifies the linear portion of slope $R_f/(R + R_f)$ for $v_{in} > V_R + V_\gamma$ in the transfer curve. Notice that the transmission characteristic is piecewise linear and continuous and has a break point at $V_R + V_\gamma$.

Let a sinusoidal signal of large amplitude be applied. The corresponding output exhibits a suppression of the positive peak of the signal. If $R_f << R$, the suppression is more and the positive excursion of the output will be sharply limited at the voltage $V_R + V_\gamma$. The positive peak of the output is clipped off, as shown in the figure.

In clipping circuit shown in Fig. 1.11(a), the diode is reversed. The corresponding piecewise linear representation of the transfer characteristic is shown in Fig. 1.11(b). In this circuit, the portion of the waveform more positive than $V_R - V_\gamma$ is transmitted without attenuation, but less positive portion is greatly suppressed. If $R_r < R$, the transmission characteristic must be modified. The

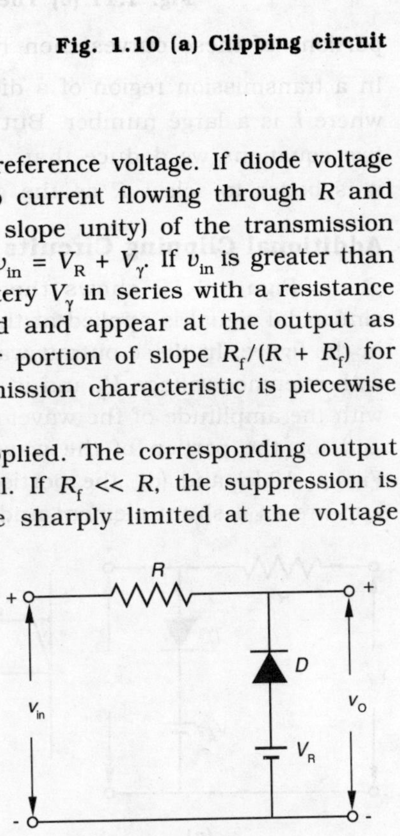


Fig. 1.11(a) A diode clipping circuit

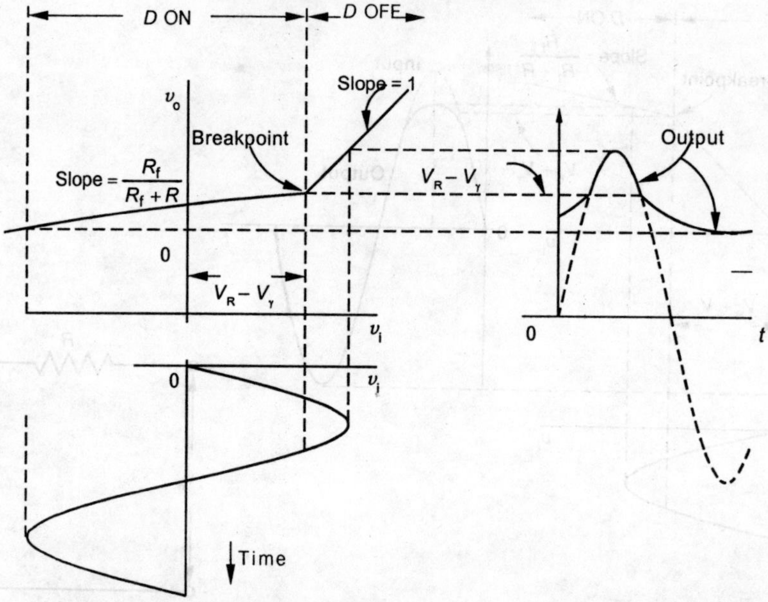


Fig. 1.11 (b) The piecewise linear transmission characteristic

portions of these curves then have a slope of $R_r / (R_r + R)$.

In a transmission region of a diode clipping circuit it is necessary that $R_r >> R$ i.e., $R_r = kR$, where $k$ is a large number. But in the attenuation region, $R >> R_f$, i.e., $R = kR_f$. From these two equations we deduce that $R = \sqrt{R_f R_r}$ and that $k = \sqrt{R_r / R_f}$. On this we conclude that it is better to select $R$ as the geometrical mean of $R_r$ and $R_f$.

## Additional Clipping Circuits

Figure 1.12 shows the diagram of four diode clipping circuits. In each case a sinusoidal signal is applied at the input and the corresponding output waveforms are shown in the figure. In these output waveforms the threshold voltage $V_\gamma$ is neglected in comparison with reference voltage $V_R$ and it is assumed that the break region is negligible in comparison with the amplitude of the waveforms. It is also assumed that $R_r >> R >> R_f$. In Figs. 1.12 (a) and (b), the portion of the waveform transmitted is that part which lies below $V_R$ and in Figs. 1.12 (c) and (d), the portion above is transmitted. In Figs. 1.12 (a) and (c), the diode appears as a shunt element and in Figs. (b) and (d), the diode appears as a series element.

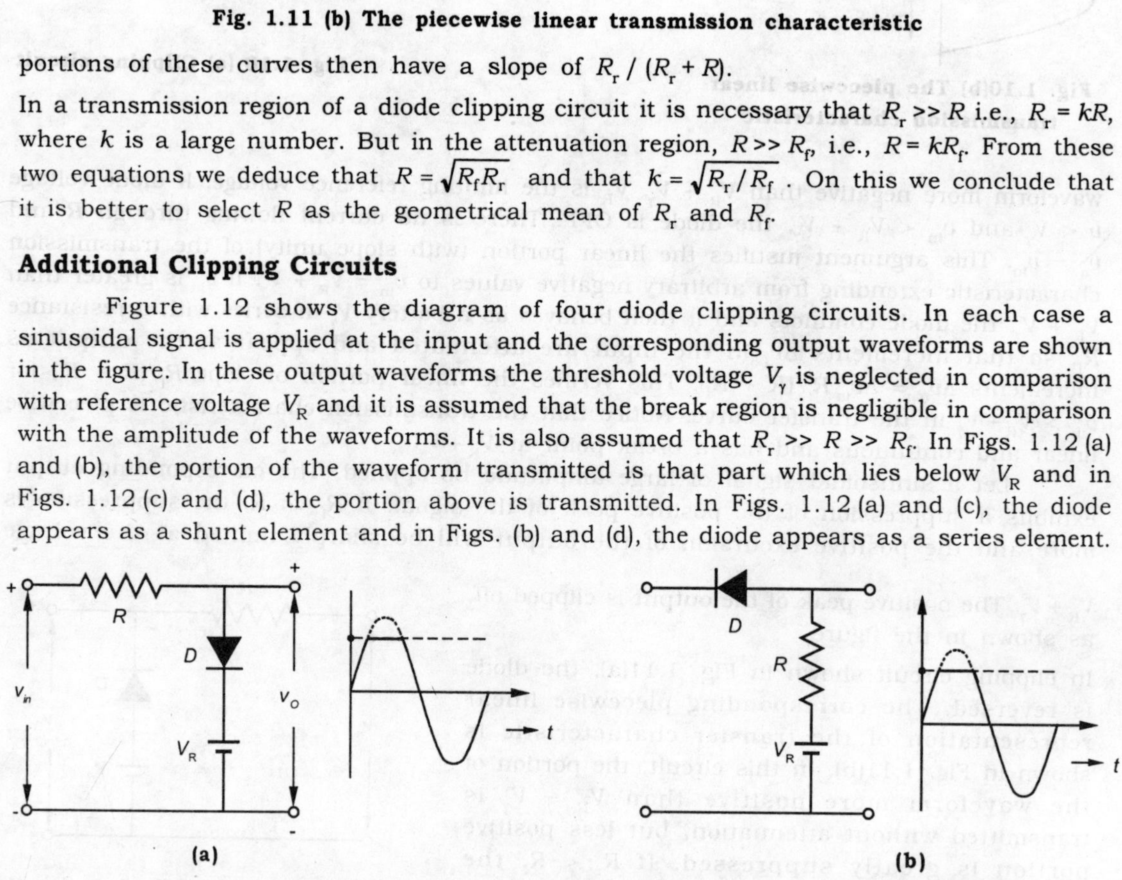


Fig. 1.12 Diode clipping circuits. In (a) the diode appears as a shunt element. In (b) the diode appears as a series element

The use of the diode as a series element has the disadvantage that when the diode is OFF and it is intended that there be transmission, fast signals or high frequency waveforms may be transmitted to the output through the diode capacitance. The use of the diode as a shunt element has the disadvantage that when the diode is open and it is intended that there be transmission, the diode capacitance, together with all other capacitances in shunt with the output terminals, will round sharp edges of input waveforms and attenuate high-frequency signals. The impedance $R_s$ of the source which supplies $V_R$ must be kept low.

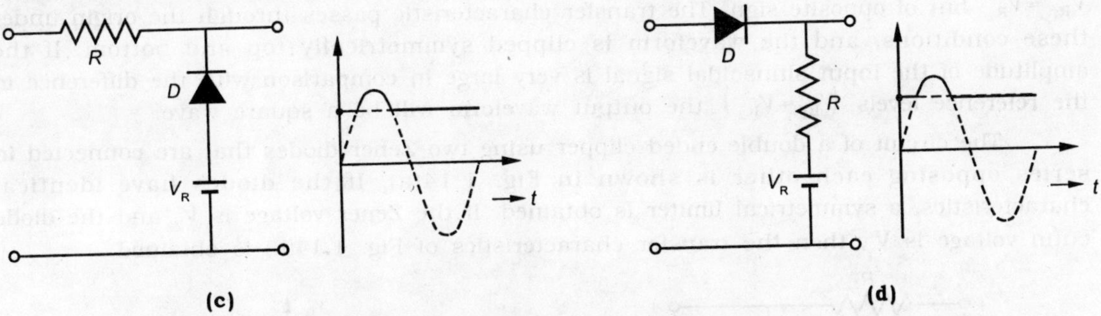


**Fig. 1.12 Diode clipping circuits. In (c) the diode appears as a shunt element. In (d) the diode appears as a series element**

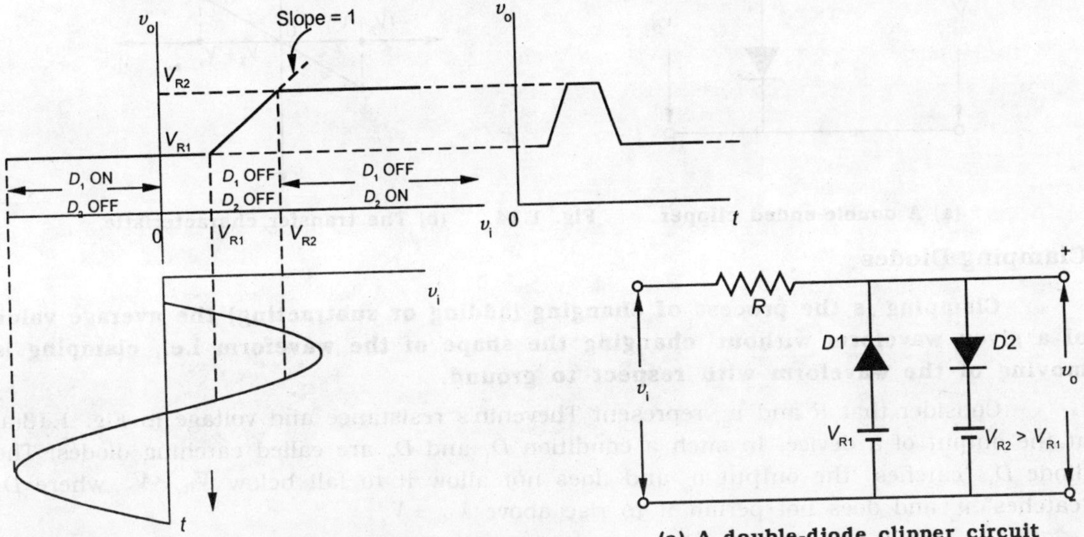


**(b) The piecewise linear transfer curve** **(a) A double-diode clipper circuit**

**Fig. 1.13**

## CLIPPING AT TWO INDEPENDENT LEVELS

Diode clippers may be used in pairs to perform double-ended limiting at independent levels. A parallel, a series, or a series-parallel arrangement may be used. A parallel arrangement of double-diode clipper is shown in Fig. 1.13(a). The piecewise linear and continuous input-output voltage curve is shown in Fig. 1.13(b). The transfer curve has two break points; one at $v_o = v_{in} = V_{R_1}$ and a second $v_o = v_{in} = V_{R_2}$ and has the following characteristics: $V_{R_2} > V_{R_1} >> V_\gamma$ and $R_f << R$.

| Input voltage | Output voltage | Diode states |
|---|---|---|
| $v_{in} \le V_{R_1}$ | $v_{in} = V_{R_1}$ | $D_1$ ON, $D_2$ OFF |
| $V_{R_1} < v_{in} < V_{R_2}$ | $v_o = v_{in}$ | $D_1$ OFF, $D_2$ OFF |
| $v_{in} \ge V_{R_2}$ | $v_o = V_{R_2}$ | $D_1$ OFF, $D_2$ ON |

The circuit shown in Fig. 1.13(a) is called the **slicer** because the output contains a slice of the input between the two reference levels $V_{R_1}$ and $V_{R_2}$ *i.e.*, it is used to convert a sinusoidal waveform into a square wave. In order to produce a symmetrical square wave, $V_{R_1} \approx V_{R_2}$ but of opposite sign. The transfer characteristic passes through the origin under these conditions, and the waveform is clipped symmetrically top and bottom. If the amplitude of the input sinusoidal signal is very large in comparison with the difference in the reference levels $\left(V_{R_1} \approx V_{R_2}\right)$ the output waveform will be a square wave.

The circuit of a double-ended clipper using two Zener diodes that are connected in series opposing each other is shown in Fig. 1.14(a). If the diodes have identical characteristics, a symmetrical limiter is obtained. If the Zener voltage is $V_Z$ and the diode cutin voltage is $V_\gamma$, then the transfer characteristics of Fig. 1.14(b) is obtained.

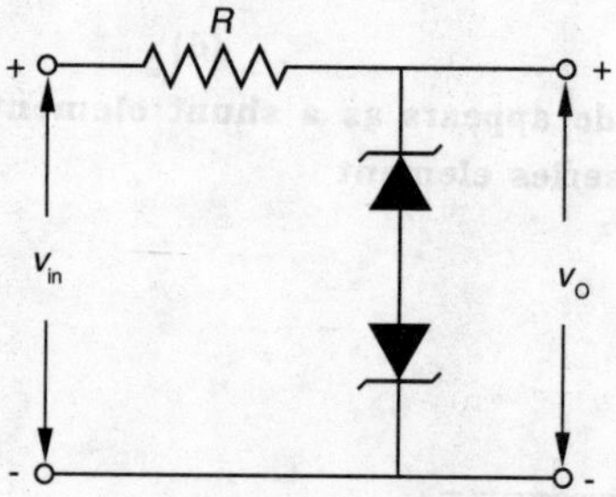


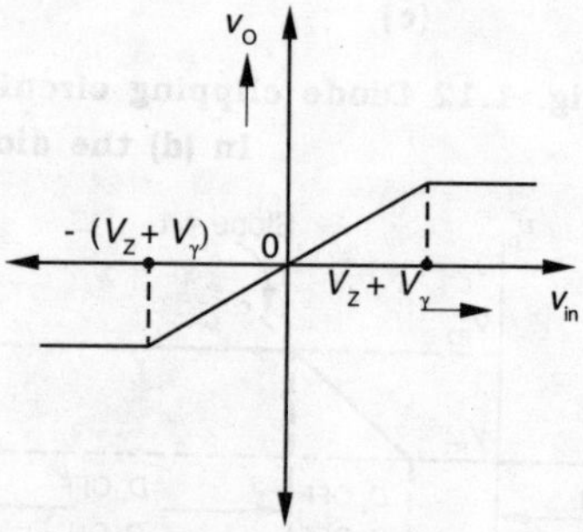


**(a) A double-ended clipper** **Fig. 1.14** **(b) The transfer characteristic**

## Clamping Diodes

**Clamping is the process of changing (adding or subtracting) the average value of a given waveform without changing the shape of the waveform i.e., clamping is moving of the waveform with respect to ground.**

Consider that $R$ and $v_{in}$ represent Thevenin's resistance and voltage in Fig. 1.13(a) at the output of a device. In such a condition $D_1$ and $D_2$ are called catching diodes. The diode $D_1$ "catches" the output $v_o$ and does not allow it to fall below $V_{R_1} - V_\gamma$ where $D_2$ "catches" $v_o$ and does not permit it to rise above $V_{R_2} + V_\gamma$.

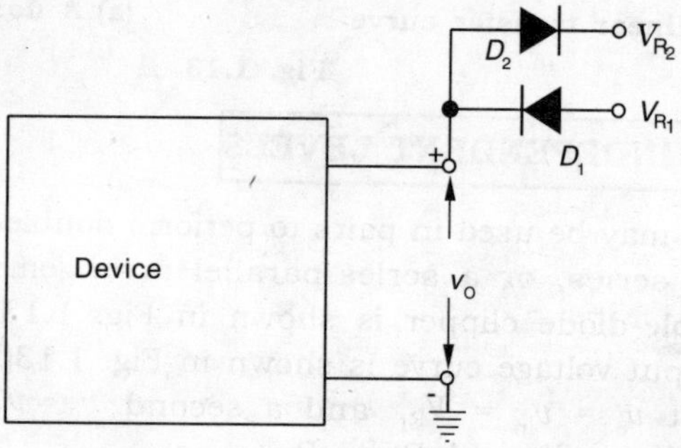


**Fig. 1.15 Clamping diodes limit the output excursion of the device between $V_{R_1}$ and $V_{R_2}$**

Whenever a node becomes connected through a forward-biased diode (low resistance) to some reference voltage $V_R$, the node is then clamped to $V_R$. Because the voltage at that node is unable to depart from $V_R$.

## COMPARATORS

A circuit that is used to mark the instant when an arbitrary waveform attains some reference level is called a **comparator.**

Figure 1.16(a) shows the circuit diagram of a diode comparator. Let the input signal be a ramp. When the input voltage crosses the voltage level $v_i = V_R + V_\gamma$ at time $t = t_1$, the output remains quiescent at $v_o = V_R$ until $t = t_1$, after which it rises with the input signal.

The device to which the comparator output is applied responds when the comparator voltage has risen to some level $V_o$ above $V_R$. However, the precise voltage at which this device responds is subject to some variability $\Delta v_o$ because of gradual changes that result from aging of components, temperature changes, etc. As a result, there will be a variability $\Delta t$ in the precise moment at which this device responds and an uncertainty $\Delta v_{in}$ in the input voltage corresponding to $\Delta t$. Furthermore, if the device responds in the range $\Delta v_o$

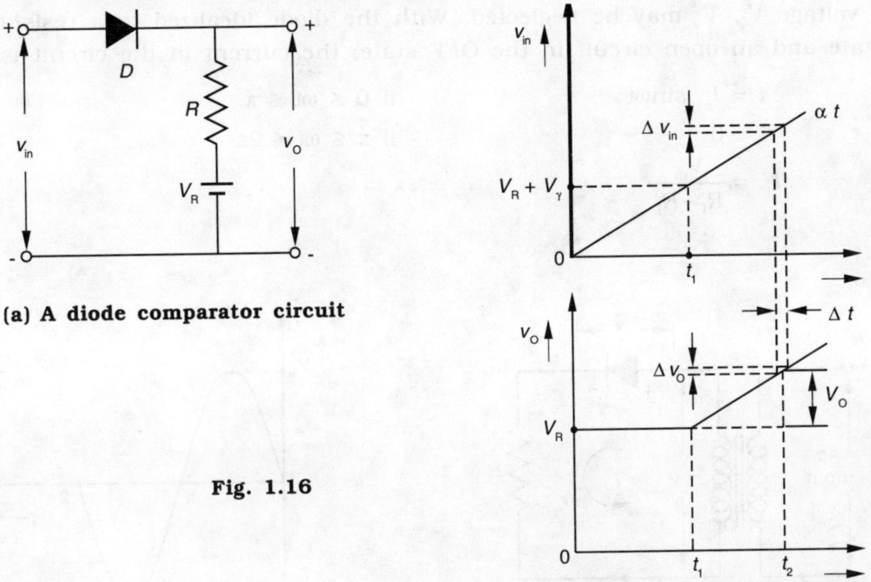


**(a) A diode comparator circuit**

**(b) Input and output waveforms**

**Fig. 1.16**

the device will respond not at time $t = t_1$, but at some later time $t_2$. The situation can be improved by increasing the slope of the rising portion of the output waveform $v_o$. For an ideal diode, it is advantageous to follow the comparator of Fig. 1.16(a) by an amplifier. In practice it is not possible to realise such an anticipated advantage because of the exponential characteristic of a practical diode.

An amplifier preceding the comparator improves the sharpness. Thus, the input signal to a diode comparator must go through a range $\Delta v_{in}$ to carry the comparator through its uncertainty region. If the amplifier's gain is $A$, the input signal then need only to go through the range $\Delta v_{in}/A$ to carry the comparator output through the same voltage range. Hence the amplifier must be direct coupled and extremely stable against drift due to aging components, temperature change, etc.

## RECTIFIERS

Almost all electronic circuits need a *dc* source of power. For portable low power systems, batteries or cells may be used. However, their voltages are low, they need frequent replacement and are expensive as compared to conventional *dc* power supplies. The most convenient way to change alternating to direct current is by means of a rectifier.

The process of converting *ac* voltage into *dc* voltage is known as **rectification**. DC power supplies are obtained with the help of (i) rectifier (ii) filter and (iii) voltage regulator circuit.

### A Half-Wave Rectifier

**A semiconductor diode which converts a sinusoidal input waveform into a unidirectional waveform, with a nonzero average component is called a rectifier.**

Figure 1.17(a) shows the basic circuit for half-wave rectifier using a diode. Let $v = V_m \sin \omega t$ be the instantaneous sinusoidal voltage of frequency $f$ appearing at the secondary coil of the transformer. Since the peak value $V_m$ is very large compared with the cutin voltage $V_\gamma$, $V_\gamma$ may be neglected. With the diode idealized to a resistance $R_f$ in the ON state and an open circuit in the OFF state, the current in the circuit is given by

$$i = I_m \sin\omega t \qquad \text{if } 0 \le \omega t \le \pi$$

$$i = 0 \qquad \text{if } \pi \le \omega t \le 2\pi \qquad \longrightarrow (1)$$

and

$$I_m = \frac{V_m}{R_f + R_L} \qquad \longrightarrow (2)$$

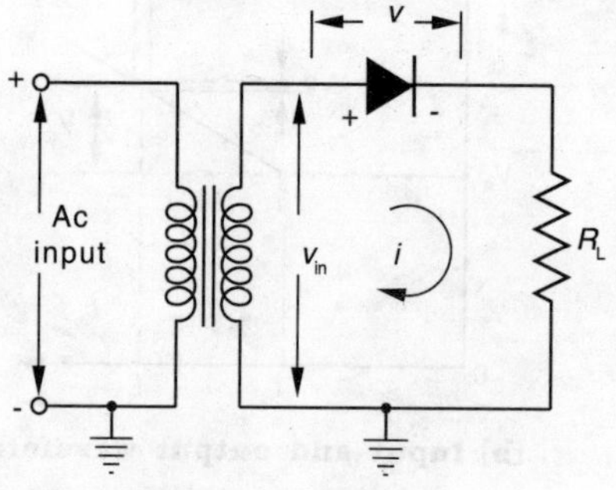


**(a) Half-wave rectifier circuit**

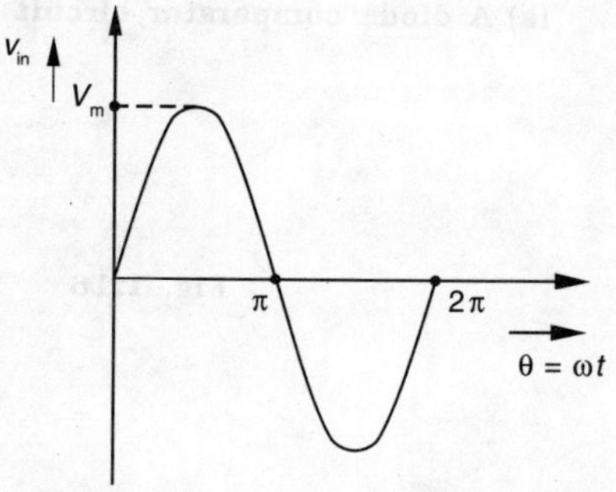


**(b) Input sinusoidal waveform**

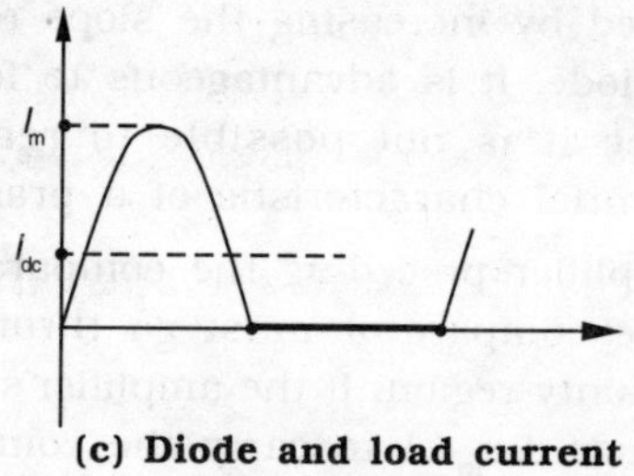


**Fig. 1.17**

**(c) Diode and load current**

The output waveform that is **unidirectional** is shown in Fig. 1.17(c). The average value of a periodic function is the area of one cycle of the curve divided by the base. Mathematically it can be expressed as

$$I_{dc} = \frac{1}{2\pi}\int_{o}^{2\pi} i\, d\theta$$

For a half-wave circuit,

$$I_{dc} = \frac{1}{2\pi}\int_{o}^{\pi} i\, d\theta = \frac{1}{2\pi}\int_{o}^{\pi} I_m \sin\theta\, d\theta$$

$$I_{dc} = \frac{I_m}{\pi} = 0.318\, I_m \qquad \longrightarrow (3)$$

The **dc output voltage** is given by

$$V_{dc} = I_{dc}\, R_L = \frac{I_m R_L}{\pi} \qquad \longrightarrow (4)$$

Since the diode has two values of resistance, i.e., $R_f$ in the ON state and $\infty$ in the OFF state, the reading of a *dc* voltmeter placed across the diode is not given by $I_{dc} R_f$. The voltmeter shows the average value. Hence, to obtain $V_{dc}$ across the diode, the instantaneous voltage must be plotted as shown in Fig. 1.18. The average value can be obtained by integrating.

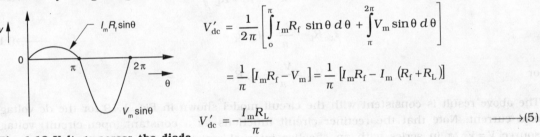

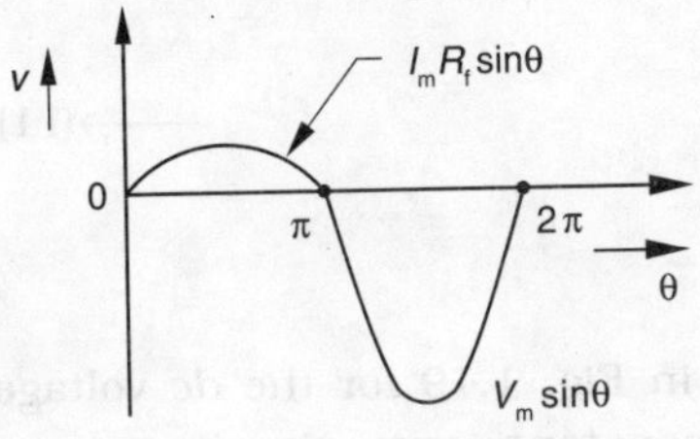


**Fig. 1.18 Voltage across the diode**

$$V'_{dc} = \frac{1}{2\pi}\left[\int_{o}^{\pi} I_m R_f \sin\theta\, d\theta + \int_{\pi}^{2\pi} V_m \sin\theta\, d\theta\right]$$

$$= \frac{1}{\pi}\left[I_m R_f - V_m\right] = \frac{1}{\pi}\left[I_m R_f - I_m\left(R_f + R_L\right)\right]$$

$$V'_{dc} = -\frac{I_m R_L}{\pi} \qquad \longrightarrow (5)$$

The above result is negative. It means that if the voltmeter is to read upscale, its positive terminal must be connected to the cathode of the diode. From Eq. (4), the *dc* diode voltage is seen to be equal to the negative of the *dc* voltage across the load resistor. This result is correct because the sum of the *dc* voltages around the circuit must add up to zero.

## The AC Current (Voltage)

The **effective** or **rms value of a periodic function of time** is defined as the ratio of area of one cycle of the curve which represents the square of the function to the base. It can be measured by a "square law" instrument, may be of the thermocouple type. A root-mean-square ammeter/voltmeter is used to measure the effective or rms current/voltage.

Mathematically,

$$I_{rms} = \sqrt{\frac{1}{2\pi}\int_{o}^{2\pi} i^2\, d\theta} \qquad \longrightarrow (6)$$

$$I_{rms} = \sqrt{\frac{1}{2\pi}\int_{o}^{2\pi} I_m^2 \sin^2\theta\, d\theta} = \frac{I_m}{2} \qquad \longrightarrow (7)$$

Applying Eq. (6) to the sinusoidal input voltage, we get

$$V_{rms} = \frac{V_m}{\sqrt{2}} = 0.707\, V_m \qquad \longrightarrow (8)$$

## Regulation

The variation of *dc* output voltage as a function of *dc* load current is called **regulation.** The percentage regulation is defined as

$$\%\ \text{Regulation} = \frac{V_{no\ load} - V_{load}}{V_{load}} \times 100\% \qquad \longrightarrow (9)$$

where **no load** represents zero current (load = ∞) and **load** indicates the normal rated load current.

For an ideal power supply the output voltage is independent of the load and the percent regulation is zero.

The variation of $V_{dc}$ with $I_{dc}$ for the half-wave rectifier is obtained as follows:

From Eqs. (2) and (3), we have

$$I_{dc} = \frac{I_m}{\pi} = \left(\frac{V_m}{R_f + R_L}\right)\Big/\pi \qquad \longrightarrow (10)$$

$$I_{dc} R_L = \frac{V_m}{\pi} - I_{dc} R_f \qquad \longrightarrow (11)$$

or

$$V_{dc} = \frac{V_m}{\pi} - I_{dc} R_f$$

The above result is consistent with the circuit model shown in Fig. 1.19 for the *dc* voltage and current. Note that the rectifier circuit functions as a constant (open-circuit) voltage source $V = V_m/\pi$ in series with an effective internal resistance $R_o = R_f$. This model shows that $V_{dc}$ equals $V_m/\pi$ at no load and that the *dc* voltage decreases linearly, with an increase in *dc* output current. In practice, the resistance $R_s$ of the secondary winding of the transformer is in series with the diode. Then

$$V_{dc} = \frac{V_m}{\pi} - I_{dc}(R_f + R_s) \qquad \longrightarrow (12)$$

**Thevenin's Theorem** states that **any two-terminal linear network may be replaced by a generator equal to the open-circuit voltage ($V_{th}$) between the terminals in series with the output impedance ($R_{th}$) seen at this port.**

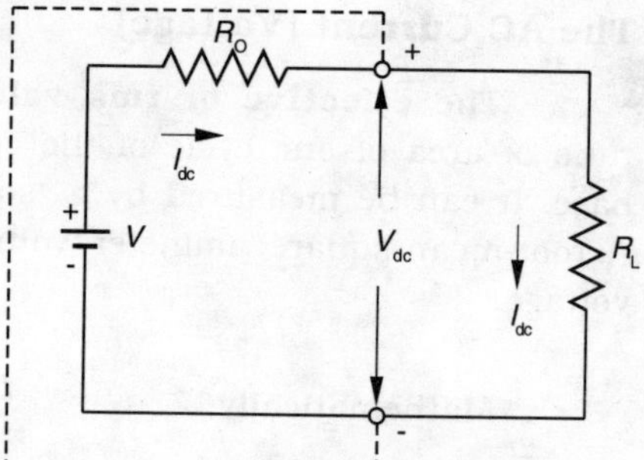

**Fig. 1.19 The Thevenin model**

## A Full-Wave Rectifier

Figure 1.20 shows the circuit of a full-wave rectifier. The circuit is seen to comprise of two half-wave circuits. During one half of the power cycle conduction takes place through one diode $D_1$ and during the second half of the cycle conduction takes place through the diode $D_2$.

The current $i(=i_1+i_2)$ to the load is sum of the diode currents $i_1$ and $i_2$. The *dc* value of load current is given by

$$I_{dc} = \frac{1}{\pi}\int_0^{\pi} i\,d\theta = \frac{1}{\pi}\int_0^{\pi} I_m \sin\theta\, d\theta$$

$$I_{dc} = \frac{2I_m}{\pi} = 0.637\,I_m \longrightarrow (1)$$

The *rms* value of current is given by

$$I_{rms} = \sqrt{\frac{1}{2\pi}\int_0^{2\pi} i^2\, d\theta} = \sqrt{\frac{1}{2\pi}\int_0^{\pi} I_m^2 \sin^2\theta\, d\theta}$$

$$I_{rms} = \frac{I_m}{\sqrt{2}} = 0.707\,I_m \longrightarrow (2)$$

DC output voltage, $I_{dc}R_L = \dfrac{2I_m R_L}{\pi} \longrightarrow (3)$

where $I_m = \dfrac{V_m}{R_f + R_L}$ and $V_m$ is the peak transformer secondary voltage from one end to the centre tap.

From Eq. (3), it is seen that the *dc* output voltage for the full-wave connection is twice that for the half-wave circuit. The output voltage is unidirectional, continuous but not constant.

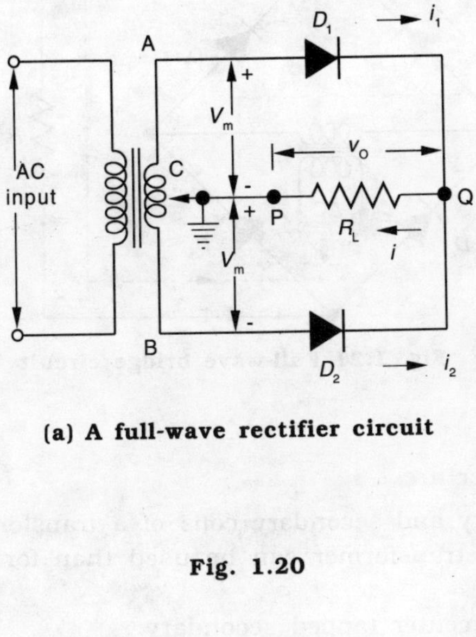


**(a) A full-wave rectifier circuit**

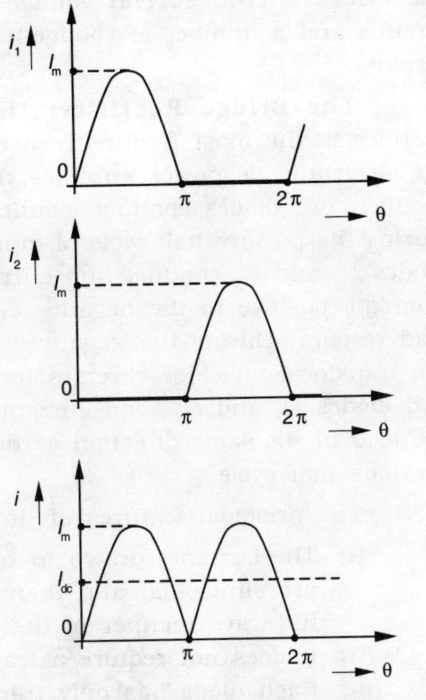


**(b) The individual currents and the load current**

**Fig. 1.20**

The variation of $V_{dc}$ with $I_{dc}$ for the full-wave rectifier is obtained as follows:

$$I_{dc} = \frac{2I_m}{\pi} = \frac{2V_m}{\pi(R_f + R_L)}$$

$$I_{dc} R_L = V_{dc} = \frac{2V_m}{\pi} - I_{dc} R_f \qquad \longrightarrow (4)$$

The Eq. (4) leads to Thevenin's *dc* model of Fig. 1.19, except that the open-circuit (internal) supply is $V = 2V_m/\pi$ instead of $V_m/\pi$.

## Peak Inverse Voltage (PIV)

The maximum reverse voltage a diode can withstand is called **peak inverse voltage.** From Fig. 1.17 it is clear that, for the half-wave rectifier, the peak inverse voltage is $V_m$. Each diode in a full-wave rectifier is alternately forward-biased and reverse-biased. When the diode $D_1$ is forward-biased the voltage across the reverse-biased diode $D_2$ is equal to sum of the voltages across the lower half of the secondary coil and the load resistor $R_L$, i.e. maximum diode voltage $V_{D_2} = V_m - (-V_m) = 2V_m$. Hence, for the full-wave rectifier, the peak inverse voltage is $2V_m$. Note that this result is obtained without reference to the nature of the load, which can be a pure resistance $R_L$ or a combination of $R_L$ and some reactive elements which may be introduced to "filter" the ripple.

## OTHER FULL-WAVE CIRCUITS

A variety of other rectifier circuits are the bridge circuit, several voltage-doubling circuits and a number of voltage-multiplying circuits.

**The Bridge Rectifier:** The bridge rectifier is the most frequently used circuit for electronic *dc* power supplies. In bridge rectifier, two diodes conduct simultaneously. During the positive half-cycle of input signal, diodes $D_1$ and $D_3$ conduct and current flows from the positive to the negative end of the load resistor. During the negative half-cycle, the transformer voltage reverses its polarity, and diodes $D_2$ and $D_4$ send current through the load in the same direction as during the previous half cycle.

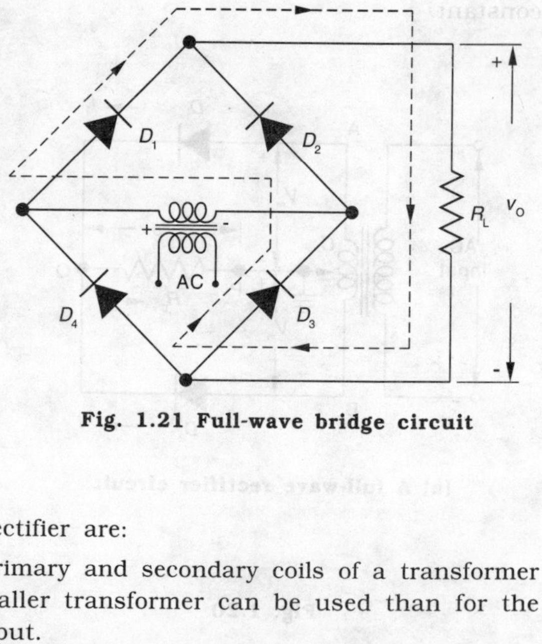


**Fig. 1.21 Full-wave bridge circuit**

The principal features of the bridge rectifier are:

(i) The currents drawn in both the primary and secondary coils of a transformer are sinusoidal and therefore a smaller transformer can be used than for the full-wave rectifier of the same output.
(ii) It does not require a transformer with center tapped secondary.
(iii) Each diode has only transformer voltage across it on the inverse cycle.
(iv) It is suitable for high-voltage applications.
(v) The PIV rating need be only half the rating required for full-wave rectifier.

## The Rectifier Meter

The rectifier meter is essentially a bridge-rectifier system, except that no transformer is required. The voltage to be measured is applied through a multiplier resistor $R$ to two corners of the bridge, a *dc* milliammeter is used as an indicating instrument across the other two corners of the bridge. Since the *dc* milliammeter reads average values of current, the meter scale is calibrated to give rms values when a sinusoidal voltage is applied to the input terminals. As a result, the instrument will not show correct values when used with waveforms which contain appreciable harmonics.

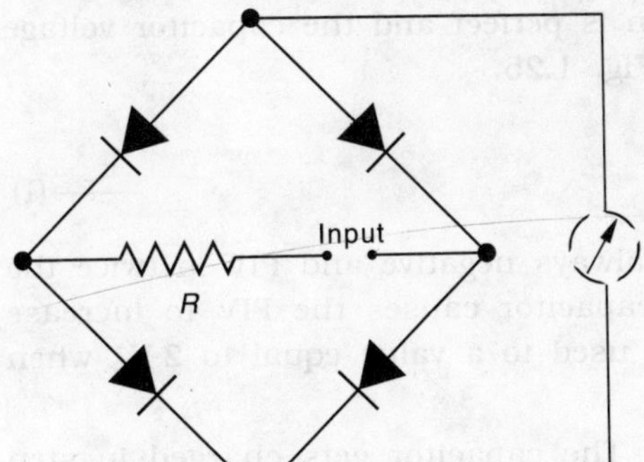


**Fig. 1.22 The rectifier voltmeter**

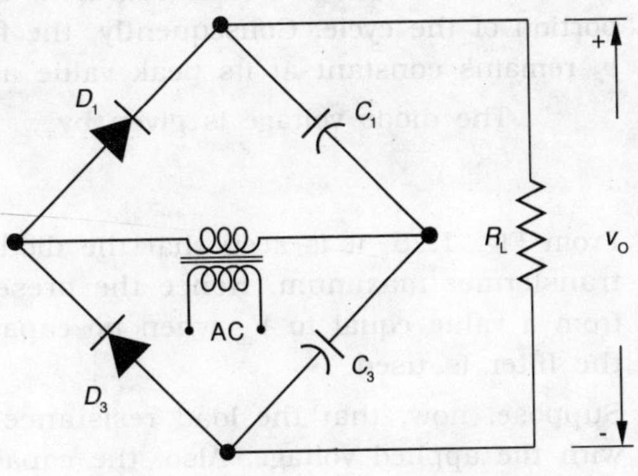


**Fig. 1.23 The bridge voltage doubler**

## Voltage Multipliers

A voltage multiplier is a circuit which produces an output *dc* voltage whose value is a multiple of peak *ac* input voltage.

Figure 1.23 shows the circuit of a voltage doubler. The output *dc* voltage is equal to twice the transformer maximum voltage at no load. This circuit is operated by alternately charging each of the two capacitors to the transformer peak voltage $V_m$. The current is continually drained from the capacitors through the load. The capacitors also smooth out the ripple in the output.

The inverse voltage across the reverse-biased diodes is twice the transformer peak voltage. The circuit has poor regulation. Hence large capacitors must be used to have good regulation.

## CAPACITOR FILTER

A circuit which removes the ripples in the rectifier output without affecting the *dc* component is called a **filter**.

Filtering is necessary because electronic circuits require a constant source of *dc* voltage and current to provide biasing for proper operation. Filtering is frequently effected by shunting the load with a capacitor. The capacitor stores

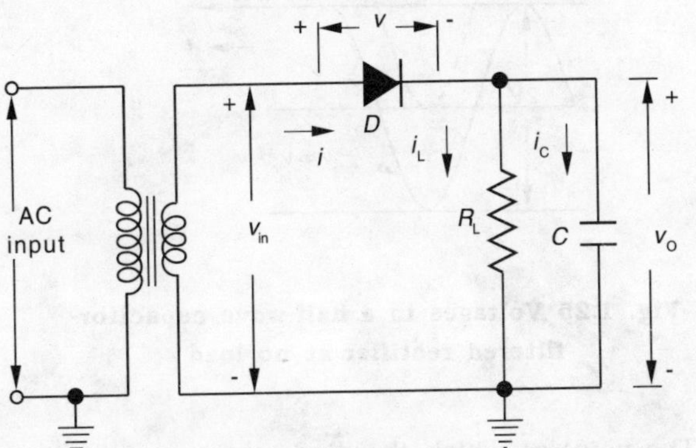


**Fig. 1.24 A half-wave capacitor-filtered rectifier**

energy during the conduction period and delivers this energy to the load during the non-conducting period. In this way, the time during which the current passes through the load is prolonged and the ripple is considerably decreased. The ripple voltage is defined as the deviation of the load voltage from its average or *dc* value.

Figure 1.24 shows the circuit of a half-wave capacitive rectifier. Assume that the load resistance $R_L = \infty$. When the rectifier output voltage increases, the capacitor charges to the peak voltage $V_m$. Further, the capacitor maintains the potential $V_m$, for no path exists by which this charge is permitted to leak off, since diode does not allow a negative current. Since the reverse-biased diode has infinite resistance, no charge flows during this portion of the cycle. Consequently, the filtering action is perfect and the capacitor voltage $v_o$ remains constant at its peak value as shown in Fig. 1.25.

The diode voltage is given by

$$v = v_i - v_o \qquad \longrightarrow (1)$$

From Fig. 1.25, it is seen that the diode voltage is always negative and PIV is twice the transformer maximum. Hence the presence of the capacitor causes the PIV to increase from a value equal to $V_m$ when no capacitor filter is used to a value equal to $2\,V_m$ when the filter is used.

Suppose, now, that the load resistance $R_L$ is finite. The capacitor gets charged in step with the applied voltage. Also, the capacitor must discharge through $R_L$, since the diode does not conduct in the negative direction. Clearly, the diode acts as a switch which permits charge to flow into the capacitor when the transformer voltage exceeds the capacitor voltage and then acts to disconnect the power source when the transformer voltage falls below that of the capacitor. Let us now understand the analysis of the circuit in two steps:

### (i) Diode Conducting

If the diode voltage $v$ is neglected, the transformer voltage is impressed directly across the load. Hence the output voltage is $v_o = V_m \sin \omega t$.

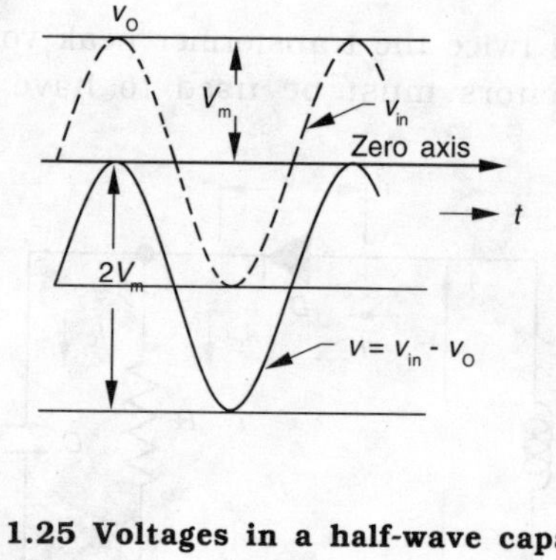


**Fig. 1.25 Voltages in a half-wave capacitor-filtered rectifier at no load**

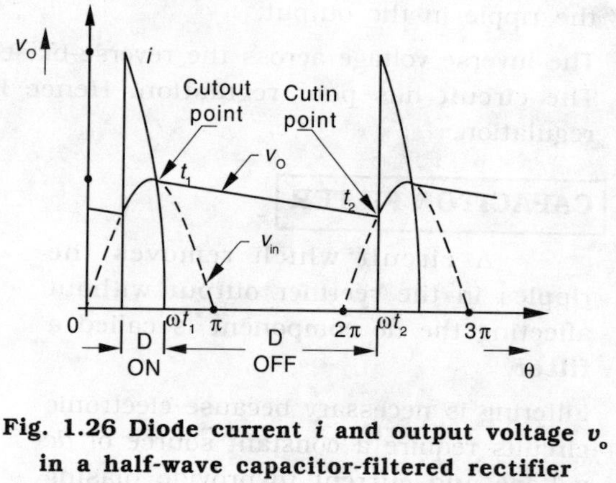


**Fig. 1.26 Diode current *i* and output voltage $v_o$ in a half-wave capacitor-filtered rectifier**

The point at which the diode starts conducting is called **cut-in point** ($t_2$) and that at which it stops conducting is called the **cutout point** ($t_1$).

Let $v = V_m \sin \omega t$ be the instantaneous sinusoidal voltage appearing across the secondary coil. When the diode conducts, the diode current $i$ is the sum of the load resistor current $i_L$ and the capacitor current $i_C$. Hence

$$i = i_L + i_C$$

$$i = \frac{v_o}{R_L} + C\frac{dv_o}{dt} = \frac{V_m \sin \omega t}{R_L} + C\frac{d(V_m \sin \omega t)}{dt} \longrightarrow (2)$$

This current is in the form of $i = I_m \sin(\omega t + \phi)$, where

$$I_m = V_m\sqrt{\frac{1}{R_L^2} + \omega^2 C^2} \quad \text{and } \phi = \text{arc tan } \omega CR_L \text{ or } \tan^{-1}(\omega CR_L) \longrightarrow (3)$$

The cutout time $t_1$ can be found by setting $i = 0$ at $t = t_1$, that leads to

$$\omega t_1 = \pi - \phi \longrightarrow (4)$$

for the cutout angle $\omega t_1$ in the first cycle. The diode current $i$ is shown in Fig. 1.26. From Eq. (2), it is clear that the use of a large capacitance to improve the filtering at a given load $R_L$ is accompanied by a high-peak diode current $I_m$. For a specified average load current, the current $i$ becomes more peaked and the conduction period decreases as $C$ is made larger. Note that the use of a capacitor filter may impose serious restrictions on the diode, since the average current may be well within the current rating of the diode, and yet the peak current may be excessive.

### (ii) Diode Nonconducting

The diode is reverse biased during the interval between the cutout time $t_1$ and the cut-in time $t_2$. The capacitor then discharges through the load resistor $R_L$ with a time constant $CR_L$. Since the capacitor voltage is equal to the load voltage, we have

$$v_o = Ae^{-t/CR_L} \longrightarrow (5)$$

where $A$ is a constant. To determine $A$, note from Fig. 1.26 that at the cutout time $t = t_1$, $v_o = v_i = V_m \sin \omega t_1$. Then Eq. (5) becomes

$$v_o = (V_m \sin \omega t_1)e^{-(t-t_1)/CR_L} \longrightarrow (6)$$

The exponential curve [Fig. 1.26] that intersects the curve $V_m \sin \omega t$ is the cut-in point $t_2$. The validity of this statement follows from the fact that at an instant of time greater than $t_2$, the transformer voltage $v_i$ (the sine curve) is greater than the capacitor voltage $v_o$ (the exponential curve).

## Full-Wave Capacitor-filtered Rectifier

Consider a full-wave rectifier with a capacitor filter obtained by connecting a capacitor $C$ across $R_L$ in Fig. 1.20. The cutout point is the same as that found for the half-wave rectifier i.e.,

$$\omega t_1 = \pi - \phi \longrightarrow (7)$$

The cut-in point $\omega t_2$ lies between $\pi$ and $2\pi$ where the exponential portion of $v_o$ intersects the sinusoid.

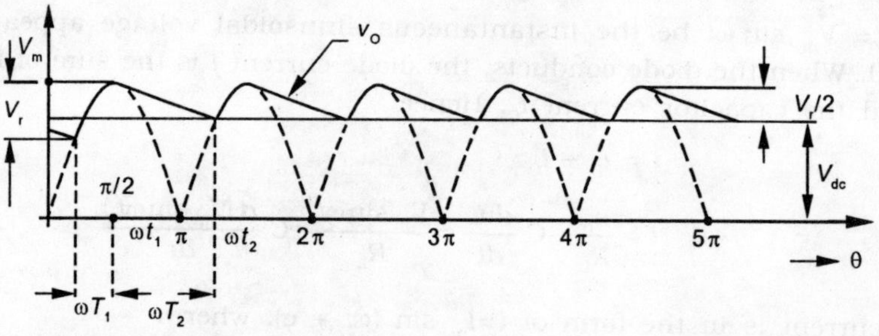


**Fig. 1.27 Output voltage waveform in a full-wave capacitor filtered rectifier**

## Approximate Analysis

Assume that the output-voltage waveform of a full-wave circuit with a capacitor filter may be represented approximately by the piecewise linear curve shown in Fig. 1.27. For large values of $C$ note that $\omega t_1 \to \pi/2$ and $v_o \to V_m$ at $t = t_1$. Also, with $C$ very large, the exponential decay in Eq. (6) can be replaced by a linear fall. If $V_r$ is the total capacitor discharge voltage (the ripple voltage), then average *dc* output voltage approximately is given by

$$V_{dc} = V_m - \frac{V_r}{2} \quad \longrightarrow (8)$$

The ripple $V_r$ is triangular in shape. The rms value of this triangular wave is

$$V_{rms} = \frac{V_r}{2\sqrt{3}} \quad \longrightarrow (9)$$

If $T_2$ is the total nonconducting time when the capacitor discharges at the constant rate $I_{dc}$, it will lose a charge of $I_{dc}\, T_2$.

Hence the change in capacitor voltage is $I_{dc}\, T_2/C$ or

$$V_r = \frac{I_{dc}\, T_2}{C} \quad \longrightarrow (10)$$

The better the filtering action, the smaller will be the conduction time $T_1$ and the closer $T_2$ will approach the time of half a cycle. Hence assume that $T_2 = T/2 = 1/2f$, where $f$ is the fundamental power-line frequency. Then

$$V_r = \frac{I_{dc}}{2fC} \quad \longrightarrow (11)$$

Eq. (8) becomes,

$$V_{dc} = V_m - \frac{I_{dc}}{4fC} \quad \longrightarrow (12)$$

The above result is consistent with Thevenin's model of Fig. 1.19, with the open-circuit voltage $V = V_m$ and the effective output resistance $R_o = 1/4fC$.

From Eq. (11), it is seen that the ripple varies directly with the load current $I_{dc}$ and also inversely with the capacitance. The ripple factor can be lowered by increasing the value of the capacitor or increasing the load resistance.

## Advantages

1. Rectifier employing capacitor input filters have the small ripple and the high voltage at light load.
2. The no-load voltage is (theoretically) equal to the maximum transformer voltages.

## Disadvantages

1. Rectifiers employing capacitor input filters have relatively poor regulation.
2. They exhibit high ripple at large loads.
3. They impose serious restrictions on the diode i.e., the peak current may be excessive.

## ADDITIONAL DIODE CIRCUITS

### 1. Peak Detector

The half-wave capacitor filtered rectifier can be used to measure the peak value of an input waveform. When $R_L = \infty$ (open-circuit), the capacitor charges to the maximum value $V_m$ of input $v_i$, the diode becomes nonconducting and output voltage $v_o$ remains at $V_m$.

In an AM radio the amplitude of the high-frequency wave (carrier signal) varies in accordance with the audio information to be transmitted. This phenomenon is known as **amplitude modulation** and such an AM waveform is shown in Fig. 1.28. The process of recovering the audio signal is called **detection** or **demodulation**. If the AM waveform is applied to the input of half-wave capacitor-filtered rectifier, the output $v_O$ is the heavy-weight curve when the time constant $CR_L$ is small. The order of magnitude of the frequency of a carrier signal in AM is 1000 *kHz* and the audio frequency ranges from 20 *Hz* to 20 *kHz*. Hence there should be at least 50 cycles of the carrier signal for each audio cycle. If Fig. 1.28 were drawn more realistically with a much higher ratio of carrier to audio frequency $(f_c/f_m)$, then the ripple amplitude of the demodulated signal will be very much smaller. This low-amplitude high-frequency ripple in $v_O$ can easily be filtered. The detected waveform is an excellent reproduction of the audio signal. Therefore, the capacitor-rectifier also acts as an **envelope demodulator.**

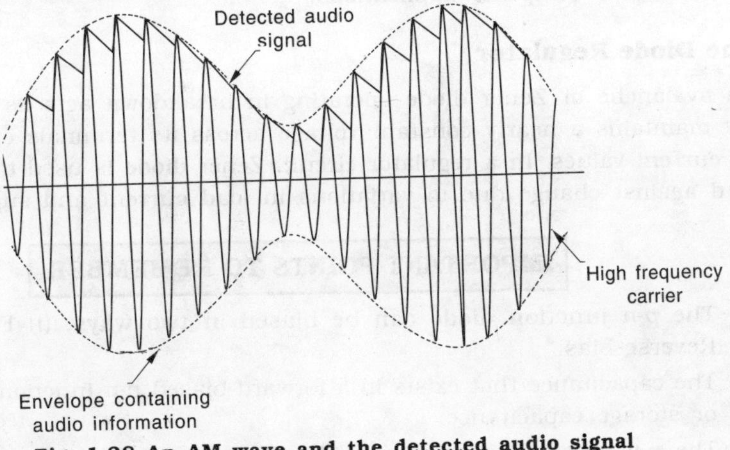


**Fig. 1.28 An AM wave and the detected audio signal**

### 2. A Clamping Circuit

The **clamping operation** is a function that must frequently be accomplished with a periodic waveform in the establishment of the recurrent positive or negative extremity at some constant reference level $V_R$. Hence, in the steady state, the circuits used to perform the clamping operation are referred to as **clamping circuits.**

Consider a clamping circuit as shown in Fig. 1.29(a). Assuming an ideal diode i.e., $R_f = 0$, $R_r = \infty$ and $V_\gamma = 0$, the voltage drop across the diode is zero when it is forward direction. Hence the output voltage cannot rise above $V_R$ and is said to be **clamped** to reference level $V_R$. If the sinusoidal voltage with a peak value $V_m$ and an average value of zero is input, as shown in Fig. 1.29(b), the output is sinusoidal with an average value of $V_R - V_m$. The waveform is obtained based on following conditions: (1) The diode must be ideal. (ii) The source impedance $R_s = 0$ and (iii) The time constant $\tau$ (= $RC$) is much larger than the period ($T$) of the signal since the clamping is not perfect.

In practice, the output voltage rises slightly above $V_R$ and slightly distorted output waveform is obtained due to forward resistance of diode.

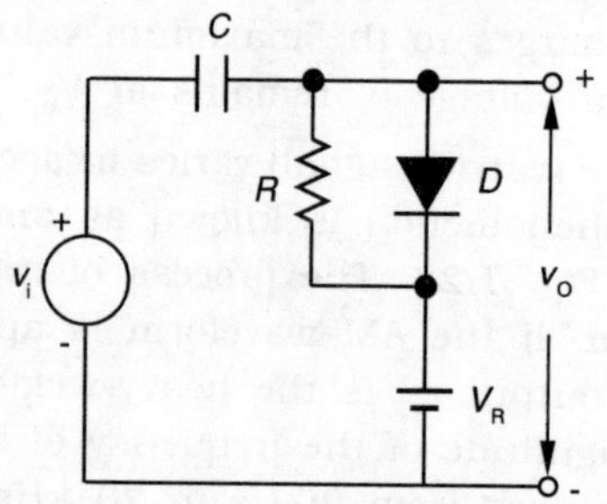


(a) A clamper circuit

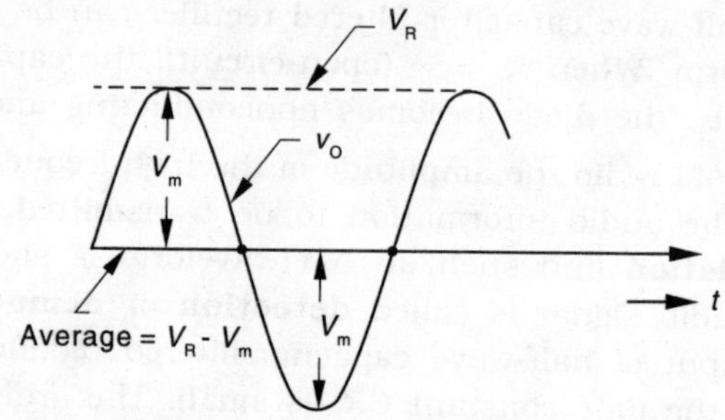


(b) Output voltage

**Fig. 1.29**

## Digital Computer Circuits

Since the diode is a binary device for a given interval of time, it is a very useful component in digital computer applications.

## Avalanche Diode Regulator

An avalanche or Zener diode operating in breakdown acts as a **voltage regulator** because it maintains a nearly constant voltage across its terminals over a specified range of reverse current values. In a regulator circuit, Zener diode is used to regulate the voltage across load against change due to variations in load current and supply voltage.

## IMPORTANT POINTS TO REMEMBER

1. The *p-n* junction diode can be biased in two ways: (i) Forward-bias and (ii) Reverse-bias.
2. The capacitance that exists in a forward-biased *p-n* junction is called a diffusion or storage capacitance.
3. The *p-n* junction diode is considered as a circuit element.
4. The piecewise linear diode model is used in applications such as clippers, comparators, diode gates, and rectifiers.
5. The curve which relates the output voltage to the input of any circuit is called the transfer characteristic.
6. The use of the load line construction allows the graphical analysis of many circuits involving devices which are more complicated than the *p-n* diode.

7. The *p-n* junction diode is a binary device as it exists in only one of two possible states, i.e., the diode is either ON or OFF at a given time.
8. The large-signal model for the ON state is represented as a battery $V_\gamma$ in series with the low forward resistance $R_f$.
9. The large-signal model for the OFF state is represented as a large resistance $R_r$.
10. Peak inverse voltage is the maximum reverse voltage that can be applied to the diode without destruction.
11. The break point is defined as the voltage $V_\gamma$, where the diode resistance changes discontinuously from the very large $R_r$ to the very small $R_f$.
12. The circuit with which the waveform is shaped by removing a particular portion of the input signal above or below a certain level is known as a clipping circuit.
13. Clipping circuits are also called voltage limiters, amplitude selectors or slicers.
14. The ratio $R_r/R_f$ serves as a figure of merit for diodes used in various applications.
15. A double-diode clipper can be used for converting a sinusoidal waveform into a square wave.
16. The circuit that is used to hold a signal at a specified *dc* level is called a *dc* restorer or clamper.
17. Whenever a node is connected through a low resistance (a conducting diode) to some reference voltage $V_R$, the node gets clamped to $V_R$.
18. The nonlinear circuits can also be used to perform the operation of comparison.
19. A comparator circuit is used to mark the instant when an arbitrary waveform attains some reference level.
20. Comparator circuit does not reproduce any part of the signal waveform.
21. The process of converting *ac* voltage into *dc* voltage is known as rectification.
22. All electronic circuits require a *dc* source of power.
23. The output voltage of a half-wave rectifier is unidirectional, pulsating and intermittent.
24. The variation of *dc* output voltage as a function of *dc* load current is called regulation.
25. Thevenin's theorem states that any linear two-terminal network may be replaced by a generator equal to the open-circuit voltage between the terminals in series with the output resistance seen at this port.
26. In a full-wave rectifier the rectifier output voltage is unidirectional, continuous but not constant.
27. In a full-wave rectifier, the PIV across each diode is twice the maximum transformer voltage measured from midpoint to either end.
28. The bridge circuit is most suitable for high-voltage applications.
29. Voltage multiplier is a circuit which produces an output *dc* voltage whose value is a multiple of peak *ac* input voltage.
30. Filter circuit removes the ripples in the rectifier output without affecting the *dc* component.
31. Filtering is frequently effected by shunting the load with a capacitor.
32. The action of the filtering depends upon the fact that the capacitor stores energy

during the conduction period and delivers the stored energy to the load during the inverse or nonconducting period.

33. The deviation of load voltage from average or *dc* voltage is called the ripple voltage.
34. In a filter circuit, the point at which the diode starts conducting is called the cut-in point and that at which it stops conducting is called the cutout point.
35. The half-wave capacitor-filtered rectifier circuit can be used to measure the peak value of an input waveform.
36. Since the diode is a binary device, it is a very useful component in digital applications.

## KEY FORMULAE

1. Diffusion capacitance of a diode, $C_D = \frac{\tau I}{\eta V_T} = \tau g = \frac{\tau}{r}$
2. At low frequencies, $C_D = \frac{1}{2}\tau g$
3. At high frequencies, $C_D = \left(\frac{\tau}{2\omega}\right)^{1/2} g$
4. Slope of the load line $= -\frac{1}{R_L}$
5. In a piecewise linear model, the current in the forward direction is $i = \frac{V_m \sin\theta - V_\gamma}{R_L + R_f}$
6. The cutin angle, $\theta = \text{arc}\sin\left(V_\gamma / V_m\right)$ or $\sin^{-1}\left(V_\gamma / V_m\right)$
7. The incremental resistance is given by $r = \eta \frac{V_T}{I_o} e^{-V/\eta V_T}$
8. For two different potentials, $\frac{r_1}{r_2} = e^{(V_2 - V_1)/\eta V_T}$
9. In a half wave rectifier, *dc* output voltage is given by $V_{dc} = \frac{I_m R_L}{\pi}$ and $I_{dc} = \frac{I_m}{\pi}$
10. In a half-wave rectifier, $V_{rms} = 0.707\, V_m$ and % regulation $= \left(\frac{V_{NL} - V_{FL}}{V_{FL}}\right) \times 100$
11. In a full-wave rectifier, $V_{dc} = \frac{2 I_m R_L}{\pi} = \frac{2V_m}{\pi} - I_{dc} R_L$
12. In a half-wave capacitor-filtered circuit, $I_m = V_m \sqrt{\frac{1}{R_L^2} + \omega^2 C^2}$
13. During nonconducting state, output voltage, $v_o = (V_m \sin\omega t_1) e^{-(t-t_1)/CR_L}$
14. In a full-wave capacitor-filtered circuit, $V_{dc} = V_m - \frac{I_{dc}}{4fC}$

## SOLVED PROBLEMS

**1. (a) Sketch the output $v_o$ and determine the *dc* level of the output for the network of Fig. 1.30.**

**(b) Repeat part (a) if the ideal diode is replaced by a *Si* diode.**

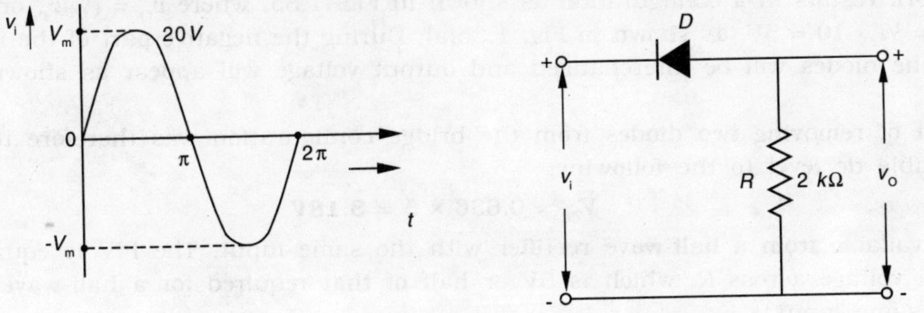


**Fig. 1.30**

**Solution:** (a) As shown in Fig. 1.30, diode conducts during the negative part of the input as shown in Fig. 1.31 and the output will appear as shown in Fig. 1.31. For the full period, the *dc* level is

$$V_{dc} = -0.318\ V_m = -0.318 \times 20 = \mathbf{-6.36V}$$

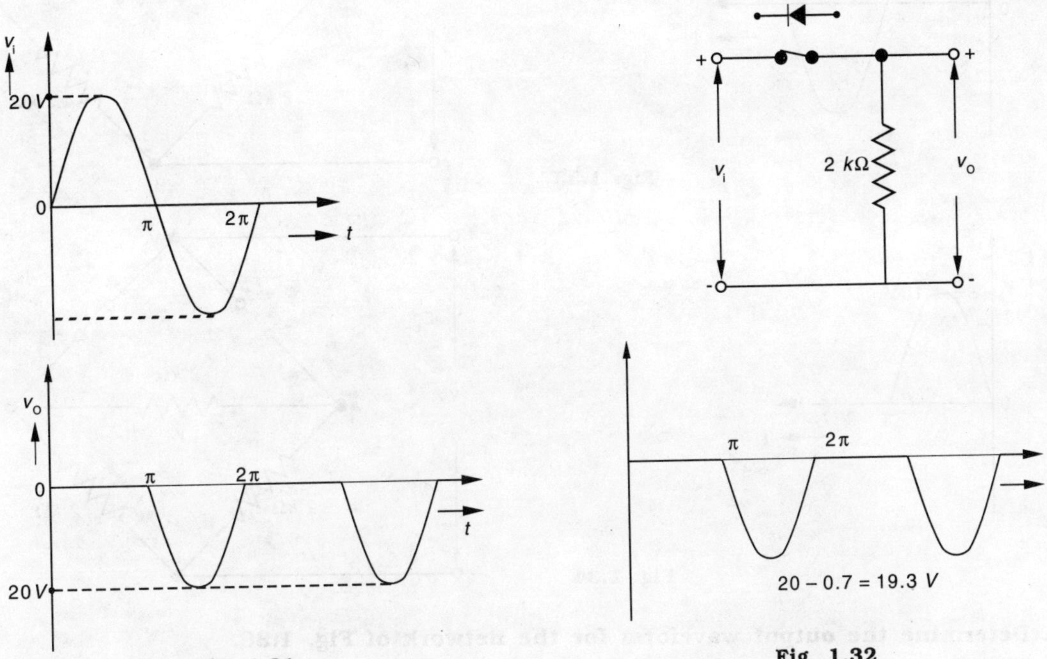


**Fig. 1.31** **Fig. 1.32**

(b) If the ideal diode is replaced by a silicon diode, the output will then appear as shown in Fig. 1.32.

$$V_{dc} = -0.318\ (V_m - 0.7) = -0.318\,(19.3) = \mathbf{-6.14\ V}$$

The resulting drop in *dc* level is 0.22*V* or about 3.5 %.

**2. Determine the output waveform for the network of Fig. 1.33 and calculate the output *dc* level and the required PIV of each diode.**

**Solution:** Since the input voltage is positive, the diode $D_1$ becomes reverse biased and diode $D_2$ gets forward-biased. Hence the network appears as shown in Fig. 1.34. Redrawing the network results in a configuration as shown in Fig. 1.35, where $v_O = (½)v_i$. or $V_{o(max)} = ½ V_{i(max)} = ½ \times 10 = 5V$, as shown in Fig. 1.35(a). During the negative part of the input the roles of the diodes will be interchanged and output voltage will appear as shown in Fig. 1.35(b).

The effect of removing two diodes from the bridge configuration was therefore to reduce the available *dc* level to the following:

$$V_{dc} = 0.636 \times 5 = \mathbf{3.18V}$$

or that available from a half-wave rectifier with the same input. The PIV is equal to the maximum voltage across *R*, which is 5*V* or half of that required for a half-wave rectifier with the same input.

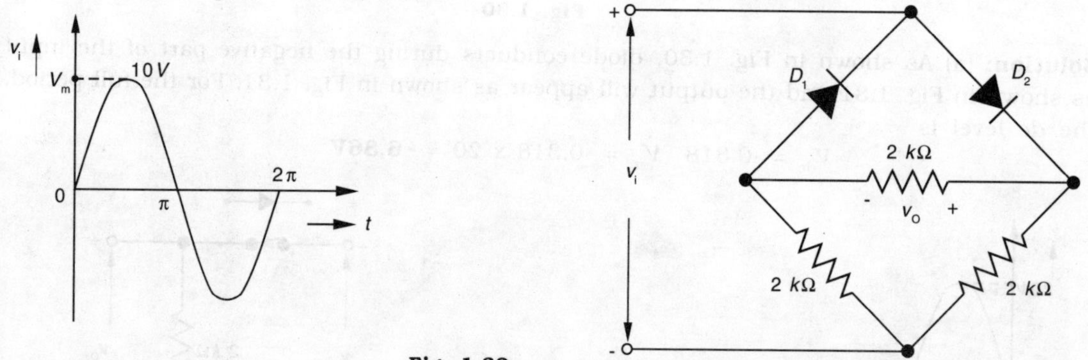


Fig. 1.33

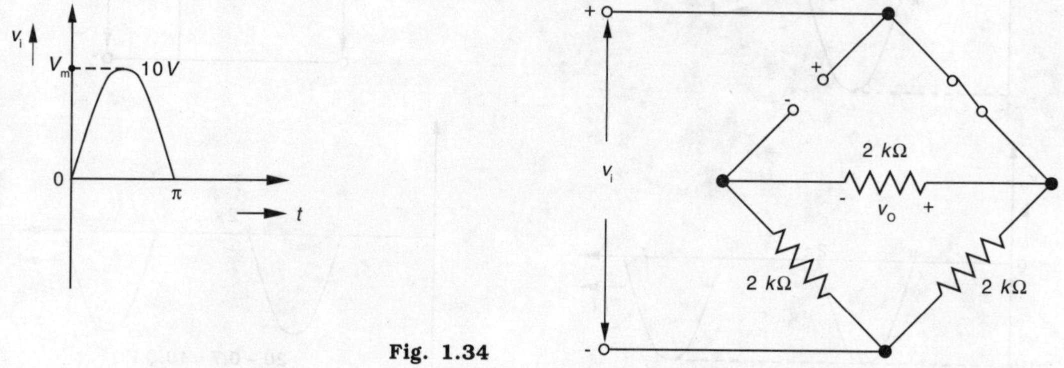


Fig. 1.34

**3. Determine the output waveform for the network of Fig. 1.36.**

**Solution:** The diode conducts only for the positive values of $v_i$ especially when we note the aiding effect of $V = 5V$. The network will then appear as shown in Fig. 1.37(a) and $v_o = v_i + 5V$. Substituting $i_d = 0$ at $v_d = 0$ for the transition voltage, we get the network of Fig. 1.37(b) and $v_i = -5V$.

For voltages more negative than -5*V* the diode does not conduct. The input and output voltages appear in Fig. 1.38.

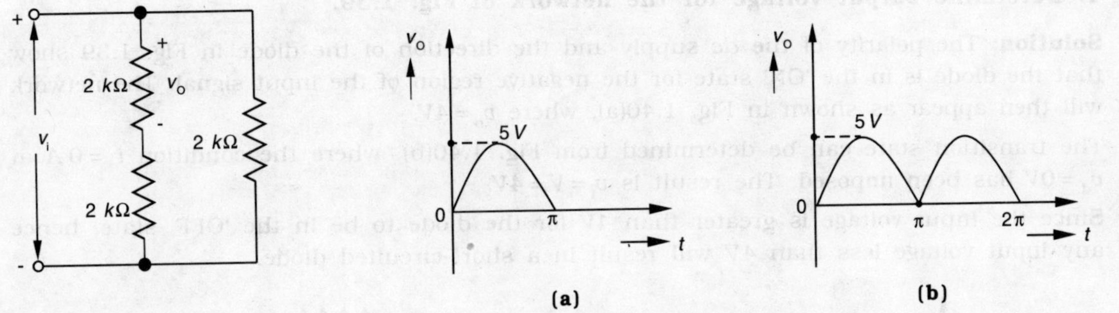

(a) (b)

Fig. 1.35

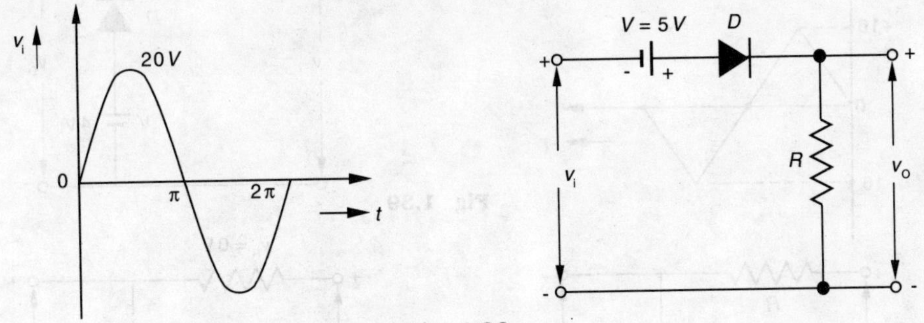

Fig. 1.36

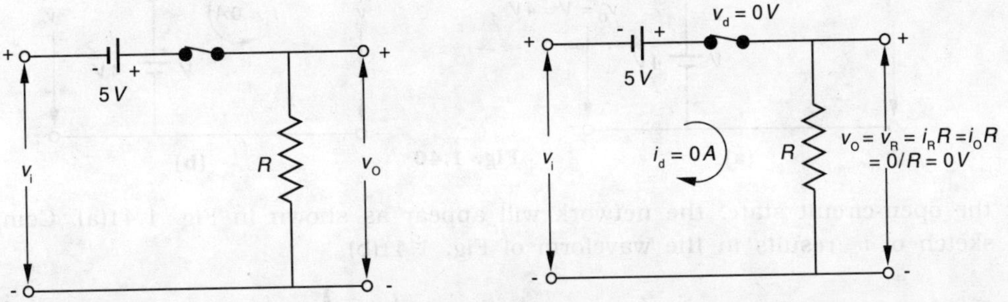

(a) Fig. 1.37 (b)

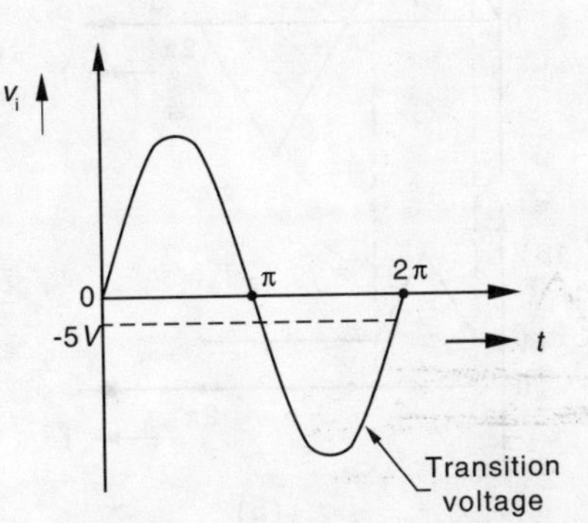

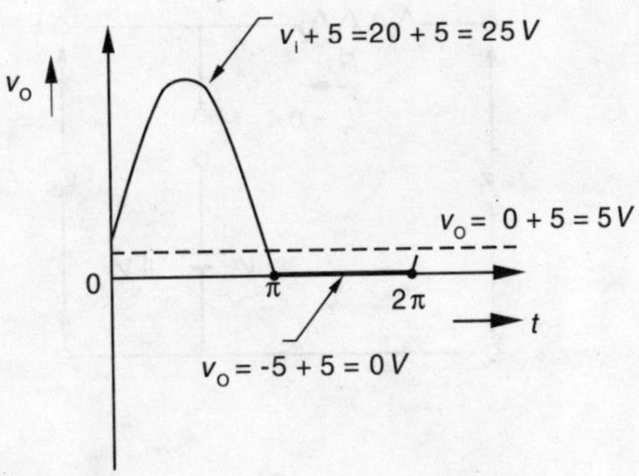

Fig. 1.38

**4. Determine output voltage for the network of Fig. 1.39.**

**Solution:** The polarity of the *dc* supply and the direction of the diode in Fig. 1.39 show that the diode is in the 'ON' state for the negative region of the input signal. The network will then appear as shown in Fig. 1.40(a), where $v_o = 4V$.

The transition state can be determined from Fig. 1.40(b), where the condition $i_d = 0\,A$ at $v_d = 0V$ has been imposed. The result is $v_i = V = 4V$.

Since the input voltage is greater than $4V$ for the diode to be in the 'OFF' state, hence any input voltage less than $4V$ will result in a short-circuited diode.

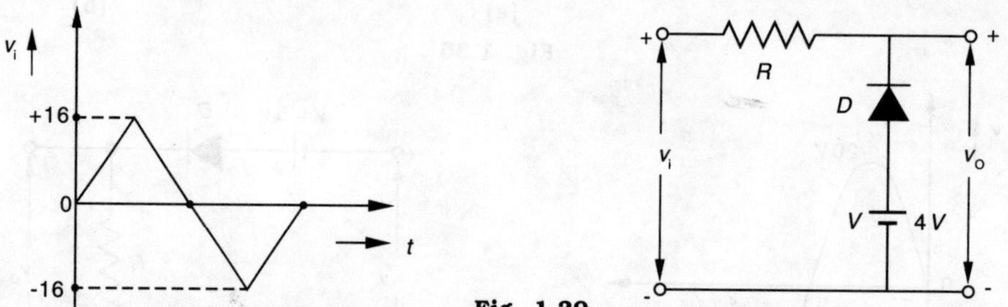


**Fig. 1.39**

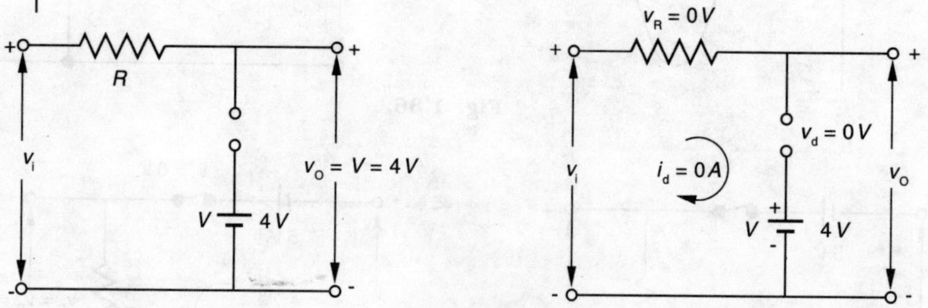


**(a)** **Fig. 1.40** **(b)**

For the open-circuit state, the network will appear as shown in Fig. 1.41(a). Completing the sketch of $v_o$ results in the waveform of Fig. 1.41(b).

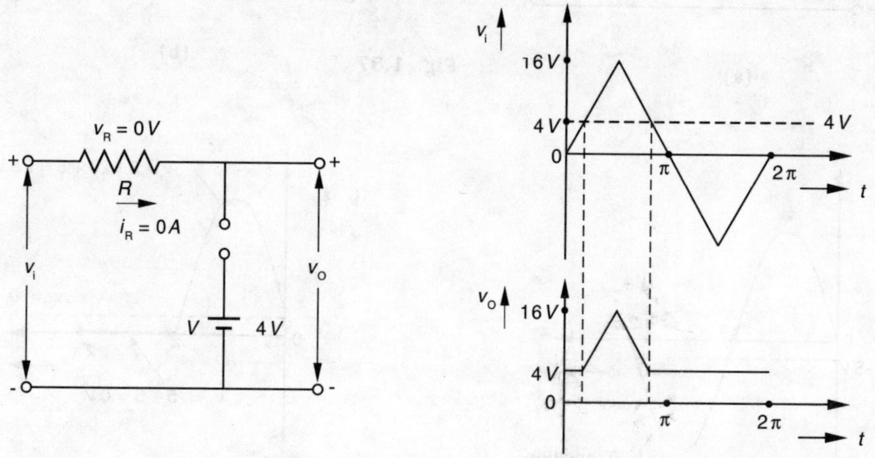


**(a)** **Fig. 1.41** **(b)**

**5. The silicon diode shown in Fig. 1.42 has a bulk resistance of 1Ω. The frequency of the 10 *mV* signal is so high that the reactance of the coupling capacitor may be taken as zero. Sketch the approximate waveform of the total voltage '*v*' across the diode.**

**Data:** Bulk resistance, $r_B = 1\Omega$ and signal amplitude $V_m = 10\ mV$

**Solution:** Apply superposition theorem to find the voltage drop *v'* across the diode.

### (a) DC equivalent circuit

From Fig. 1.42 it is seen that the circuit to the left of the point A is open to *dc* source, because *C* does not allow *dc* to pass through it. Thus the *dc* equivalent circuit is as shown in Fig. 1.43(a). In this circuit, the diode is forward-biased. Under this condition, the voltage drop across the diode is equal to 0.7*V dc* and current through it is given by

$$I = \frac{20 - 0.7}{20} = \mathbf{1}\boldsymbol{mA}$$

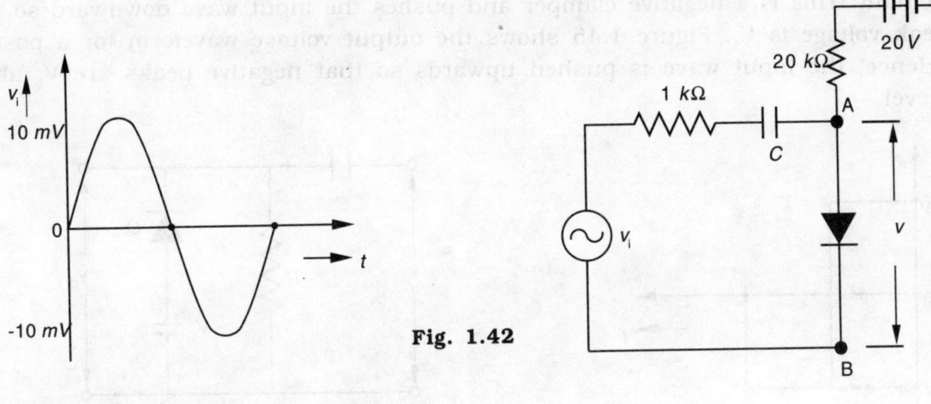


Fig. 1.42

### (b) AC equivalent circuit

In this case, the capacitor *C* and 20*V* battery are treated as shorts. The *ac* equivalent circuit is as shown in Fig. 1.43(b). For a given silicon diode, the junction resistance,

$$r_j = \frac{50}{1} = 50\ \Omega$$

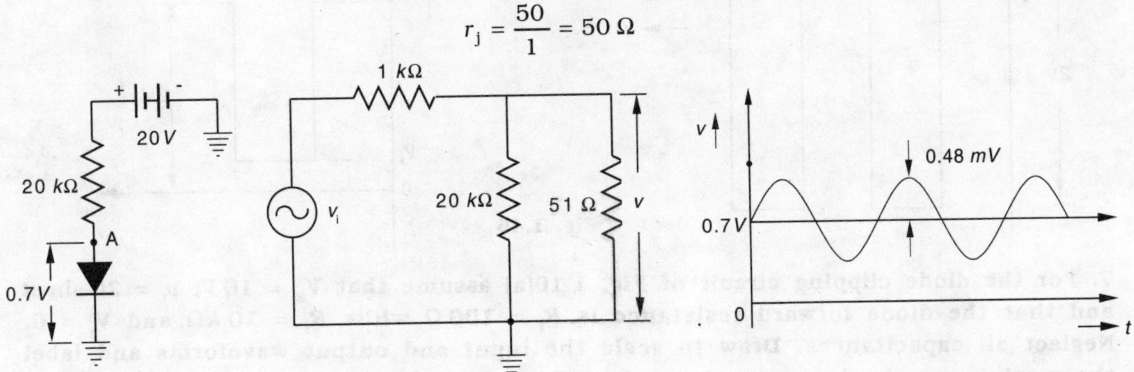


(a) (b) (c)

Fig. 1.43

AC resistance $r'_e = r_B + r_j = 1 + 50 = 51\Omega$

From Fig. 1.43(b), it is seen that resistances 20 $k\Omega$ and 51$\Omega$ are parallel for an *ac* signal. The equivalent resistance is given by

$$20\ k\Omega \,||\, 51\Omega \cong 51\Omega$$

Hence, 1 $k\Omega$ and 51$\Omega$ are put in series across the *ac* signal source of peak value 10 $mV$. The peak value of the *ac* voltage drop across 51$\Omega$ resistance

$$V_{51\Omega} = 10 \times \frac{51}{51 + 1000} = \mathbf{0.48\ mV}$$

The total voltage drop across the diode is equal to the sum of *dc* and *ac* voltage drops and is as shown in Fig. 1.43(c). The resulting waveform consists of a *dc* voltage of 0.7$V$ over which an *ac* voltage of peak value of 0.48 $mV$ is superimposed.

**6. Draw the output voltage waveform for the input wave shown in Fig. 1.44.**

**Solution:** Figure 1.44, shows the output voltage waveform for the input wave to the clamping circuit. This is a negative clamper and pushes the input wave downward so that positive peak voltage is $V_1$. Figure 1.45 shows the output voltage waveform for a positive clamper. Hence, the input wave is pushed upwards so that negative peaks are $V_1$ above the zero level.

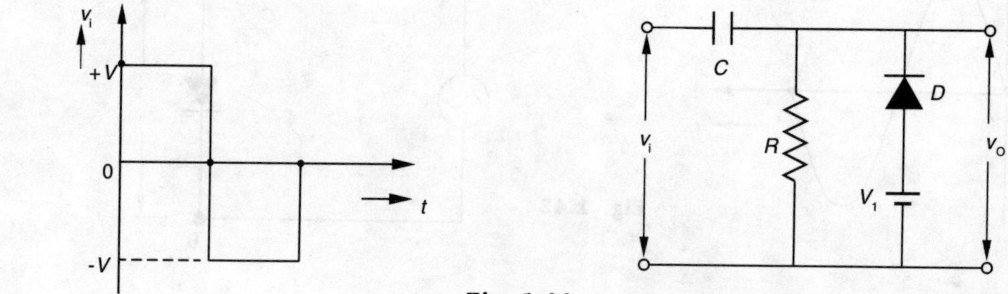


**Fig. 1.44**

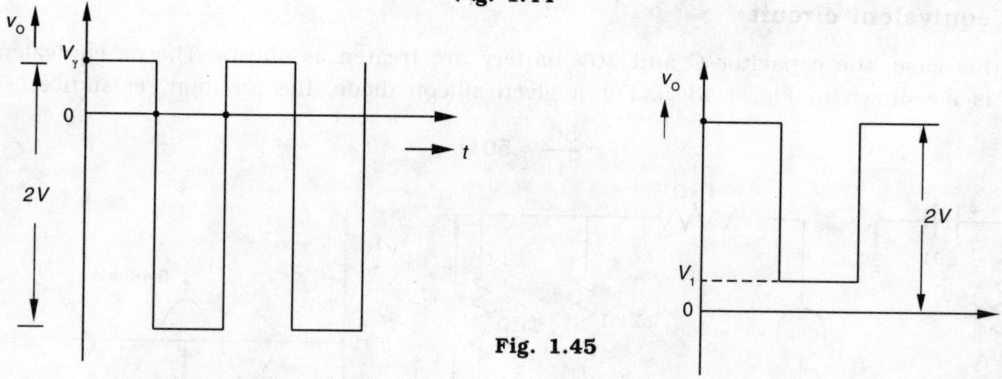


**Fig. 1.45**

**7. For the diode clipping circuit of Fig. 1.10(a) assume that $V_R$ = 10 $V$, $v_i$ = 20 sin$\omega t$ and that the diode forward resistance is $R_f$ = 100 $\Omega$ while $R_r$ = 10 $k\Omega$ and $V_\gamma$ = 0. Neglect all capacitances. Draw to scale the input and output waveforms and label the maximum and minimum values of (a) $R$ = 100 $\Omega$ (b) $R$ = 1 $k\Omega$ and (c) $R$ = 10 $k\Omega$.**

**Data:** $V_R$ = 10 $V$, $v_i$ = 20 sin$\omega t$, $R_f$ = 100 $\Omega$, $R_r$ = 10 $k\Omega$, $V_\gamma$ = 0

**To find:** $v_{o\,(max)}$ and $v_{o(min)}$

**Solution:** The diode will start conducting when $v_i$ = 20 sin$\omega t$ = 10, or $\omega t$ = 30°. When diode conducts

$$v_o = (v_i - 10)\frac{R_f}{R+R_f} + 10$$

$$v_{o\,(max)} = 10 \cdot \left(\frac{R+2R_f}{R+R_f}\right) \qquad \text{since } v_{i\,(max)} = 20V$$

when $v_i < 10\,V$, $$v_o = (v_i - 10)\frac{R_r}{R+R_f} + 10$$

or $$v_{i\,(min)} = 10\left(\frac{R-2R_r}{R+R_r}\right) \qquad \text{since } v_{i\,(min)} = 20V$$

(a) When $R$ = 100, $v_{o\,(min)}$ = **– 19.7 V**

(b) When $R$ = 1 $k\Omega$, $v_{o\,(min)}$ = **– 17.3 V**

(c) When $R$ = 10 $k\Omega$, $v_{o\,(min)}$ = **– 5 V**

**8. The clipping circuit employs temperature compensation. The *dc* voltage source $V_\gamma$ represents the diode offset voltage; otherwise the diodes are assumed to be ideal with $R_f$ = 0 and $R_r$ = ∞.**

**(a) Sketch the transfer curve $v_o$ versus $v_i$.**

**(b) Show that the maximum value of the input voltage $v_i$ so that the current in $D_2$ is always in the forward direction is $v_{i(max)} = V_R + \frac{R}{R'}(V_R - V_\gamma)$.**

**(c) What is the temperature dependent of the point on the input waveform at which clipping occurs?**

**Solution:**

(a) If $v_i \le V_R$, $v_o = V_R - V_\gamma$

If $v_i \ge V_R$, $v_o = v_i - V_\gamma$

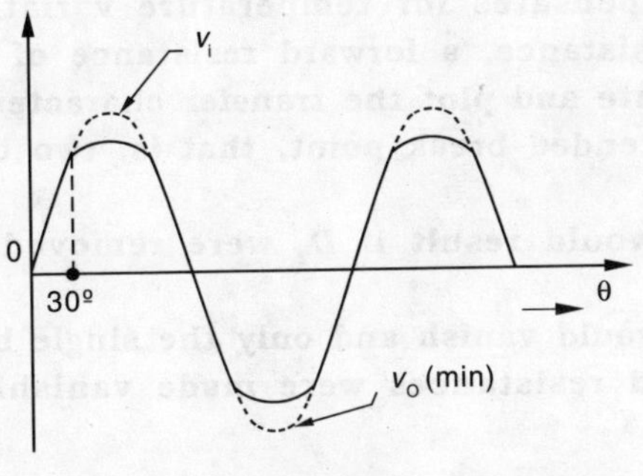


Fig. 1.46

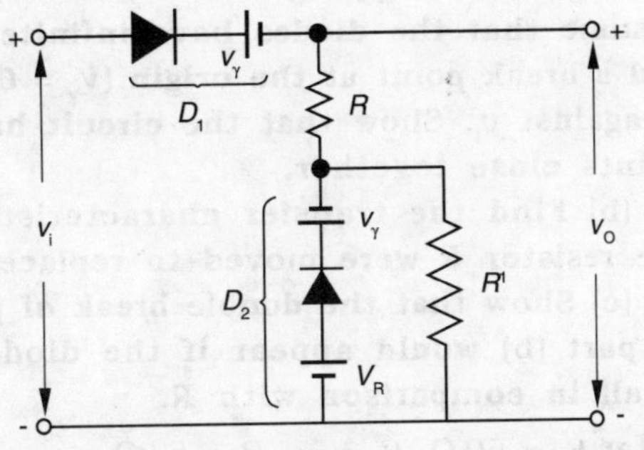


Fig. 1.47

(b) Assume both diodes conducting.

$$v_i = Ri_1 + V_R + V_\gamma - V_\gamma$$

$$V_R - V_\gamma = R'i_2$$

But $$i_1 = i_{D_1} = \frac{v_i - V_R}{R}$$

and $$i_{D_2} = i_2 - i_1 \qquad \therefore \frac{V_R - V_\gamma}{R'} = \frac{v_i - V_R}{R}$$

The current $i_{D_2}$ becomes negative when $v_i \geq \left(V_R + \frac{R}{R'}(V_R - V_\gamma)\right)$. Then $D_1$ is ON and $D_2$ is OFF. Therefore the maximum value of the input voltage $v_i$, so that the current in $D_2$ is always in forward direction is

$$v_{i(max)} = V_R + \frac{R}{R'}(V_R - V_\gamma)$$

Under the previous conditions $v_o = v_i - V_\gamma$

If $v_i < v_{i(max)}$ when both $D_1$ and $D_2$ are ON and $v_o = v_i - V_\gamma$

(c) The break point is $v_i = V_R$ and it is independent of temperature.

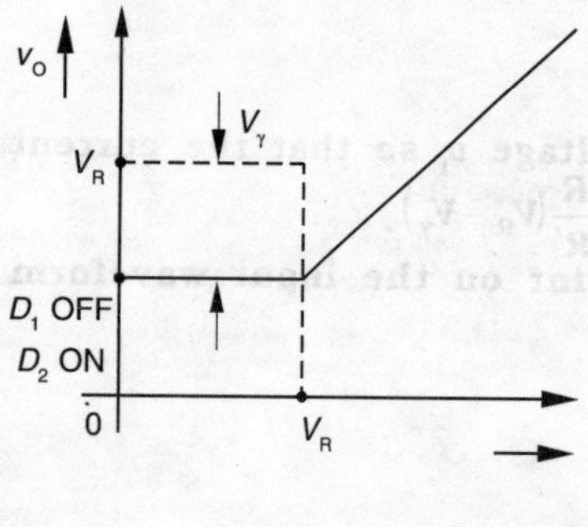


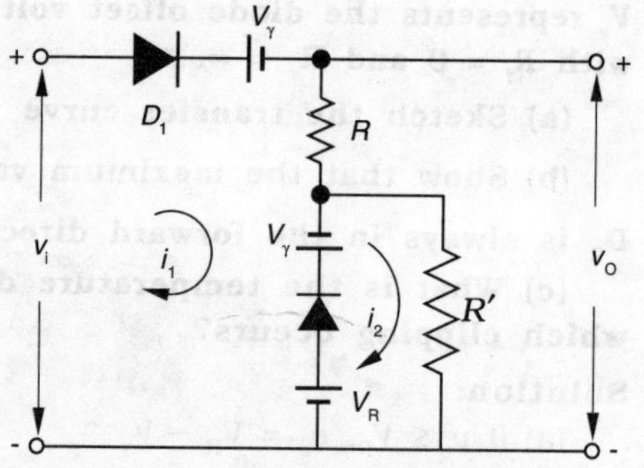


Fig. 1.48

**9. (a) In the clipping circuit shown, $D_2$ compensates for temperature variations. Assume that the diodes have infinite back resistance, a forward resistance of 50 Ω and a break point at the origin ($V_\gamma$ = 0). Calculate and plot the transfer characteristic $v_o$ against $v_i$. Show that the circuit has an extended break point, that is, two break points close together.**

**(b) Find the transfer characteristic that would result if $D_2$ were removed and the resistor $R$ were moved to replace $D_2$.**

**(c) Show that the double break of part (a) would vanish and only the single break of part (b) would appear if the diode forward resistances were made vanishingly small in comparison with $R$.**

**Data:** $R_f$ = 50 Ω, $R_r$ = ∞, $R$ = 5 kΩ

**To find:** (a) Two break points close together, (b) Transfer characteristic

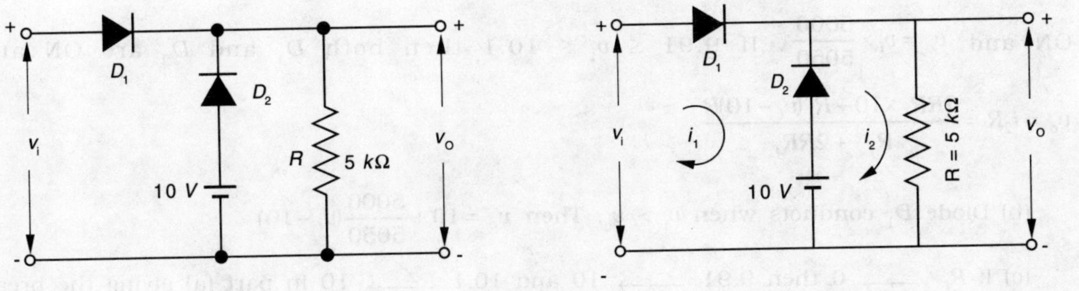


(a) **Fig. 1.49** (b)

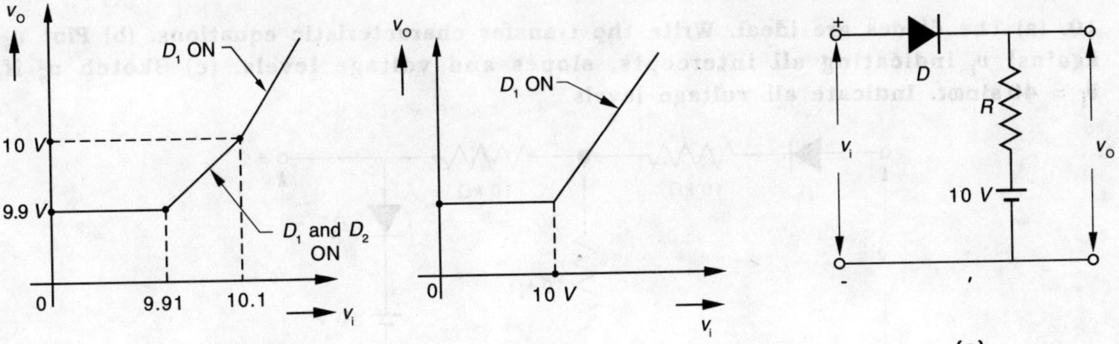


(a) (b) (c)

**Fig. 1.50**

**Solution:** Assume both diodes conduct. Then

$$v_i = 2R_f i_1 - i_2 R_f + 10 + V_\gamma - V_\gamma \longrightarrow (1)$$

$$10 - V_\gamma = i_2 (R + R_f) - i_1 R_f \longrightarrow (2)$$

Since $V_\gamma = 0$ and solving for $i_1$ and $i_2$, we have

$$i_1 = i_{D_1} = \frac{(v_i - 10)(R_f + R) + 10R_f}{R_f^{\ 2} + 2RR_f}$$

and

$$i_2 - i_1 = i_{D_2} = \frac{10R_f - R(v_i - 10)}{R_f^{\ 2} + 2RR_f}$$

If $i_{D_2} \geq 0$ then $D_2$ is ON when $v_i \leq 10 + \frac{R_f}{R} \times 10$

If $i_{D_1} \geq 0$ then $D_1$ is ON when $v_i \geq 10 - \left(\frac{R_f}{R + R_f} \times 10\right)$

Hence one break point occurs at $v_i$ = 10. 1$V$ and another at $v_i = 9.91\,V$. If $v_i < 9.91\,V$ then $D_1$ is OFF and $D_2$ is ON and $v_o = 10 \times \frac{5000}{5050}$. If $v_i > 10.1\,V$ then $D_2$ is OFF and $D_1$ is

ON and $v_o = v_i \times \frac{5000}{5050}$. If $9.91 \le v_i \le 10.1$ then both $D_1$ and $D_2$ are ON and

$$v_o = i_2 R = \frac{2RR_f \times 10 + R_f(v_i - 10)R}{R_f^2 + 2RR_f}$$

(b) Diode $D_1$ conducts when $v_i > v_o$. Then $v_o = 10 + \frac{5000}{5050}(v_i - 10)$

(c) If $R_f \longrightarrow 0$ then $9.91 \longrightarrow 10$ and $10.1 \longrightarrow 10$ in part (a) giving the break point of part (b).

**10. (a) The diodes are ideal. Write the transfer characteristic equations. (b) Plot $v_o$ against $v_i$ indicating all intercepts, slopes and voltage levels. (c) Sketch $v_o$ if $v_i$ = 40 sin$\omega t$. Indicate all voltage levels.**

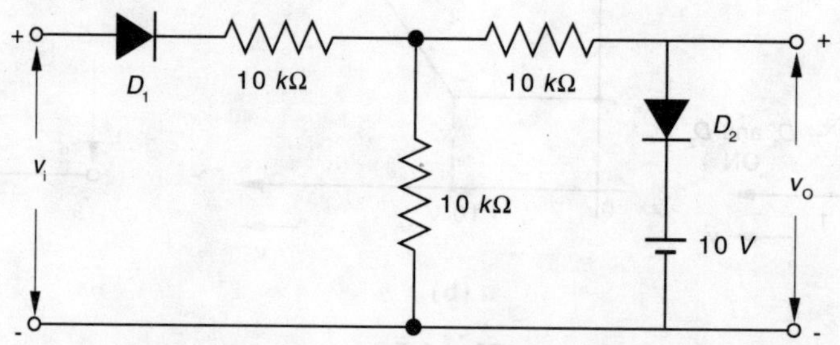


**Fig. 1.51**

**Solution:** (a) Diode $D_1$ conducts when $v_i > 0$. $D_2$ conducts when $v_o > 10\,V$. Thus

(i) $v_o = 0$ for $v_i \le 0$

(ii) $v_o = v_i/2$ for $0 \le v_i \le 20$ V

(iii) $v_o = 10\,V$ for $20\,V \le v_i$

(b)

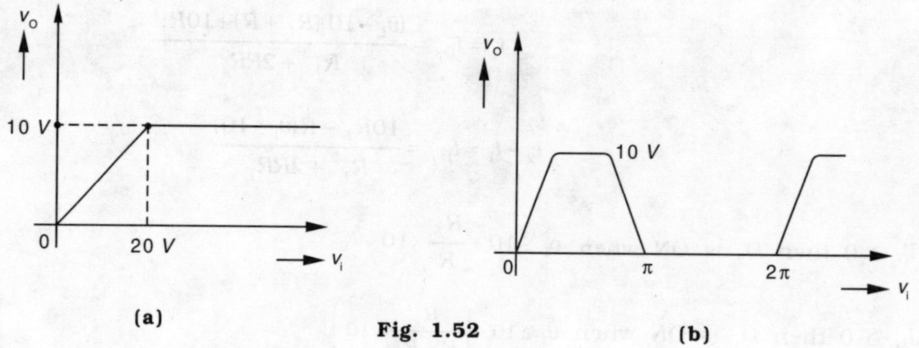


**Fig. 1.52**

(c) When $0 \le v_i \le 20\,V$ then $v_o = 20 \sin\omega t$. For $v_i = 20\,V$ that is at $\sin\omega t = \frac{20}{40}$ or $\omega t = 30^o$. $D_2$ conducts and $v_o = 10\,V$. For $v_i < 0$, $v_o = 0$.

**11. A full-wave single-phase rectifier consists of a double-diode vacuum tube, the internal resistance of each element of which may be considered to be constant and equal to 500 Ω. These feed into a pure resistance load of 2 kΩ. The secondary transfer voltage to centre tap is 280V. Calculate (a) the *dc* load current, (b) the direct current in each tube, (c) the *ac* voltage across each diode, (d) the *dc* output power, (e) the percentage regulation.**

**Data:** $R_f = 500\ \Omega,\ V = 280V,\ R_L = 2\ k\Omega$

**To find:** $I_{dc}$, $I_{dc(tube)}$, $V_{rms(tube)}$, $P_{dc}$ and % regulation.

**Solution:**

(a) $$I_{dc} = \frac{2I_m}{\pi} = \frac{2V_m}{\pi(R_L + R_f)}$$

$$= \frac{2 \times 280 \times \sqrt{2}}{\pi(2000 + 500)} = \mathbf{101\ mA}$$

(b) $$I_{dc(tube)} = \frac{1}{2} I_{dc} = \frac{101}{2} = \mathbf{50.5\ mA}$$

(c) The voltage to centre tap is impressed across $R_f$ in series with $R_L$. Hence the voltage across the conducting tube is sinusoidal with a peak value $V_m \frac{500}{2500} = 0.2 V_m$.

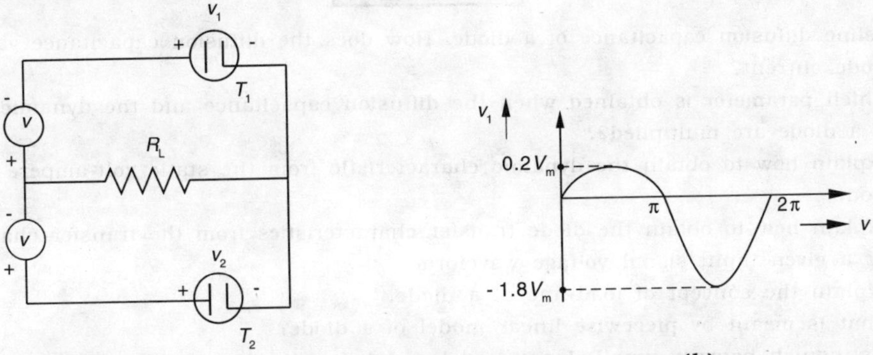


(a) **Fig. 1.53** (b)

During non-conducting of $V_1$ we see by transversing the outside path of the circuit sketched that $v_i = -2V - V_2$. Since $V_2$ is now conducting its peak value is $0.2\ V_m$ and that of $v_i$ is $-2\ V_m + 0.2\ V_m = -1.8\ V_m$. Thus, the voltage $V_1$ across $T_1$ is as shown, and its *ac* value is given by

$$V_{rms}^2 = \frac{1}{2\pi}\int_0^{\pi} (0.2V_m)^2 \sin^2\theta\, d\theta + \frac{1}{2\pi}\int_{\pi}^{2\pi} (-1.8V_m)^2 \sin^2\theta\, d\theta = \frac{V_{rms}^2}{4}[(0.2)^2 + (1.8)^2]$$

$$V_{rms} = 0.905\ V_m = 0.905 \times 280 \times \sqrt{2} = \mathbf{358\ V}$$

(d) $$P_{dc} = I_{dc}^2\ R_L = (0.101)^2\,(2000) = \mathbf{20.4\ W}$$

(e) $$\text{Percentage regulation} = \frac{R_f}{R_L} \times 100\ \% = \frac{500}{2000} \times 100\ \% = \mathbf{25\ \%}$$

**12. A 1-*mA dc* meter whose resistance is 10 Ω is calibrated to read rms volts when used in a bridge circuit with semiconductor diodes. The effective resistance of each element may be considered to be zero in the forward direction and infinite in the inverse direction. The sinusoidal input voltage is applied in series with a 5kΩ resistance. What is the full-scale reading of this meter?**

**Data:** $R_m = 10\,\Omega$, $R_L = 5\,k\Omega$, $I_{dc} = 1\,mA$

**To find:** $V_{rms}$

**Solution:**

$$I_{dc} = \frac{2I_m}{\pi} = \frac{2V_m}{\pi R_L}$$

$$I_{dc} = \frac{2\sqrt{2}V_{rms}}{\pi(5010)} = 1mA = 1\times10^{-3}$$

$$V_{rms} = \frac{5010\pi}{2\sqrt{2}} = \mathbf{5.56V}$$

with the reverse polarity, (c) Repeat parts (a) and (b) if the Zener breakdown voltage is 10*V*.

## QUESTIONS

1. Define diffusion capacitance of a diode. How does the diffusion capacitance vary with *dc* diode current?
2. Which parameter is obtained when the diffusion capacitance and the dynamic resistance of a diode are multiplied?
3. Explain how to obtain the dynamic characteristic from the static volt-ampere curve of a diode.
4. Explain how to obtain the diode transfer characteristics from the transfer characteristics for a given input signal voltage waveform.
5. Explain the concept of load-line of a diode.
6. What is meant by piecewise linear model of a diode?
7. Distinguish between small-signal model and large-signal model.
8. The diode is a binary device. Justify.
9. You are given the *V-I* output characteristics in graphical form of a new device (a) Sketch the circuit using this device which will require a load-line construction to determine *i* and *v*. (b) Is the load line vertical, horizontal at 135° or 45° for infinite load resistance? (c) For zero load resistance?
10. (a) Explain and draw the piecewise linear volt-ampere characteristics of a *p-n* diode. (b) What is the circuit model for the ON state? (c) The OFF state?
11. Explain briefly the analysis of diode circuits using the piecewise linear diode model.
12. For a circuit consisting of a diode, a resistance and a signal source in series, define (a) static characteristics (b) dynamic characteristics and (c) transfer characteristics. (d) What is the correlation between (b) and (c)?
13. What is meant by breakdown region of a diode?
14. In analysing a circuit containing several diodes by the piecewise linear method, assume that certain of the diodes are ON and others are OFF. Explain how to determine whether or not the assumed state of each diode is correct.

15. Consider a series circuit consisting of a diode, a resistance, $R$, a reference battery $V_R$ and an input signal $v_i$. The output is taken across $R$ and $V_R$ in series. Using the piecewise linear model, draw the transfer characteristic if the diode is connected to the positive terminal of the battery.
16. Repeat question no. 15, if the diode is connected to the negative terminal of the battery.
17. What is a clipper circuit?
18. Explain with relevant diagrams a diode clipping circuit that transmits the part of the waveform more negative than $V_R + V_\gamma$ using piecewise linear transmission characteristics.
19. Explain with relevant diagrams a diode clipping circuit that transmits the part of the waveform more positive than $V_R - V_\gamma$ using piecewise linear transmission characteristics.
20. What is a comparator circuit? How does a comparator circuit differ from a clipping circuit?
21. What is a rectifier circuit?
22. Define in words and as an equation (a) *dc* current $I_{ac}$ (b) *dc* voltage $V_{dc}$ (c) *ac* current $I_{rms}$ for a half-wave/full-wave rectifier.
23. Explain with relevant diagrams the circuit of a half-wave rectifier. Derive the expression for (a) the *dc* current (b) the *dc* load voltage (c) the *ac* rms current.
24. Explain with relevant diagrams the circuit of a full-wave rectifier. Derive the expressions for (a) the *dc* current (b) the *dc* load voltage (c) the *ac* rms current and (d) the *dc* diode voltage.
25. Define regulation. Derive the expression for regulation of a full wave rectifier circuit.
26. Define Thevenin's theorem. Draw the Thevenin's model for a full-wave rectifier.
27. Define peak inverse voltage. What is value of peak inverse voltage for a half-wave/full-wave/bridge rectifier circuit using ideal diodes.
28. Explain with circuit diagram the operation of a bridge rectifier.
29. Explain with circuit diagram the operation of a rectifier meter (voltmeter).
30. Explain with circuit diagram the operation of a voltage multiplier.
31. (a) Explain multiplier with circuit diagram, the half-wave capacitor rectifier. (b) At no-load draw the steady-state voltage across the capacitor and also across the load?
32. (a) Explain with circuit diagram the operation of a full-wave capacitive rectifier. (b) Draw the output voltage under load. Indicate over what period of time the diode conducts. (c) Indicate the diode current waveform superimposed upon the output waveform.
33. (a) Derive an expression for (a) the diode current and (b) How is the cut-in angle found?
34. (a) Explain with circuit diagram the operation of full-wave capacitor filter. Draw the approximate output waveform if the circuit uses a large value of $C$.
    (b) Derive an expression for the peak ripple voltage.
    (c) Derive the Thevenin's model for this rectifier.
35. Is it possible to use half-wave capacitor filtered rectifier as a peak detector. Justify.
36. What is a clamping circuit?
37. Write the advantages and disadvantages of a full-wave capacitor filter.
38. Define amplitude modulation and detection.

## EXERCISES

1. A *p-n* Ge junction diode at room temperature has a reverse saturation current of 10 $\mu A$, negligible ohmic resistance and a Zener breakdown voltage of 100$V$. A 1 $k\Omega$ resistor is in series with this diode, and a 30 $V$ battery is impressed across this combination. Find the current (a) if the diode is forward-biased, (b) if the battery is inserted into the circuit. **[Ans: (a) 29.8 *mA* (b) -10 $\mu$A (c) 20 *mA*]**

2. Each diode is described by a linearized volt-ampere characteristic, with incremental resistance $r$ and offset voltage $V_\gamma$. Diode $D_1$ is Ge with $V_\gamma = 0.2V$ and $r = 20\ \Omega$, whereas $D_2$ is Si with $V_\gamma = 0.6V$ and $r = 15\ \Omega$. Find the diode currents if (a) $R = 10\ k\Omega$ and (b) $R = 1 k\Omega$.
**[Ans: (a) 9.97 *mA* and (b) 56.5 *mA* and 42.3 *mA*]**

3. Calculate the breakdown region over which the dynamic resistance of a diode is multiplied by factor of 1000. **[Ans: 359 *mV* for Si and 180 *mV* for Ge at room temperature]**

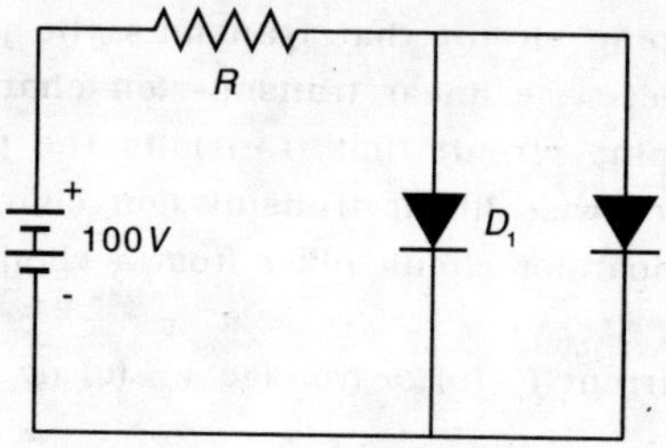


**Fig. 1.54**

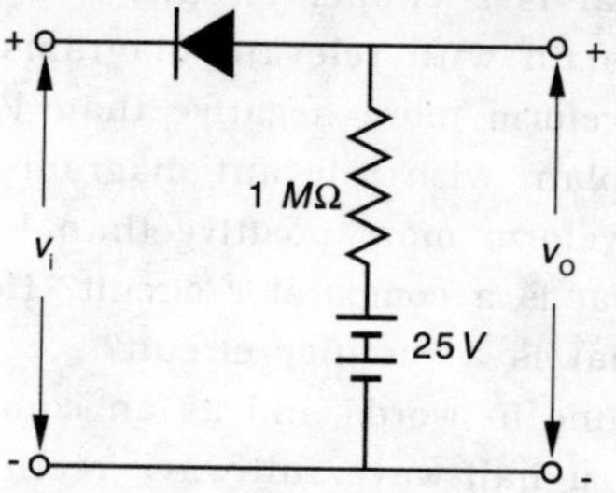


**Fig. 1.55**

4. For the diode clipping circuit of 1.10(a) assume that $V_R = 10V$, $v_i = 20$ sinωt and that the diode forward resistance is $R_f = 100\ \Omega$ while $R_\gamma = \infty$ and $V_\gamma = 0$. Neglect capacitances. Draw to scale the input and output waveforms and label the maximum and minimum values if (a) $R = 100\ \Omega$ (b) $R = 1\ k\Omega$ and (c) $R = 10\ k\Omega$.
**[Ans: (a) $V_{O(max)} = 15V$, $V_{O(min)} = -20V$ (b) $V_{O(max)} = 10.9V$, $V_{O(min)} = -20V$ and (c) $V_{O(max)} = 10.1V$ and $V_{O(min)} = -20V$]**

5. A symmetrical 5 *kHz* square wave whose output varies between +10 and -10*V* is impressed upon the clipping circuit shown. Assume $R_f = 0$, $R_r = 2\ M\Omega$ and $V_\gamma = 0$. Sketch the steady state output waveform, indicating numerical values of the maximum, minimum and constant portions.
**[Ans: $v_i = v_o < 2.5V$]**

6. For the clipping circuits shown in 1.12(b) and (d) draw the transfer characteristic $v_o$ versus $v_i$ taking into account $R_f$ and $V_\gamma$ and considering $R_r = \infty$.
**[Ans: See Fig. 1.56(a) and (b)]**

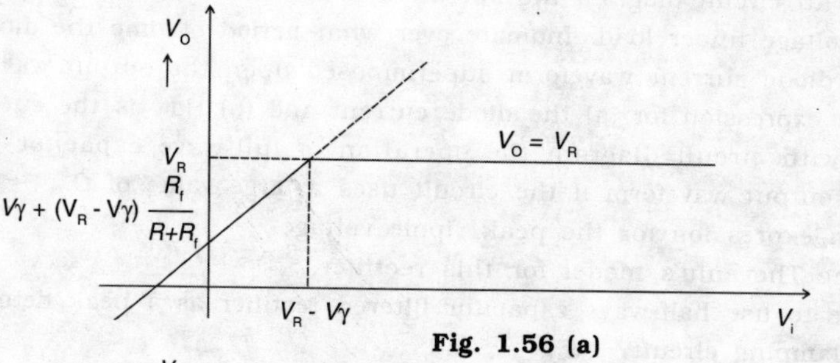


**Fig. 1.56 (a)**

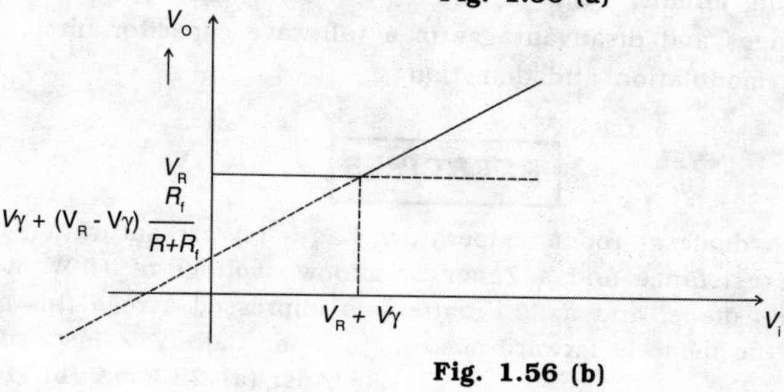


**Fig. 1.56 (b)**

7. (a) In the peak clipping circuit shown, add anode diode $D_2$ and a resistor $R'$ in a manner that will compensate for drift with temperature.

(b) Show that the break point of the transmission curve occurs at $V_{R'}$. Assume $R_r >> R >> R_f$.

(c) Show that if $D_2$ is always to remain in conduction it is necessary that $v_i < v_{i,max} = V_R + \frac{R}{R'}(V_R - V_\gamma)$

**[Ans: (a) see Fig.1.57(b)]**

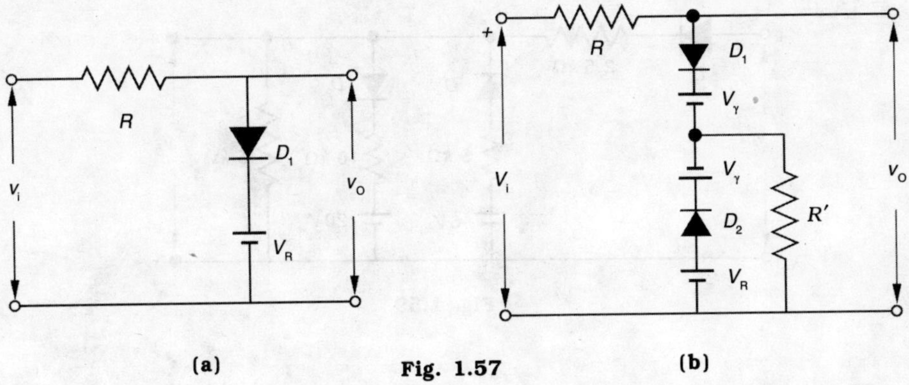


**Fig. 1.57**

8. (a) The input voltage $v_i$ to the two-level clipper shown in part (a) Fig. 1.58 varies linearly from 0V to 150V. Sketch the output voltage $v_o$ to the same time scale as the input voltage. Assume ideal diodes. (b) Repeat (a) for the circuit shown in part (b) of Fig. 1.58.

9. The circuit shown in Fig. 1.11(a) is used to "square" a 10 kHz input sinewave whose peak value is 50V. It is desired that the output voltage waveform be flat for 90 per cent of the time. Diodes are used having a forward resistance of 100 Ω and a backward resistance of 100 kΩ.

(a) Find the values of $V_{R_1}$ and $V_{R_2}$.

(b) What is a reasonable value to use R? **[Ans: (a) $V_{R_1} = V_{R_2}$ = 7.8V (b) 2.5kΩ]**

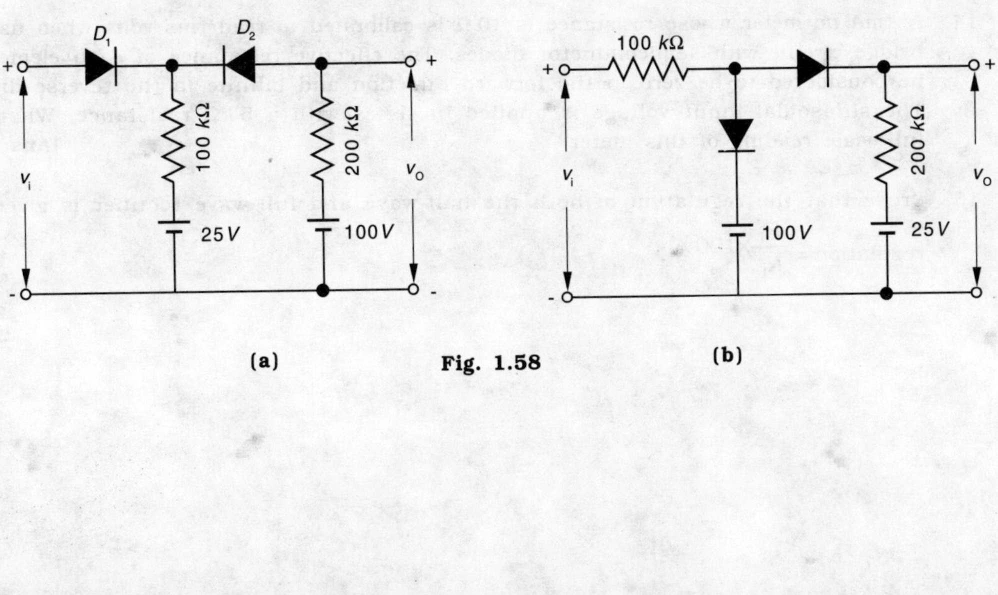


**Fig. 1.58**

10. Assume that the diodes are ideal. Make a plot of $v_o$ against $v_i$ for the range of $v_i$ from 0 to 50V. Indicate all slopes and voltage levels. Indicate, for each region, which diodes are conducting.

**[Ans: (i) $D_1$, $v_o = 3V$ (ii) $D_1$ and $D_3$, $\frac{1}{2}(v_i - 3) + 3$ (iii) $D_3$, $2v_i/3$ (iv) $D_2$ and $D_3$, $\frac{4}{7}\left(v_i - \frac{20}{3}\right) + \frac{20}{3}$]**

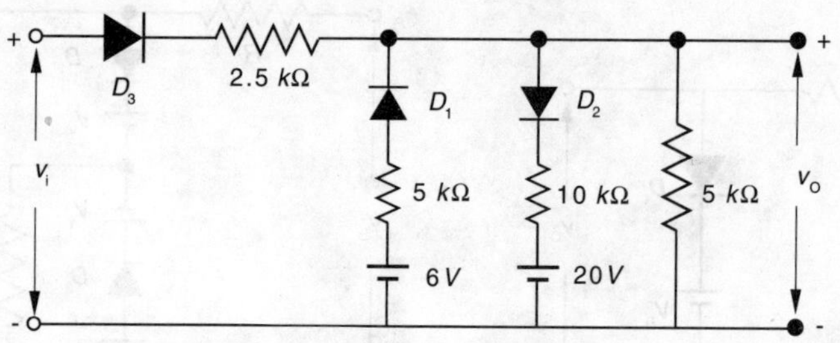


**Fig. 1.59**

11. The diode resistor comparator circuit of Fig. 1.14 is connected to a device which responds when the comparator output attains a level of 0.1V. The input is a ramp which rises at the rate 10V/μs. The Ge diode has a reverse saturation current of 1μA. Initially $R = 1k\Omega$ and the 0.1V output level is attained at a time $t = t_1$. If we now set $R = 100\ k\Omega$, what will be the corresponding change in $t_1$? $V_R = 0$. **[Ans: $\Delta t = 10.2\ ns$]**

12. A diode whose internal resistance is 20 Ω is to supply power to a 1000 Ω load from a 110V (rms) source of supply. Calculate (a) the peak load current, (b) the *dc* load current (c) the *ac* load current (d) the *dc* diode voltage (e) the total input power to the circuit (f) the percentage regulation from no load to the given load.
**[Ans: (a) 152.5 *mA* (b) 48.5 *mA* (c) 76.2 *mA* (d) -48.5V (e) 5.92 *W* (f) 2.06%]**

13. Show that the maximum of *dc* output power $P_{dc} = V_{dc} I_{dc}$ in a half-wave single-phase circuit occurs when the load resistance equals the diode resistance $R_f$.

14. A 1*mA* *dc* meter whose resistance is 10 Ω is calibrated to read rms volts when used in a bridge circuit with semiconductor diodes. The effective resistance of each element may be considered to be zero in the forward direction and infinite in the reverse direction. The sinusoidal input voltage is applied in series with a 5 kΩ resistance. What is the full-scale reading of this meter? **[Ans: 5.56V]**

15. Prove that the regulation of both the half-wave and full-wave rectifier is given by % regulation $= \frac{R_f}{R_L} \times 100\%$.

# 2 TRANSISTOR BIASING

## Chapter Outline

- Introduction
- The Operating Point
- Bias Stability
- Self Bias Or Emitter Bias
- Stabilization Against Variations In $I_{co}$, $V_{be}$ And $\beta$
- General Remarks On Collector-Current Stability
- Bias Compensation
- Biasing Techniques For Linear Integrated Circuits

## INTRODUCTION

The most important amplifying device is the **bipolar junction transistor.** The bipolar junction transistor consists of three doped semiconductor regions (emitter, base and collector) separated by two *p-n* junctions. There are two types of junction transistor. One of them consists of two *p*-regions separated by an *n*-region (*p-n-p*) and the other type consists of two *n*-regions separated by a *p*-region (*n-p-n*).

Apart from the transistor being used as an amplifying device it can be used as a switch, a current source, linear circuit and also used in radio, TV and communication circuits. To use devices for amplification of voltage or current, or as control (ON or OFF) elements it is necessary **to bias** the device.

DC biasing is a static operation. It deals with setting a fixed level of the current which should flow through the transistor with a desired fixed voltage drop across the transistor junctions. Usually, the currents $I_B$ and $I_C$ and the voltages $V_{BE}$ and $V_{CE}$ are required to be set by the biasing circuit. The proper values of these currents and voltages allow a transistor to amplify the weak signals faithfully. If an amplifier is not biased with correct *dc* voltages on the input and output, it can go into saturation or cutoff when an input signal is applied.

**The proper flow of zero signal collector current and the maintenance of proper collector emitter voltage during the passage of signal is known as transistor biasing.**

The aim of biasing is to achieve a certain condition of current and voltage called the **operating point (Quiescent** point or $Q$-point). The method of establishing the operating point is called **biasing.** However, the operating point shifts with changes in temperature $T$ because the transistor parameters such as $\beta$, $I_{CO}$ and $V_{BE}$ are functions of $T$.

## THE OPERATING POINT

The transistor functions linearly when it is made to operate in its active region. To establish an operating point in the active region it is necessary to provide appropriate direct potentials and currents using external sources. Once an operating point $Q$ is established, time-varying excursions of the input signal should cause an output signal of the same waveform.

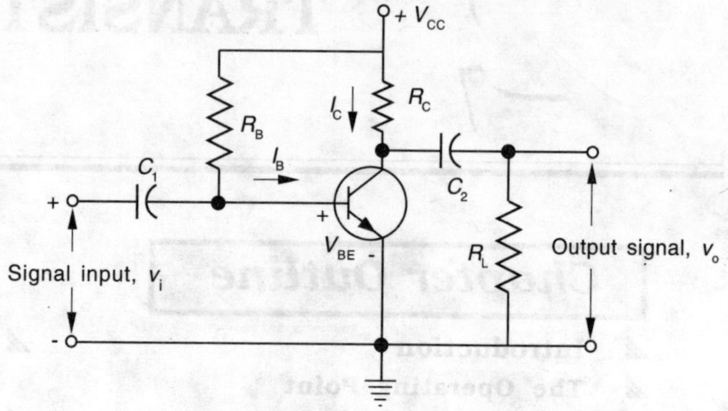

**Fig. 2.1 The fixed bias circuit**

Consider a common-emitter circuit as shown in Fig. 2.1 The transistor is biased with $V_{CC}$ to obtain certain values of $I_C$, $I_B$, $I_E$, $V_{BE}$ and $V_{CE}$. Figure 2.2 is the family output characteristics. The transistor cannot be operated everywhere in the active region because the various transistor ratings such as $P_{C\,(max)}$, $V_{C\,(max)}$, $I_{C\,(max)}$ and $V_{EB\,(max)}$ limit the range of useful operation. Figure 2.2 shows the three of these bounds on typical collector characteristics, $I_{C\,(max)}$, $V_{C\,(max)}$, and $P_{C\,(max)}$.

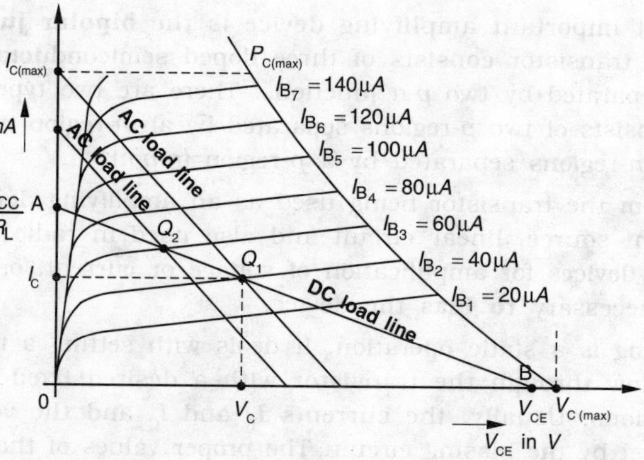

**Fig. 2.2 Common-emitter collector characteristics**

**Capacitive Coupling:** The capacitances $C_1$ and $C_2$ are called **coupling** or **blocking** capacitors as they block *dc* voltages and pass signal voltages. The capacitances are chosen large enough so that for the lowest frequency of signal they act as short.

**The Static and Dynamic Load Lines:** Under *dc* condition $C_2$ acts as an open-circuit. Applying Kirchhoff's voltage law to the collector circuit ($R_L = \infty$), we have

$$V_{CC} = I_C R_C + V_{CE} \longrightarrow (1)$$

where $$I_C R_C = V_{CC} - V_{CE}$$

or $$I_C = \frac{V_{CC}}{R_C} - \frac{V_{CE}}{R_C} = \left(-\frac{1}{R_C}\right)V_{CE} + \frac{V_{CC}}{R_C} \longrightarrow (2)$$

The above equation is in the form of $y = mx + c$ which is the equation of a straight line. Here $m = -\frac{1}{R_C}$ is the slope of the line and $c = V_{CC}/R_C$ is the intercept on Y-axis. As $V_{CC}$ and $R_C$ are constants, it is a first degree equation line called the ***DC* load line** equation. The load line determines the locus of $V_{CC}$ and $I_C$ for any given value of $R_C$. The two points needed to draw the *dc* load line are located as follows:

(i) When $V_{CE} = V_{CC}$ ; $I_C = 0$ and

(ii) When $V_{CE} = 0$ ; $I_C = \frac{V_{CC}}{R_C}$

These two points can be located on the output characteristics. By joining them *dc* load line is obtained. If $R_L = \infty$ and if the input signal is large and symmetrical, we must locate the operating point $Q_1$, at the centre of the load line. Hence the collector voltage and current vary approximately symmetrically around the quiescent values $V_C$ and $I_C$ respectively. If $R_L \neq \infty$, a dynamic (*ac*) load line must be drawn. Since $C_2$ acts as a short at the signal frequency, the effective load $R_L'$ at the collector is given by $R_L' = R_C \parallel R_L$. The dynamic load line must be drawn through $Q_1$ and must have a slope corresponding to $R_L'$ (Fig. 2.2). It is seen that the input signal may swing a maximum of 40 μ*A* around $Q_1$. If the base current decreases by more than 40 μ*A*, the transistor is driven to cutoff.

In order to avoid cutoff, the quiescent point must be located at a higher current $I_{B_2}$. This choice of $Q_2$ allows an input peak current swing of about 60 μ*A* and an output without distortion.

## The Fixed Bias Circuit

Applying KVL to base circuit, we have

$$V_{CC} = I_B R_B + V_{BE}$$

or $$I_B = \frac{V_{CC} - V_{BE}}{R_B} = I_{B_2} \longrightarrow (3)$$

The operating point $Q_2$ can be established by noting the required current $I_{B_2}$ and choosing the resistance $R_B$ so that the base current $I_B$ is equal to $I_{B_2}$.

Since $V_{CC}$ is much larger than $V_{BE}$, we have

$$I_B = \frac{V_{CC}}{R_B} \longrightarrow (4)$$

The supply voltage $V_{CC}$, base-emitter voltage $V_{BE}$ and a resistor $R_B$ are fixed, $I_B$ is also constant. Hence this circuit is called **fixed bias circuit.**

In summary, it is clear that the selection of an operating point $Q$ depends upon the *ac* and *dc* loads, the available power supply, the maximum transistor ratings, the peak signal excursions to be handled by the stage and the tolerable distortion.

## BIAS STABILITY

We have examined the problem of selecting an operating point $Q$ on the load line of the transistor. Now consider some of the problems of maintaining the operating point stable.

Two reasons can be attributed to the shift of operating point. Firstly, the transistor parameters such as $\beta \approx h_{FE}$, $V_{BE}$ are not the same for every transistor even though they are of the same type. Secondly, the transistor parameters are temperature dependent.

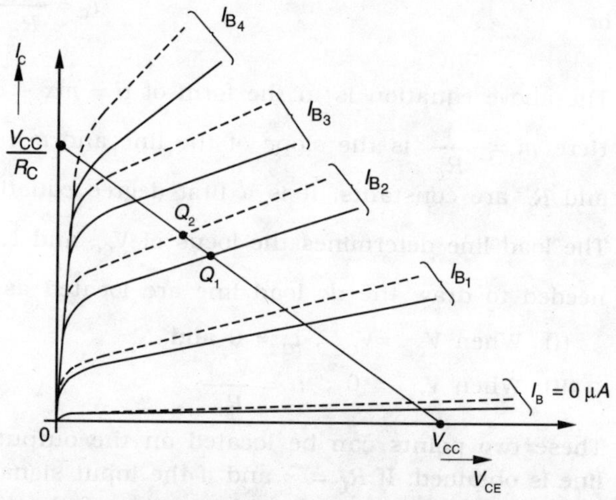


**Fig. 2.3 Graphs showing the collector characteristics for two transistors of the same type.**

For equal changes in $I_B$ the spacing of the output characteristics may increase or decrease as $\beta$ varies. In Fig. 2.3, it is assumed that $\beta$ is greater for the replacement transistor of Fig. 1.1. Since $I_B$ is maintained constant at $I_{B_2}$ by the external biasing circuit, it follows that the operating point will shift to $Q_2$. This new operating point may be completely unsatisfactory. It is clear that maintaining $I_B$ constant will not provide operating point stability as $\beta$ changes. Hence $I_B$ should be allowed to change so as to maintain $I_C$ and $V_{CE}$ constant as $\beta$ changes.

The inherent variations of transistor parameters may change the operating point, resulting in unfaithful amplification. It is therefore, very important that biasing network be so designed that operating should be independent of transistor parameter's variations, and must not drift.

**The process of making operating point independent of temperature changes or variation in transistor parameters is called the stabilization.**

Once stabilization is properly done, the zero signal $I_C$ and $V_{CE}$ become independent of temperature variations or replacement of transistor i.e., the operating point is fixed.

**Thermal Instability:** The reverse saturation current $I_{CO}$ increases greatly with temperature. Specifically, $I_{CO}$ doubles for every 10° *C* rise in temperature. The collector current causes the collector-junction temperature to rise, which in turn increases $I_{CO}$. The collector current is given by

$$I_C = \beta I_B + (\beta + 1)\, I_{CO} \qquad \longrightarrow (1)$$

As $I_{CO}$ increases, $I_C$ will increase which in turn further the junction temperature and consequently $I_{CO}$. It is possible for this succession of events to become cumulative, thereby the ratings of the transistor exceed and the transistor may burn out. This situation is referred to as **thermal runaway** of the transistor.

If $I_B = 0$, then Eq. (1) becomes $I_C = (\beta + 1)\, I_{CO}$ ⟶(2)

As the temperature increases, $I_{CO}$ increases. Assuming β remains constant (actually, it also increases), it is clear that the $I_B = 0$ line in the C-E output characteristics will move upward. The output characteristics for other values of $I_B$ will also move upward by the same amount and consequently the operating point $Q$ will move if $I_B$ is forced to remain constant. The output characteristics of the 2N708 *n-p-n* transistor at temperatures of + 25° C and + 100° C are shown in Fig. 2.4.

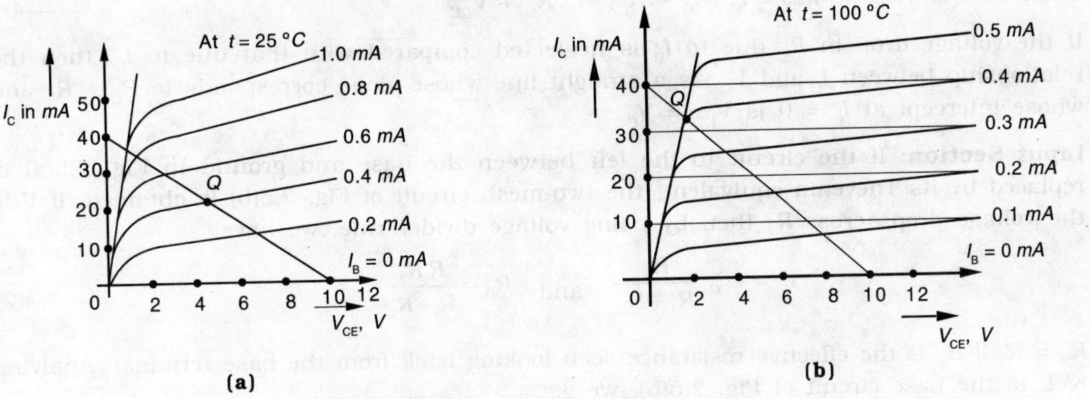


**Fig. 2.4 Output characteristics of an *n-p-n* transistor (2N708) for (a) 25 °C (b) 100 °C**

## SELF BIAS OR EMITTER BIAS

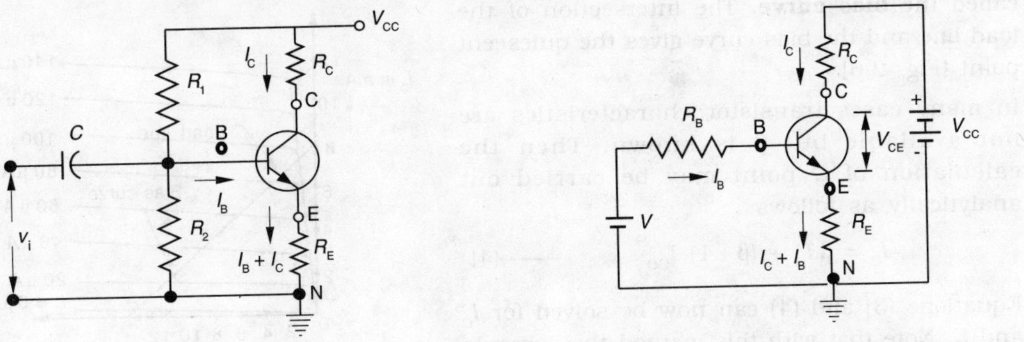


**(a) A self-biasing circuit** **Fig. 2.5** **(b) Simplification of the base circuit**

The most widely used biasing circuit is self-bias. It is used to establish a stable operating point. Figure 2.5 shows a self-biasing circuit using an *n-p-n* transistor. In this circuit, the biasing is provided by three resistors $R_1$, $R_2$ and $R_E$. The resistors $R_1$ and $R_2$

act as a voltage divider giving a fixed voltage to the base. The current in the resistance $R_E$ causes a voltage drop which reverse biases the emitter junction. Since this junction must be forward-biased, the base voltage is obtained from the supply through the $R_1$ $R_2$ network.

The physical reason for an improvement in stability with this circuit is as follows: If $I_C$ increases due to change in temperature or changes in β, the emitter current $I_E$ also increases. As a consequence of the increase in voltage drop across $R_E$, the voltage $V_{BE}$ gets decreased. Due to reduction in $V_{BE}$, base current $I_B$ and hence $I_c$ also decreases. This reduction in $I_C$ compensates for the original increase in $I_C$.

**Circuit Analysis**

**Output Section:** Applying KVL to the collector circuit, we have

$$V_{CC} = I_C R_C + V_{CE} + I_E R_E$$

Since $I_E = I_C + I_B$, $\quad V_{CC} = I_C (R_C + R_E) + I_B R_E + V_{CE}$ ⟶(1)

If the voltage drop in $R_E$ due to $I_B$ is neglected compared with that due to $I_C$, then the relationship between $I_C$ and $V_{CE}$ is a straight line whose slope corresponds to $R_C + R_E$ and whose intercept at $I_C = 0$ is $V_{CE} = V_{CC}$.

**Input Section:** If the circuit to the left between the base and ground in Fig. 2.5(a) is replaced by its Thevenin equivalent, the two-mesh circuit of Fig. 2.5(b) is obtained. If $V$ is the voltage drop across $R_2$ then by using voltage divider rule, we have

$$V = V_{CC} \frac{R_2}{R_1 + R_2} \quad \text{and} \quad R_B = \frac{R_1 R_2}{R_1 + R_2} \qquad \longrightarrow (2)$$

$R_B = R_1 \parallel R_2$ is the effective resistance seen looking back from the base terminal. Applying KVL to the base circuit of Fig. 2.5(b), we get

$$V = I_B R_B + V_{BE} + I_E R_E = I_B R_B + V_{BE} + (I_C + I_B) R_E \qquad \longrightarrow (3)$$

Substituting the value of $I_C$ from Eq. (3) in Eq. (1), a relationship between $I_B$ and $V_{CE}$ is obtained. For each value of $I_B$ given on the collector curves, $V_{CE}$ can be calculated. The locus of these corresponding points $V_{CE}$ and $I_B$ plotted on the output characteristics is called the **bias curve.** The intersection of the load line and the bias curve gives the quiescent point [Fig. 2.6].

In many cases transistor characteristics are not available but β is known. Then the calculation of $Q$ point may be carried out analytically as follows:

$$I_C = \beta I_B + (\beta + 1) I_{CO} \qquad \longrightarrow (4)$$

Equations (3) and (4) can now be solved for $I_C$ and $I_B$. Note that with this method the currents in the active region can be determined by the base circuits and the values of β and $I_{CO}$.

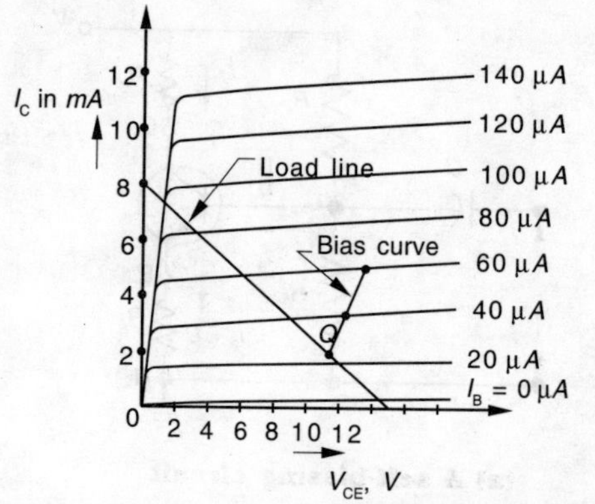


**Fig. 2.6 The intersection of the load line and the bias curve determines the Q point**

## STABILIZATION AGAINST VARIATIONS IN $I_{CO}$, $V_{BE}$ AND β

The sources of instability of $I_C$ are: (i) the reverse saturation current $I_{CO}$ doubles for every 10° C increase in temperature, (ii) the base to emitter voltage $V_{BE}$ decreases at the rate of 2.5 *mV/°C* for both *Ge* and *Si* transistors and (iii) β, increases with temperature [Tables 2.1 and 2.2].

Since the effect of the change of $V_{CE}$ with temperature is very small, it may be neglected. Assumes that the transistor operates in the active region, where $I_C$ is approximately independent of $V_{CE}$.

**The Transfer Characteristic:** The transfer characteristics i.e., output current $I_C$ versus input voltage $V_{BE}$ respectively for *Ge* and *Si* transistors are shown in Fig. 2.7. Each curve shifts to the left at the rate of 2.5 *mV/°C* (at constant $I_C$) for increasing temperature.

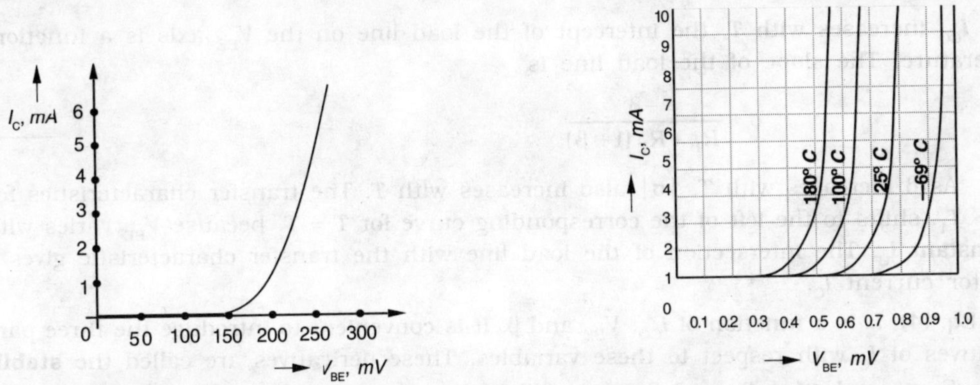


**(a) Transfer characteristics for Ge *p-n-p* transistor** **Fig. 2.7** **(b) Transfer characteristic for Si transistor**

Now let us examine the effect of shift in transfer characteristics and the variation of β and $I_{CO}$ with temperature. Applying Kirchhoff's law to the base circuit, we have

$$V = I_B R_B + V_{BE} + (I_B + I_C) R_E \longrightarrow (1)$$

$$V_{BE} = V - I_B R_B - (I_B + I_C) R_E \longrightarrow (2)$$

We know that $I_C = \beta I_B + (\beta + 1) I_{CO}$

or

$$I_B = \frac{I_C}{\beta} - \left(\frac{\beta+1}{\beta}\right) I_{CO} \longrightarrow (3)$$

Substituting $I_B$ in Eq. (2), we get

$$V_{BE} = V - \left[\frac{I_C}{\beta} - \left(\frac{\beta+1}{\beta}\right) I_{CO}\right](R_B + R_E) - I_C R_E$$

Rearranging,

$$V_{BE} = V + (R_B + R_E)\left(\frac{\beta+1}{\beta}\right) I_{CO} - \frac{R_B + R_E(1+\beta)}{\beta} I_C \longrightarrow (4)$$

The above equation represents a load line in $I_C - V_{BE}$ plane and is shown in Fig. 2.8. The intercept on the $V_{BE}$ axis is $V + V'$, where

$$V' = (R_B + R_E)\left(\frac{\beta+1}{\beta}\right)I_{CO} \approx (R_B + R_E)\,I_{CO} \quad \text{(because } \beta >> 1) \qquad \longrightarrow (5)$$

If $I_{CO_1}$ and $I_{CO_2}$ are reverse saturation currents, and $\beta_1$ and $\beta_2$ current gain at temperatures $T_1$ and $T_2$ respectively, then

$$V_1' \approx (R_B + R_E)\,I_{CO_1}$$

and $$V_2' \approx (R_B + R_E)\,I_{CO_2}$$

Since $I_{CO}$ increases with $T$, the intercept of the load line on the $V_{BE}$ axis is a function of temperature. The slope of the load line is

$$\sigma = \frac{-\beta}{R_B + R_E(1+\beta)} \qquad \longrightarrow (6)$$

As $\beta$ increases with $T$, $|\sigma|$ also increases with $T$. The transfer characteristics for $T = T_2 > T_1$, shifts to the left of the corresponding curve for $T = T_1$ because $V_{BE}$ varies with $T$ at constant $I_C$. The intersection of the load line with the transfer characteristic gives the collector current $I_C$.

From Eq. (4), $I_C$ is a function of $I_{CO}$, $V_{BE}$ and $\beta$, it is convenient to introduce the three partial derivatives of $I_C$ with respect to these variables. These derivatives are called the **stability** factors $S$, $S'$ and $S''$ and are defined as follows:

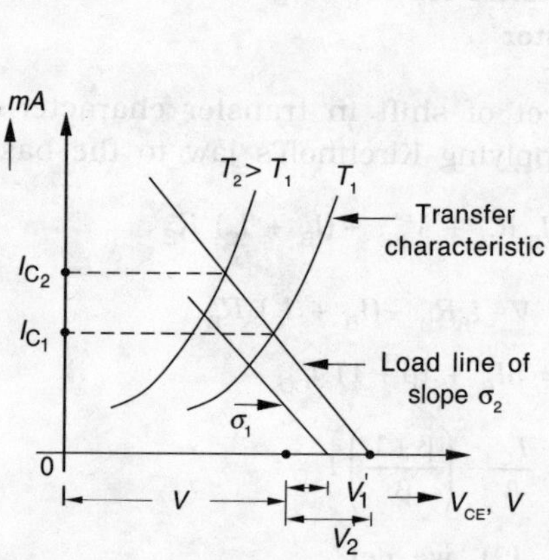


**Fig. 2.8 Illustrating that the collector current varies with temperature $T$ because $V_{BE}$, $I_{CO}$ and $\beta$ change with $T$**

**1. The Stabilization Factor(S): The stabilization factor is defined as the ratio of change of collector current with respect to the change in reverse saturation current, with $\beta$ and $V_{BE}$ constant.**

$$\textbf{Stabilization factor,} \quad S = \frac{\partial I_c}{\partial I_{co}} \approx \frac{\Delta I_c}{\Delta I_{co}} \qquad \longrightarrow (7)$$

'S' is a measure of bias stability of a transistor circuit. The larger the value of S, better is the thermal instability.

Differentiating Eq. (4) with $I_{co}$, we get

$$o = (R_B + R_E)\left(\frac{\beta+1}{\beta}\right) - \frac{R_B + R_E(\beta+1)}{\beta}\frac{\Delta I_c}{\Delta I_{co}}$$

or

$$S = \frac{\Delta I_c}{\Delta I_{co}} = \frac{(R_B + R_E)(\beta+1)}{R_B + R_E(\beta+1)}$$

$$S = (\beta+1)\frac{1 + R_B/R_E}{1 + \beta + R_B/R_E} \longrightarrow (8)$$

Note that S varies between 1 for small $R_B/R_E$ and $(1 + \beta)$ for $R_B/R_E \to \infty$. If $(\beta+1) >> R_B/R_E$, then Eq. (8) reduces to

$$S \approx 1 + \frac{R_B}{R_E} \longrightarrow (9)$$

Thus, for constant β, $V_{BE}$ and small S, we have

$$\frac{\Delta I_c}{I_c} \approx S\frac{\Delta I_{co}}{I_c} \approx \frac{\Delta I_{co}}{I_c} + \frac{R_B}{R_E}\frac{\Delta I_{co}}{I_c} \longrightarrow (10)$$

**2. The Stability Factor (S′):** The variation of $I_c$ with $V_{BE}$ is given by the stability factor $S'$, defined by

$$\boxed{S' = \frac{\partial I_c}{\partial V_{BE}} \approx \frac{\Delta I_c}{\Delta V_{BE}}} \longrightarrow (11)$$

keeping both $I_{co}$ and β are constant.

Differentiating Eq. (4) with $V_{BE}$, we get.

$$1 = -\frac{R_B + R_E(1+\beta)}{\beta}\frac{\Delta I_c}{\Delta V_{BE}}$$

or

$$\frac{\Delta I_c}{V_{BE}} = S' = \frac{-\beta}{R_B + R_E(1+\beta)} = \frac{-\beta/R_E}{1 + \beta + R_B/R_E} \longrightarrow (12)$$

Assuming that $(\beta+1) >> R_B/R_E$ and also $\beta >> 1$, Eq. (12) reduces to

$$S'' \approx \frac{\Delta I_c}{\Delta V_{BE}} \approx -\frac{1}{R_E} \longrightarrow (13)$$

or

$$\frac{\Delta I_c}{I_c} \approx \frac{S'\Delta V_{BE}}{I_c} \approx -\frac{\Delta V_{BE}}{I_c R_E} \longrightarrow (14)$$

From Eqs. (10) and (14), it is seen that the dominant factor in stabilizing against $I_{co}$ and $V_{BE}$ is the quiescent voltage drop across the emitter resistance $R_E$. The larger the voltage drop, the smaller is the percentage change in collector current due to $\Delta I_{co}$ and $\Delta V_{BE}$.

**3. The Stability Factor ( S″ ):** The variation of $I_c$ with respect to β is given by the stability factor S", defined by

$$S'' \approx \frac{\partial I_c}{\partial \beta} \approx \frac{\Delta I_c}{\Delta \beta} \qquad (15)$$

keeping both $I_{co}$ and $V_{BE}$ constant.

Eq. (4) $\Rightarrow V_{BE} = V + V'\left(\frac{\beta+1}{\beta}\right) - \frac{R_B + R_E(1+\beta)}{\beta} I_c$

or $I_c\left(\frac{R_B + R_E(1+\beta)}{\beta}\right) = V + V'\left(\frac{\beta+1}{\beta}\right) - V_{BE}$

or
$$I_c = \frac{\beta(V + V' - V_{BE})}{R_B + R_E(1+\beta)} \qquad (16)$$

Differentiating the above equation with β, we get

$$\frac{\Delta I_c}{\Delta \beta} = \frac{(R_B + R_E)(V + V' - V_{BE})}{[R_B + R_E(1+\beta)]^2}$$

After doing some algebraic manipulation, we get

$$S'' = \frac{\Delta I_c}{\Delta \beta} = \frac{I_c S}{\beta(1+\beta)} \qquad (17)$$

The change in collector current due to a change in β is

$$\Delta I_c \approx S'' \Delta \beta = \frac{I_c S}{\beta(1+\beta)} \Delta \beta \qquad (18)$$

where $\Delta\beta = \beta_2 - \beta_1$ may represent a large change in β. Hence it is not clear to use $\beta_1$, $\beta_2$ or perhaps some average value of β in the expressions for $S''$. This difficulty is avoided if $S''$ is obtained by taking finite differences rather than by evaluating a derivative. Thus

$$S'' \approx \frac{I_{c_2} - I_{c_1}}{\beta_2 - \beta_1} = \frac{\Delta I_c}{\Delta \beta} \qquad (19)$$

From Eq. (16), we have

$$\frac{I_{c_2}}{I_{c_1}} = \frac{\beta_2}{\beta_1} \frac{R_B + R_E(1+\beta_1)}{R_B + R_E(1+\beta_2)} \qquad (20)$$

Subtracting 1 from both sides of Eq. (20), we get

$$\frac{I_{c_2}}{I_{c_1}} - 1 = \left(\frac{\beta_2}{\beta_1} - 1\right) \frac{R_B + R_E}{R_B + R_E(1+\beta_2)} \qquad (21)$$

or
$$S'' = \frac{\Delta I_c}{\Delta \beta} = \frac{I_{c_1} S_2}{\beta_1(1+\beta_2)} \qquad (22)$$

where $S_2$ is the value of the stabilizing factor $S$ when $\beta = \beta_2$ as given by Eq. (8). Note that the equation reduces to Eq. (17) as $\Delta\beta = \beta_2 - \beta_1 \rightarrow 0$.

If we assume that $S_2$ is small so that the approximate value given in Eq. (9) is valid, then from Eq. (22) with $\beta >> 1$ we find

$$\frac{\Delta I_c}{I_{c_1}} \approx \left(1+\frac{R_B}{R_E}\right)\frac{\Delta\beta}{\beta_1\beta_2} = \left(1+\frac{R_B}{R_E}\right)\left[\frac{\beta_2/\beta_1 - 1}{\beta_2}\right] \longrightarrow (23)$$

From the above discussion it is clear that $R_B/R_E$ should be kept small. Also, for a given variation in the value of $\beta$, a high $\beta$ circuit will be more stable compared to lower-$\beta$ transistor.

## GENERAL REMARKS ON COLLECTOR-CURRENT STABILITY

The stability factors are defined as

$$S = \frac{\Delta I_c}{\Delta I_{co}}, \quad S' = \frac{\Delta I_c}{\Delta V_{BE}}, \quad \text{and} \quad S'' = \frac{\Delta I_c}{\Delta \beta} \longrightarrow (1)$$

In order to obtain the total change in $I_c$ over a specified temperature range, it is better to express $\Delta I_c$ as the sum of the individual changes due to the three stability factors. Hence, the total differential of $I_c = f(I_{co}, V_{BE}, \beta)$, we get

$$\Delta I_c = \frac{\partial I_c}{\partial I_{co}}\Delta I_{co} + \frac{\partial I_c}{\partial V_{BE}}\Delta V_{BE} + \frac{\partial I_c}{\partial \beta}\Delta\beta$$

$$\Delta I_c = S\,\Delta I_{co} + S'\,\Delta V_{BE} + S''\,\Delta\beta \longrightarrow (2)$$

The stability factors may also be expressed in terms of the parameter $M$ defined as

$$M = \frac{1}{1 + R_B/[R_E(1+\beta)]} \approx \frac{1}{1 + R_B/\beta R_E} \longrightarrow (3)$$

If $\beta R_E >> R_B$, then $M = 1$.

For the fractional change in collector current

$$\frac{\Delta I_c}{I_{c_1}} = \left(1+\frac{R_B}{R_E}\right)\frac{M_1 \Delta I_{co}}{I_{c_1}} - \frac{M_1 \Delta V_{BE}}{I_{c_1} R_E} + \left(1+\frac{R_B}{R_E}\right)\frac{M_2 \Delta\beta}{\beta_1\beta_2} \longrightarrow (4)$$

As $T$ increases, $\Delta I_{co}/\Delta I_{c_1}$ and $\Delta\beta$ increase, whereas $\Delta V_{BE}/I_{c_1}$ decreases. Tables 2.1 and 2.2 show typical parameters of Si and Ge transistors.

**Table 2.1 Typical Silicon transistor parameters**

| | | | |
|---|---|---|---|
| $t\,^\circ C$ | - 65 | + 25 | + 175 |
| $I_{co}$, $nA$ | $1.95 \times 10^{-3}$ | 1.0 | 33,000 |
| $\beta$ | 25 | 55 | 100 |
| $V_{BE}$, $V$ | 0.78 | 0.60 | 0.225 |

**Table 2.2 Typical germanium transistor parameters**

| | | | |
|---|---|---|---|
| $t\,^{\circ}C$ | -65 | +25 | +175 |
| $I_{co}$, $nA$ | $1.95 \times 10^{-3}$ | 1.0 | 32 |
| $\beta$ | 20 | 55 | 90 |
| $V_{BE}$, $V$ | 0.38 | 0.20 | 0.10 |

Note that the tolerance in bias resistors and supply voltages must be taken into account, in addition to the variation of $\beta$, $I_{co}$ and $V_{BE}$.

## BIAS COMPENSATION

In biasing a transistor in the active region it is difficult to maintain the operating point stable by keeping $I_c$ and $V_{CE}$ constant. There are two techniques to do so. (i) **Stabilization techniques** and (ii) **Compensation techniques**. Stabilization techniques refer to the use of resistive biasing circuits which allow $I_B$ to vary so as to keep $I_{CO}$, $\beta$ and $V_{BE}$ constant. Compensation techniques refer to the use of temperature-sensitive devices such as diodes, transistors, thermistors, etc., which provide compensating voltages and currents to maintain the operating point stable.

**Diode Compensation for $V_{BE}$:** Figure 2.9 shows a circuit utilizing the self-bias stabilization technique and diode compensation. The diode $D$ is kept forward-biased by the source $V_{DD}$ and resistance $R_D$. If the diode and the transistor are of the same material and type, the voltage $V_D$ across the diode will have the same temperature coefficient ($-2.5\ mV/^{\circ}C$) as the base-to-emitter voltage $V_{BE}$.

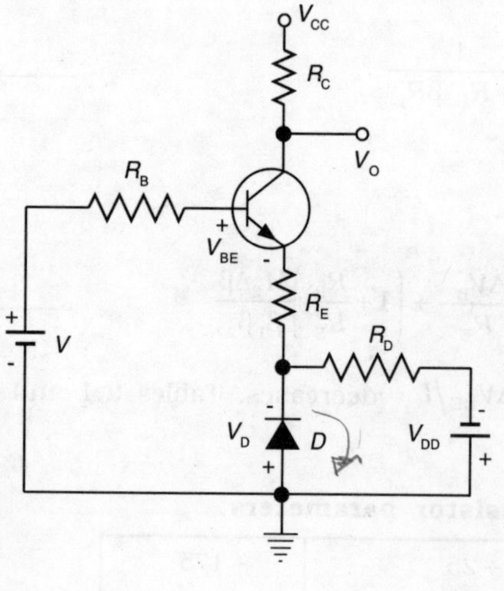


**Fig. 2.9 Stabilization by means of self-bias and diode-compensation techniques**

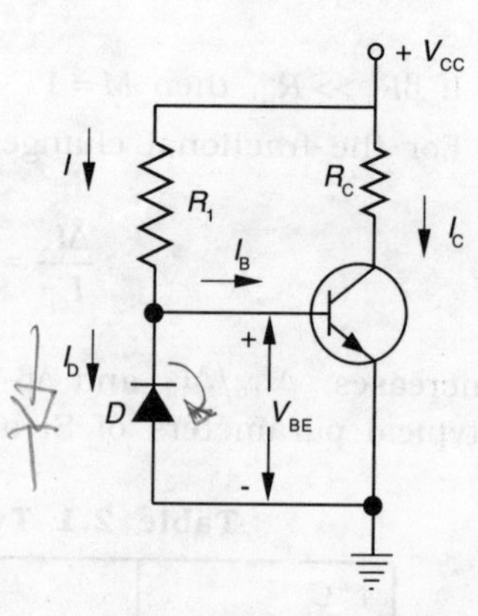


**Fig. 2.10 Diode compensation for a germanium transistor**

Applying KVL to the base circuit, we have

$$V = I_B R_B + V_{BE} + I_E R_E - V_D = I_B R_B + V_{BE} + (I_C + I_B) R_E - V_D \quad \longrightarrow (1)$$

We know that $$I_c = \frac{\beta(V + V' - V_{BE})}{R_B + R_E(1+\beta)} \longrightarrow (2)$$

Solving, $$I_c = \frac{\beta[V - (V_{BE} - V_D)] + (R_B + R_E)(\beta+1)I_{co}}{R_B + R_E(1+\beta)} \longrightarrow (3)$$

Since $V_{BE}$ tracks $V_D$ with respect to temperature, it is clear from Eq. (3) that $I_c$ is not sensitive to variations in $V_{BE}$. In practice, the compensation of $V_{BE}$ is not exact, but it is sufficiently effective to take care of a great part of transistor drift due to variations in $V_{BE}$.

**Diode Compensation for $I_{co}$:** We know that changes in $V_{BE}$ with temperature contribute significantly to changes in $I_c$ of silicon transistors. On the other hand changes in $I_{co}$ with temperature play an important role in collector current stability for germanium transistors. Figure 2.10 shows the diode compensation circuit that offers stabilization against variations in $I_{co}$.

If the diode and the transistor are of the same material and type, the reverse saturation current $I_o$ of the diode increases with temperature at the same rate as the transistor collector reverse saturation current $I_{co}$. From Fig. 2.10, we have

$$V_{cc} = IR_1 + V_{BE}$$

or $$I = \frac{V_{cc} - V_{BE}}{R_1} \approx \frac{V_{cc}}{R_1} = \text{constant} \longrightarrow (4)$$

Since the diode $D$ is reverse-biased by an amount $V_{BE} \approx 0.2V$ for Ge devices, the current through $D$ is reverse saturation current $I_o$. Hence the current $I_B = I - I_o$.

We know $$I_c = \beta I_B + (\beta+1)I_{co}$$

$$I_c = \beta(I - I_o) + (\beta+1)I_{co} \longrightarrow (5)$$

From Eq. (5), it is seen that if $\beta >> 1$ then $I_o$ of diode and $I_{co}$ of transistor track each other over the desired temperature range, and $I_c$ remains essentially constant.

## BIASING TECHNIQUES FOR LINEAR INTEGRATED CIRCUITS

The self-bias circuit requires a capacitor across $R_E$ since otherwise the negative feedback, due to $R_E$, reduces the signal gain drastically. This bypass capacitance is too large to be fabricated by integrated-circuit technology. Hence the biasing technique shown in Fig. 2.11 has been developed for monolithic circuits. The transistor $Q_1$ is connected as a diode across the base-emitter junction of $Q_2$ whose collector current is to be temperature stabilized [Fig. 2.11(a)]. The collector current of $Q_1$ is given by

$$I_{c_1} = \frac{V_{cc} - V_{BE}}{R_1} - I_{B_1} - I_{B_2} \longrightarrow (1)$$

Since $V_{BE} << V_{cc}$ and $(I_{B_1} + I_{B_2}) << I_{c_1}$, then Eq. (1) reduces to

$$I_{c_1} \approx \frac{V_{cc}}{R_1} = \text{constant} \longrightarrow (2)$$

If transistors $Q_1$ and $Q_2$ are identical and have the same value of $V_{BE}$, then their collector currents $I_{c_1} = I_{c_2}$ = constant. Even if the two transistors are not identical, experiments have shown that this biasing scheme gives collector current matching between the biasing and operating transistors typically better than 5 percent and is also stable over a wide temperature range.

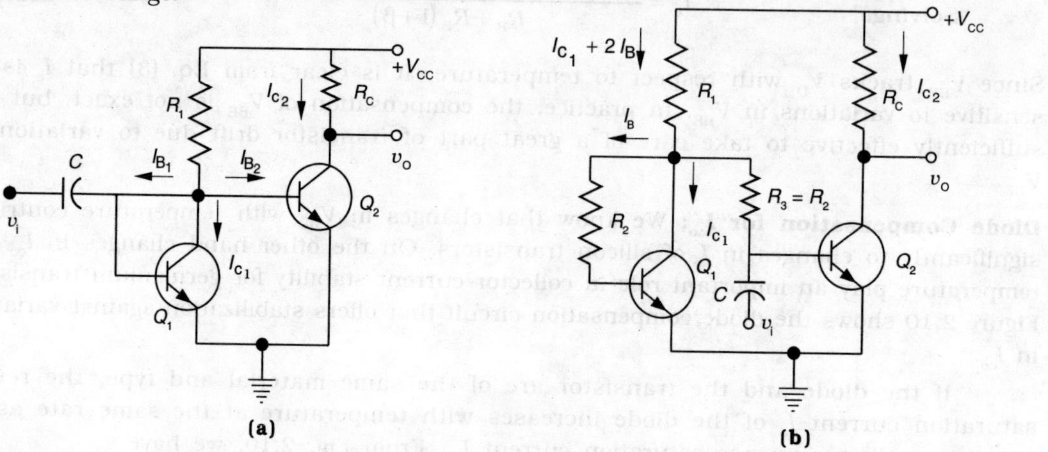


**Fig. 2.11 Biasing techniques for linear integrated circuits**

The circuit of Fig. 2.11(b) is a modified version of Fig. 2.11(a) i.e., where the transistors are driven by equal base currents rather than the same base voltage. Since the collector current in the active region varies linearly with $I_B$, but exponentially with $V_{BE}$, improved matching of collector currents results. Since the two bases are driven from a common node voltage through equal resistances, $I_{B_1} = I_{B_2} = I_B$ and thereby the collector currents are well matched for identical transistors.

From Fig. 2.11(b), the collector current of $Q_1$ is given by

$$I_{c_1} = \frac{V_{cc} - V_{BE}}{R_1} - \left(2 + \frac{R_2}{R_1}\right) I_B \longrightarrow (3)$$

Since $V_{BE} << V_{cc}$ and $(2 + R_2/R_1)I_B << V_{cc}/R_1$, then

$$I_{c_1} = I_{c_2} = \frac{V_{cc}}{R_1} \longrightarrow (4)$$

If $R_c = \frac{R_1}{2}$, then $$V_{CE} = V_{cc} - I_{c_2} R_c \approx \frac{V_{cc}}{2}, \longrightarrow (5)$$

Hence, the amplifier will be biased at $V_{cc}/2$, independent of the supply voltage as well as temperature and dependent only on the matching components within the integrated circuit.

## IMPORTANT POINTS TO REMEMBER

1. The purpose of biasing a circuit is to establish a proper stable *dc* operating point ($Q$ point).
2. The $Q$-point of a circuit is defined by specific values for $I_c$ and $V_{CE}$. These values are called the coordinates of the $Q$-point.

3. A *dc* load line passes through the *Q*-point on a transistor's collector curves intersecting vertical axis at approximately $I_{c(sat)}$ and the horizontal axis at $V_{CE(off)}$.
4. The active operating region of a transistor lies along the load line below saturation and above cutoff.
5. A coupling capacitor transmits an *ac* voltage from one point to another. It acts like a short for low frequency *ac*.
6. Make the reactance at least 10 times smaller than the total resistance in series with the capacitor i.e. $X_c < 0.1\,R$., in designing a circuit.
7. In a fixed-bias circuit, the supply voltage $V_{cc}$, $V_{BE}$ and $I_B$ are constant.
8. The two reasons for the shift of operating points are (i) the transistor parameters such as β, $V_{BE}$ are not the same for every transistor even of the same type. (ii) The transistor parameters are temperature dependent.
9. The process of making operating point independent of temperature changes or variation in transistor parameters is called the stabilization.
10. Self-bias provides good *Q*-point stability with a single polarity supply voltage.
11. The intersection of the load line and the bias curve gives the quiescent point.
12. The reverse saturation current $I_{co}$ doubles for every $10°C$ increase in temperature.
13. The base to emitter voltage $V_{BE}$ decreases at the rate of $2.5\ mV/°C$ for both Ge and Si transistors.
14. β increases with temperature.
15. The stabilization factor is defined as the rate of change of collector current with respect to the reverse saturation current, keeping β and $V_{BE}$ constant.
16. The two techniques called stabilization techniques and compensation techniques maintain the operating point stable.
17. Stabilization techniques refer to the use of resistive biasing circuits which allow $I_B$ to vary so as to keep $I_c$ relatively constant, with variations in $I_{co}$, β and $V_{BE}$.
18. Compensation techniques refer to the use of temperature sensitive devices such as diodes, transistors, thermistor, etc., which provide compensation voltages and currents to maintain the operating point stable.
19. Bypass capacitor should be connected in parallel across a resistor. It bypasses or shunts *ac* current away from the resistor.
20. At high frequencies, the bypass capacitor acts like a short.

## KEY FORMULAE

1. In a fixed-bias circuit, $I_B = \dfrac{V_{cc} - V_{BE}}{R_B} \approx \dfrac{V_{cc}}{R_B}$
2. Collector current, $I_c = \beta I_B + (\beta+1) I_{co}$
3. In self-bias circuit, $V = \dfrac{V_{cc} R_2}{R_1 + R_2}$, $R_B = \dfrac{R_1 R_2}{R_1 + R_2}$ and $V = I_B R_B + V_{BE} + (I_B + I_C) R_E$
4. The slope of the load line is $\sigma = \dfrac{-\beta}{R_B + R_E(1+\beta)}$
5. Stabilization factor, $S = \dfrac{\Delta I_c}{\Delta I_{co}} = 1 + \dfrac{R_B}{R_E}$

6. Stability factor, $S' = \dfrac{\Delta I_c}{\Delta V_{BE}} = \dfrac{-\beta/R_E}{1+\beta+R_B/R_E}$

7. Stability factor, $S'' = \dfrac{\Delta I_c}{\Delta\beta} = \dfrac{I_c S}{\beta(1+\beta)}$

8. $\Delta I_c = S\Delta I_{co} + S'\Delta V_{BE} + S''\Delta\beta$

9. $M = \dfrac{1}{1+R_B/R_E(1+\beta)} \approx \dfrac{1}{1+R_B/\beta R_E}$

10. Diode compensation for $V_{BE}$, $I_c = \dfrac{\beta[V-(V_{BE}-V_D)]+(R_B+R_E)(\beta+1)I_{co}}{R_B+R_E(1+\beta)}$

11. Diode compensation for $I_{co}$, $I_c = \beta(I - I_o) + (\beta+1)I_{co}$

## SOLVED PROBLEMS

**1. (a) Determine the quiescent currents and the collector-to-emitter voltage for a Si transistor with β = 50 in the self-bias circuit. The circuit component values are $V_{cc}$ = 20 V, $R_c$ = 2 kΩ, $R_E$ = 0.1 kΩ, $R_1$ = 100 kΩ and $R_2$ = 5 kΩ. (b) Repeat (a) for Ge transistor.**

**Data:** $V_{cc} = 20\,V$, $R_1 = 100\,k\Omega$, $R_2 = 5\,k\Omega$, $R_C = 2\,k\Omega$, $R_E = 100\,\Omega$, and $\beta = 50$

**To find:** Quiescent currents and $V_{CE}$,

**Solution:** We know $V = \dfrac{V_{CC}R_2}{R_1+R_2}$

$$V = \frac{20\times5}{100+5} = \mathbf{0.952\,V}$$

and $R_B = \dfrac{R_1R_2}{R_1+R_2} = \dfrac{100\times5}{105} = \mathbf{4.76\,k\Omega}$

Applying KVL to the base-emitter loop, we have

$$V = I_BR_B + V_{BE} + (I_B + I_C)R_E$$

$$V = I_BR_B + V_{BE} + (1+\beta)I_B R_E$$

$$0.952 = I_B\,[4.76 + 51\times0.1] + 0.7 = 9.86\,I_B + 0.7$$

or $I_B = \dfrac{0.952-0.7}{9.86} = \mathbf{0.0256\,mA}$

or $I_C = \beta I_B = 50\times0.0256 = \mathbf{1.26\,mA}$

and $I_E = I_B + I_C = 1.26+0.02 = \mathbf{1.28\,mA}$

We know $V_{CE} = V_{CC} - I_CR_C - I_ER_E$

$$V_{CE} = 20 - 1.26\times2 - 1.28\times0.1 = \mathbf{17.35\,V}$$

(b) $R_B$ and $V$ has the same value as in part (a) but for Ge transistor $V_{BE}$ = 0.2 V
Applying KVL to the base-emitter loop, we get

$$V = I_B R_B + V_{BE} + (I_B + I_C)R_E = I_B R_B + V_{BE} + (1+\beta)I_B R_E$$

$$0.952 = I_B[4.76+51\times0.1]+0.2 = 9.86 I_B + 0.2$$

$$I_B = \frac{0.952-0.2}{9.86} = \mathbf{76.2\ \mu A}$$

$$I_C = \beta I_B = 0.5\times76.2\mu A = \mathbf{3.81\ mA}$$

$$I_E = 3.81+0.0762 = \mathbf{3.89\ mA}$$

$$V_{CE} = V_{CC} - I_C R_C - I_E R_E$$

$$= 20-3.81\times2-3.89\times0.1 = 20-7.62-0.39$$

$$V_{CE} = \mathbf{12\ V}$$

**2. A *p-n-p* silicon transistor is used in a C-C circuit [Fig. 2.5 with $R_C$ = 0]. The circuit component values are $V_{CC}$ = 3.0 V. $R_E$ = 1 kΩ, $R_1$ = $R_2$ = 5 kΩ. If β = 44 (a) find the quiescent point, (b) recalculate these values, taking the base-spreading resistance of 690 Ω into account.**

**Data:** $V_{cc}$ = 3 V, β = 44, $R_E$ = 1 kΩ, $R_1$ = $R_2$ = 5 kΩ, Base spreading resistance = 690 Ω

**To find:** Quiescent point

**Solution:** (a) We know $V = V_{CC}\frac{R_2}{R_1+R_2} = 3\times\frac{5}{5+5} = \frac{3}{2} = 1.5V$

$$R_B = \frac{R_1 R_2}{R_1+R_2} = \frac{5\times5}{5+5} = 2.5\ k\Omega$$

Applying KVL to the collector-emitter circuit, we have

$$V_{CC} = V_{CE} + I_E R_E = V_{CE} + (I_B + I_C)R_E$$

$$-3 = V_{CE} + I_B + I_C \longrightarrow (1)$$

Applying KVL to the base-emitter circuit, we have

$$V = I_B R_B + V_{BE} + (I_B + I_C)R_E$$

$$1.5 = -I_B\times2.5+0.70-(I_B+I_C)$$

Since $I_C = \beta I_B = 44\,I_B$, we have

$$-0.8 = 3.5I_B + 44I_B = 47.5I_B$$

or $$I_B = \frac{-0.8}{47.5} = \mathbf{-0.0168\ mA}$$

$$I_C = -44\times0.0168 = \mathbf{-0.74\ mA}$$

Then $$V_{CE} = -3+0.74+0.0168 = \mathbf{-2.24\ V}$$

(b) $R_B$ is increased to 2.5 + 0.69 = 3.19 $k\Omega$.

Applying KVL to the base-emitter circuit, we obtain

$$-0.80 = [3.19+1+44]I_B = 48.19 I_B$$

or $$I_B = \frac{-0.80}{48.19} = \mathbf{-0.0166\ mA}$$

and $$I_C = \beta I_B = 44 \times -0.0166 = \mathbf{-0.73\ mA}$$

$$V_{CE} = -3 + 0.746 = \mathbf{-2.254\ V}$$

**3. For the circuit shown in Fig. 2.12, (a) calculate $I_B$, $I_C$ and $V_{CE}$ if a silicon transistor is used with β = 50. (b) Specify a value for $R_B$ so that $V_{CE}$ = 7V.**

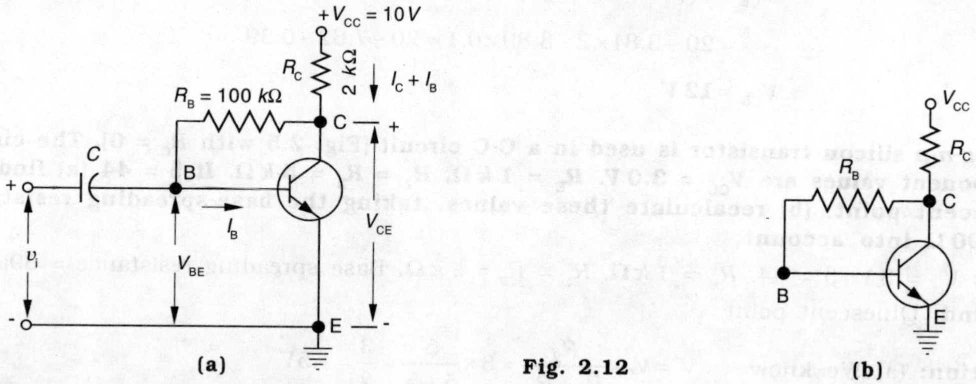


**(a)** **Fig. 2.12** **(b)**

**Data:** $V_{CC}$ = 10V, $R_C$ = 2 kΩ, $R_B$ = 100 kΩ, and β = 50

**To find:** $I_B$, $I_C$, $V_{CE}$ and $R_B$ if $V_{CE}$ = 7V

**Solution:** (a) Applying KVL around the loop $V_{CC}$-C-B-E, we obtain

$$V_{CC} = R_C(I_B + I_C) + I_B R_B + V_{BE}$$

Since the transistor is in the active region, $I_C = 50\, I_B$ and $V_{BE} = 0.7V$

∴ $$10 = 51 \times 2 \times I_B + 100\, I_B + 0.7$$

or $$I_B = \frac{9.3V}{202\, k\Omega} = \mathbf{46\ \mu A}$$

$$I_C = \beta I_B = 50 \times 46\ \mu A = \mathbf{2.30\ mA}$$

Then $$V_{CE} = V_{CC} - (I_C + I_B) R_C$$

$$V_{CE} = 10 - (2.346)2 = \mathbf{5.32\ V}$$

(b) Given $V_{CE}$ = 7 V. Applying KVL in the collector-emitter circuit,

$$I_C + I_B = \frac{V_{CC} - V_{CE}}{R_C} = \frac{10-7}{2} = \frac{3}{2} = 1.5 mA$$

or $$I_B = \frac{1.5}{51} = \mathbf{0.0294\ mA}$$

$$I_C = 50 \times 0.0294 = \mathbf{1.47\ mA}$$

Applying KVL around the loop $V_{CC}$-C-B-E, we have

$$10 = 2 \times 1.5 + 0.70 + 0.0294\, R_B$$

or

$$R_B = \frac{6.30}{0.0294} = \mathbf{214\ k\Omega}$$

**4. Calculate the values of $R_E$, $V_{CE}$ and stability factor for the self-bias circuit. Given $V_{CC}$ = 12 V, $R_1 = 50\ k\Omega$, $R_2 = 5\ k\Omega$, $R_C = 2\ k\Omega$, $I_C$ = 2 mA and $\beta = 100$.**

**Data:** $V_{CC} = 12V$, $R_1 = 50\ k\Omega$, $R_2 = 5\ k\Omega$, $R_C = 2\ k\Omega$, $I_C = 2\ mA$ and $\beta = 100$

**To find:** $R_E$, $V_{CE}$ and $S$

**Solution:** We know

$$I_B = \frac{I_C}{\beta} = \frac{2mA}{100} = \mathbf{0.020\ mA}$$

Applying KVL to the base circuit, we have

$$V_{CC} = (I + I_B)\, R_1 + IR_2$$

$$12 = (I + 0.02)50 + I \times 5$$

or

$$I = 0.2\ mA$$

We know

$$V = IR_2 = 0.2 \times 5 = 1V$$

$$V = I_B R_B + V_{BE} + I_E R_E$$

where

$$R_B = \frac{R_1 R_2}{R_1 + R_2} = \frac{50 \times 5}{55} = \frac{50}{11} = 4.545 k\Omega$$

Then

$$1 = 0.02 \times 4.55 + 0.7 + 2.02 \times R_E$$

$$1 = 0.091 + 0.7 + 0.22\, R_E$$

or

$$R_E = \frac{0.219}{2.22} \approx 100\,\Omega$$

Applying KVL to the collector circuit, we obtain

$$V_{CC} = I_C R_C + V_{CE} + I_E R_E$$

$$V_{CE} = V_{CC} - I_C R_C - I_E R_E = 12 - 2 \times 2 - 2.02 \times 0.1$$

$$V_{CE} = \mathbf{7.8V}$$

Stability factor,

$$S = (1+\beta)\left[\frac{1 + (R_B/R_E)}{1 + \beta + (R_B/R_E)}\right]$$

$$= (101)\left[\frac{1 + 4.55/0.1}{101 + 4.55/0.1}\right] = 101\left[\frac{46.5}{146.5}\right] = \mathbf{32.06}$$

**5. For the two-battery transistor circuit shown, prove that the stabilization factor S is given by $\mathbf{S} = \dfrac{1+\beta}{1 + \beta R_E/(R_B + R_E)}$.**

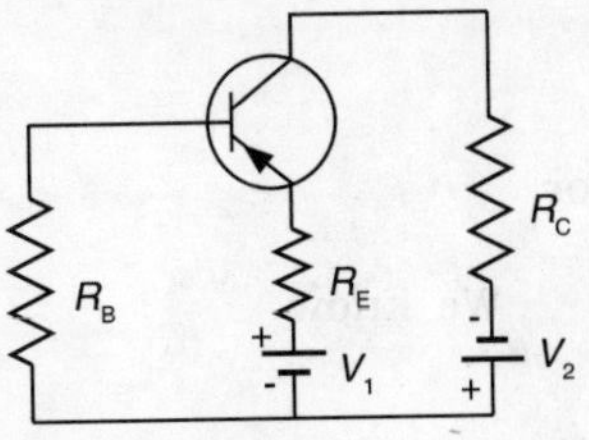

**Fig. 2.13**

**Solution:** Neglecting $V_{BE}$, we obtain from the circuit

$$V_1 = I_E R_E - I_B R_B$$

Since $I_E = -(I_C + I_B)$,

$$V_1 = -(I_B + I_C)R_E - I_B R_B$$

or

$$I_B = -\frac{I_C R_E + V_1}{R_E + R_B}$$

We know $$I_C = (1+\beta)I_{CO} + \beta I_B$$

$$I_C = (1+\beta)I_{CO} - \left(\frac{\beta}{R_E + R_B}\right)(I_C R_E + V_1)$$

or $$I_C\left(1+\frac{\beta R_E}{R_E+R_B}\right) = (1+\beta)I_{CO} - \frac{\beta V_1}{R_E+R_B}$$

We know $$S = \frac{\Delta I_C}{\Delta I_{CO}}$$

Differentiating we have

$$S\left(1+\frac{\beta R_E}{R_E+R_B}\right) = (1+\beta)$$

or $$S = \frac{1+\beta}{1+\left(\beta R_E / R_E + R_B\right)}$$

**6. In the circuit shown in Fig. 2.14(a), $V_{CC} = 24$ V, $R_C = 10$ kΩ, and $R_E = 270$ Ω. If a silicon transistor is used with $\beta = 45$ and if under quiescent conditions $V_{CE} = 5V$, determine (a) R (b) the stability factor S.**

**Data:** $V_{CC} = 24$ V, $R_C = 10$ kΩ, $R_E = 270$ Ω, $V_{CE} = 5V$ and $\beta = 45$

**To find:** R and S

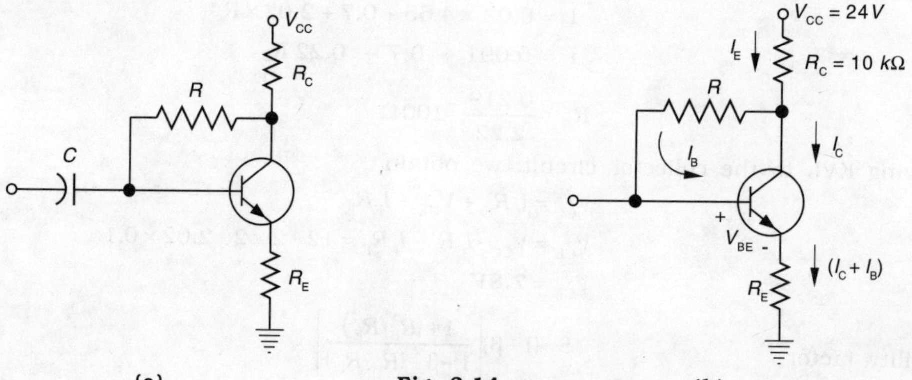


**Fig. 2.14**

**Solution:** (a) Applying KVL to the collector circuit, we have

$$V_{CC} = (I_C + I_B)R_C + V_{CE} + (I_C + I_B)R_E$$

$$24 = (I_C + I_B)10 + 5 + (I_C + I_B)\times 0.27$$

or $$I_C + I_B = \frac{19}{10.27} = \mathbf{1.85\ mA}$$

We know $$I_C = \beta I_B = 45 I_B$$

∴ $$46 I_B = 1.85 \quad \text{or} \quad I_B = \frac{1.85}{46} = \mathbf{0.0403\ mA}$$

and $$I_C = 45 \times 0.0403 = \mathbf{1.81\ mA}$$

From the figure, $$V_{CE} = V_{BE} + I_B \times R$$

or $$5 = 0.7 + I_B R$$

$$R = \frac{4.3}{0.0403} = \mathbf{107\ k\Omega}$$

(b) Applying KVL to the collector-base circuit

$$V_{CC} = (I_B + I_C)(R_C + R_E) + I_B R + V_{BE} \longrightarrow (1)$$

or $$I_B = \frac{V_{CC} - V_{BE} - I_C(R_C + R_E)}{R_C + R_E + R} \longrightarrow (2)$$

We know $$I_C = \beta I_B + (\beta + 1)I_{CO}$$

Substituting $I_B$ in Eq.(2), we get

$$I_C = \beta\left[\frac{V_{CC} - V_B - I_C(R_C + R_E)}{(R_C + R_E + R)}\right] + (\beta + 1)I_{CO}$$

or $$I_C\left[1 + \frac{\beta(R_C + R_E)}{R_C + R_E + R}\right] = (\beta + 1)I_{CO} + \beta\left(\frac{V_{CC} - V_{BE}}{R_C + R_E + R}\right)$$

Stability factor, $$S = \frac{\Delta I_C}{\Delta I_{CO}}$$

$$S = \frac{\beta + 1}{1 + [\beta(R_C + R_E)/(R_C + R_E + R)]}$$

$$= \frac{46}{1 + (45 \times 10.27/117.27)} = \mathbf{9.31}$$

**7. In the transformer-coupled amplifier shown in Fig. 2.15, $V_{BE} = 0.7$ V, and the quiescent voltage is $V_{CE} = 4V$. Determine (a) $R_E$ (b) S.**

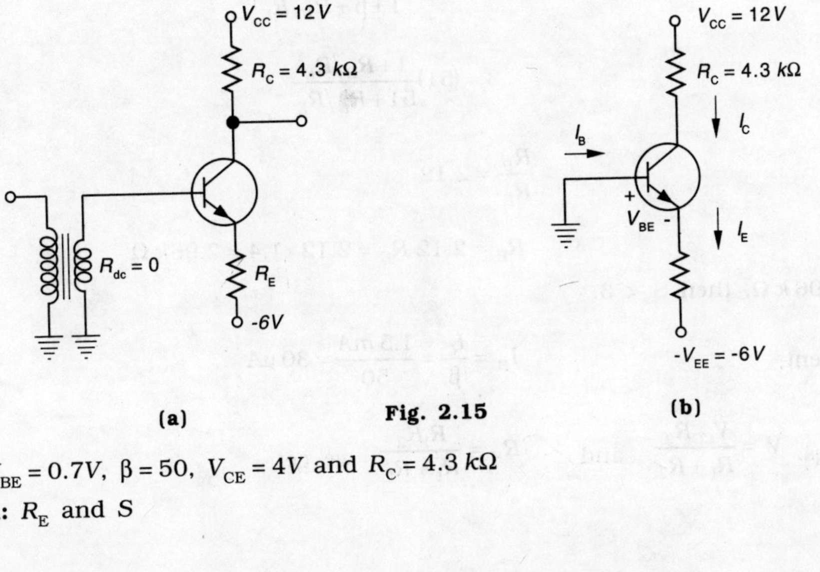


**Fig. 2.15**

**Data:** $V_{BE} = 0.7V$, $\beta = 50$, $V_{CE} = 4V$ and $R_C = 4.3\ k\Omega$

**To find:** $R_E$ and S

**Solution:** (a) Applying KVL to the collector-omitter circuit, we have

$$V_{CC} + V_{EE} = I_C R_C + (I_C + I_B)R_E + V_{CE}$$

$$12+6=4.3I_C + R_E\left(1+\frac{1}{50}\right)I_C + 4 \qquad \longrightarrow(1)$$

For base-emitter circuit, $V_{EE} = (I_C + I_B)R_E + V_{BE}$

$$6 = \left(1+\frac{1}{50}\right)I_C R_E + 0.7 \qquad \longrightarrow(2)$$

Solving Eqs. (1) and (2), we have

$$I_C = 2\ mA \text{ and } R_E = 2.74\ k\Omega$$

(b) Comparing the circuit with that of Fig. 2.5(b), we see that they are similar except here $R_B = 0$.

Under this condition, $S = \mathbf{1}$

**8. Assume that a silicon transistor with $\beta = 50$, $V_{BE,\ active} = 0.7$, $V_{CC} = 22.5V$, and $R_c = 5.6\ k\Omega$ is used in Fig. 2.5(a). It is desired to establish a $Q$ point at $V_{CE} = 12V$, $I_C = 1.5\ mA$ and stability factor $S \le 3$. Find $R_1$, $R_2$ and $R_E$.**

**Data:** $V_{CC} = 22.5V$, $V_{BE,active} = 0.7V$, $\beta = 50$, $R_C = 5.6\ k\Omega$, $V_{CE} = 12V$ and $I_C = 1.5\ mA$

**To find:** $R_1$, $R_2$ and $R_E$

**Solution:** Since the current in $R_E$ is $(I_C + I_B)$, from the collector circuit of Fig. 2.5(b) we have

$$R_E + R_C = \frac{V_{CC} - V_{CE}}{I_C}$$

$$R_E + R_C = \frac{22.5-12}{1.5} = \frac{10.5}{1.5} = 7k\,\Omega$$

and $R_E = 7k\,\Omega - R_C = 7 - 5.6 = \mathbf{1.4\,k\,\Omega}$

We know, $$S = (1+\beta)\frac{1 + R_B/R_E}{1+\beta+R_B/R_E}$$

$$3 = (51)\frac{1+R_B/R_E}{51+R_B/R_E}$$

Solving, $$\frac{R_B}{R_E} = 2.12$$

$\therefore$ $$R_B = 2.12\ R_E = 2.12 \times 1.4 = 2.96k\,\Omega$$

If $R_B < 2.96\,k\,\Omega$, then $S < 3$.

Base current, $$I_B = \frac{I_C}{\beta} = \frac{1.5\ mA}{50} = 30\ \mu A$$

Solving Eqs. $V = \dfrac{V_{CC}R_2}{R_1 + R_2}$ and $R_B = \dfrac{R_1 R_2}{R_1 + R_2}$, we get

$$R_1 = R_B \frac{V_{CC}}{V} \quad \text{and} \quad R_2 = \frac{R_1 V}{V_{CC} - V}$$

We know $\quad V = I_B R_B + V_{BE} + (I_C + I_B) R_E$

$$V = 0.03 \times 2.96 + 0.7 + (0.03 + 1.5)1.4 = \mathbf{2.93\,V}$$

$$R_1 = R_B \times \frac{V_{CC}}{V} = 2.96 \times \frac{22.5}{2.93} = \mathbf{22.8\,k\,\Omega}$$

and $\quad R_2 = \dfrac{R_1 V}{V_{CC} - V} = \dfrac{22.8 \times 2.93}{22.5 - 2.93} = \mathbf{3.4\,k\,\Omega}$

**9. Prove that for the circuit of Fig. 2.5(b) the stability factor S′ is given by $S'' = \dfrac{I_C S}{\beta(1+\beta)}$.**

**Solution:** We know $\quad I_C = \beta \dfrac{(V + V' - V_{BE})}{R_B + R_E(1+\beta)}$

Differentiating the above equation with respect to β, we have

$$\frac{\Delta I_C}{\Delta \beta} = \frac{[R_B + R_E(1+\beta)][V + V' - V_{BE}] - \beta[V + V' - V_{BE}][R_E]}{[R_B + R_E(1+\beta)]^2}$$

$$= \frac{(R_B + R_E)(V + V' - V_{BE})}{[R_B + R_E(1+\beta)]^2} = \frac{(R_B + R_E)}{R_B + R_E(1+\beta)} \cdot \frac{I_C}{\beta}$$

$$\frac{\Delta I_C}{\Delta I_B} = \frac{I_C S}{(1+\beta)\beta}$$

**10. Determine the values of $I_C$ and $V_{CE}$ in self-bias circuit if $R_1$ = 10 kΩ, $R_2$ = 5.6 kΩ, $R_E$ = 560 Ω, $R_C$ = 1 kΩ, $V_{CC}$ = 10 V and β = 100.**

**Data:** $R_1 = 10k\,\Omega,\ R_2 = 5.6k\,\Omega,\ R_E = 560\Omega,\ R_C = 1k\,\Omega,\ V_{CC} = 1$

**To find:** $I_C$ and $V_{CE}$

**Solution:** We know that $\quad V = V_{CC} \dfrac{R_2}{R_1 + R_2}$

$$V = 10 \times \frac{5.6}{10 + 5.6} = 3.59\,V$$

We know that $\quad R_B = \dfrac{R_1 R_2}{R_1 + R_2} = \dfrac{10 \times 5.6}{15.6} = 3.59k\,\Omega$

$$V = I_B R_B + V_{BE} + I_E R_E$$

$$3.59 = I_B \times 3.59 + 0.7 + (I_C + I_B)0.56$$

$$2.89 = I_B(3.59 + 0.56) + 0.56 I_C$$

$$2.89 = 4.15 I_B + 0.56 \times 100 I_B = 60.15 I_B$$

or
$$I_B = \frac{2.89}{60.15} = 0.048\,mA$$

$$I_C = \beta I_B = 100 \times 0.048 = \mathbf{4.8\,mA}$$

$$I_E = I_C + I_B = 4.8 + 0.048 = \mathbf{4.848\,mA}$$

Applying KVL to the collector circuit, we have

$$V_{CC} = I_C R_C + V_{CE} + I_E R_E$$

$$10 = 4.8 \times 1 + V_{CE} + 4.848 \times 0.56$$

$$\therefore \quad V_{CE} = \mathbf{2.49\,V}$$

## QUESTIONS

1. Explain the concept of *dc* bias in a linear amplifier.
2. What is meant by transistor biasing? Explain why a transistor should be biased.
3. Draw a *dc* load line for a given biased transistor circuit.
4. What ratings limit the range of operation of a transistor?
5. Why is capacitive coupling used to connect a signal source to an amplifier?
6. Explain quiescent operating point.
7. Explain the conditions for linear operation.
8. Explain the factors against which an amplifier needs to be stabilized.
9. For a capacitively coupled load, is the *dc* load larger or smaller than the *ac* load? Explain.
10. Explain with a diagram the fixed-bias circuit is unsatisfactory if the transistor is replaced by another of the same type.
11. What is meant by stabilization? Explain thermal instability.
12. Explain with a diagram qualitatively why a self-bias circuit is an improvement on the fixed-bias circuit, as far as stability is concerned.
13. How is the load line drawn for a self-bias circuit? Justify your answer.
14. Define bias curve. Explain how bias curve is used to obtain the quiescent point for self-bias circuit.
15. Mention the three sources of instability of collector current. Define the three stability factors.
16. Explain properly how to minimize the percentage variations in $I_C$ (i) due to variations in $I_{CO}$ and $V_{BE}$ and (ii) due to variations in $\beta$.
17. Over what temperature range can a transistor be used if it is silicon/germanium?
18. The collector-current variation is usually greater due to which parameter change $I_{CO}$ or $V_{BE}$ for (a) silicon and (b) germanium?
19. Define (a) stabilization techniques and (b) compensation techniques.
20. Explain with a circuit which uses a diode to compensate for changes (a) in $V_{BE}$ and (b) in $I_{CO}$.
21. Explain with circuit diagrams a properly biased integrated circuit linear amplifier. How are the parameter values chosen so that the quiescent output voltage is $V_{CC}/2$?

## EXERCISES

1. Determine the values of $I_B$, $I_C$ and $V_{CE}$ in a fixed bias circuit. Given $R_B$ = 180 kΩ, $V_{CC}$ = 25 V and β = 80. **[Ans: 0.14 *mA*; 11.2 *mA*; and 15.8 V]**
2. For the circuit of self-bias, draw the *dc* load line and mark the Q-point of the circuit. Assume $V_{BE}$ = 0.7 V. Given $R_1$= 5kΩ, $R_2$ = 5 kΩ, $R_C$= 2 kΩ, $R_E$= 3kΩ. **[Ans: 143 *mA*; 12.85 V]**
3. Find the values of $I_C$ and $V_{CE}$ for self-bias circuit if $R_1$ = 68 kΩ, $R_2$ = 47 kΩ, $R_C$= 1.8 kΩ, $R_E$= 2.2 kΩ, $V_{CC}$ = - 6 V and β = 75. **[Ans: -636 μA; -3.46 V]**
4. Transistor type 2N335, used in fixed-bias circuit, may have any value between 36 and 90 at a temperature of 25°C and the leakage current has negligible effect on $I_C$ at room temperature. Find $R_1$, $R_2$ and $R_E$ subject to the following specifications: $R_C$ = 4 kΩ, $V_{CC}$ = 20 V; the nominal bias point is to be at $V_{CE}$ = 10 V, $I_C$ = 2 *mA* and $I_C$ should be in the range 1.75 *mA* to 2.25 *mA* as β varies from 36 to 90. **[Ans: 117 kΩ; 24.2 kΩ and 1 kΩ]**
5. For the self-bias circuit, $R_C$ = 4.7 kΩ, $R_B$ = 7.75 kΩ and $R_B/R_E$ = 1.65. The collector supply voltage and $R_C$ are adjusted to establish a collector current of 1.5 *mA*, at 25°C.
   (a) Determine the variation of $I_C$ in the temperature range of -65 to +175°C when Si transistor of Table 2.1 is used.
   (b) Repeat (a) for the range -65 to +75°C when Ge transistor of Table 2.2 is used.
   **[Ans: (a) -0.118 *mA* to 0.199 *mA*; (b) -0.159 *mA* to 0.131 *mA*]**
6. A *p-n-p* Ge transistor is used in the self-biasing circuit. The circuit component values are $V_{CC}$ = 4.5 V, $R_C$ = 1.5 kΩ, $R_E$ = 0.27 kΩ, $R_2$ = 2.7 kΩ and $R_1$ = 27 kΩ. If β = 44.
   (a) Find the quiescent point. (b) Recalculate these values if the base spreading resistance of 690 Ω is taken into account.
   **[Ans: -0.63 *mA*; -3.38V and -0.60 *mA*; -3.43V]**
7. (a) A Ge transistor is used in self-bias circuit with $V_{CC}$ = 16V and $R_C$ = 1.5 kΩ. The quiescent point is chosen to be $V_{CE}$ = 8V and $I_C$ = 4 *mA*. A stability factor S = 12 is desired. If β = 50, find $R_1$, $R_2$ and $R_E$. (b) Repeat part (a) for S = 3.
   **[Ans: (a) $R_E$ = 0.49 kΩ; $R_1$ = 41 kΩ; $R_2$ = 8.56 kΩ and (b) $R_E$ = 0.49 kΩ; $R_1$ = 7.3 kΩ and $R_2$ = 1.21 kΩ]**
8. Determine the stability factor S for the circuit shown in Fig. 2.16.

   **[Ans:** $S = \frac{(\beta+1)[R_2(R_C+R_1)+R_E(R_1+R_2+R_C)]}{R_1R_2+(\beta+1)[R_2R_C+R_E(R_1+R_2+R_C)]}$ **]**
9. In the two stage circuit shown in Fig. 2.17, assume β = 100 for each transistor. (a) Determine R so that the quiescent conditions are $V_{CE_1}$ = −4V and $V_{CE_2}$ = −6V.
   (b) Explain how quiescent point stabilization is obtained. Assume $V_{BE}$ = 0.2V.
   **[Ans: (a) 210 kΩ; (b) By negative *dc* feedback through R as well as 1 kΩ and 3 kΩ]**

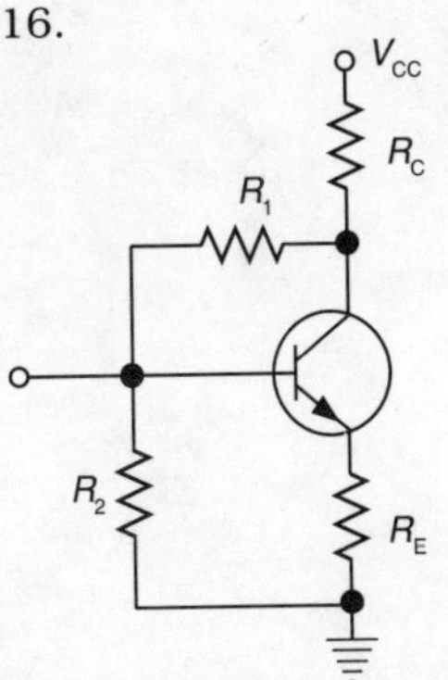

**Fig. 2.16**

10. In the Darlington stage shown in Fig. 2.18, $V_{CC} = 24V$, $\beta_1 = 24$, $\beta_2 = 39$, $V_{BE} = 0.7V$, $R_C = 330\ \Omega$ and $R_E = 120\ \Omega$. If at the quiescent point $V_{CE_2} = 6V$, determine (a) $R$ (b) the stability factor defined as $S = dI_C/dI_{CO_1}$. **[Ans: (a) 115 kΩ; (b) 198]**

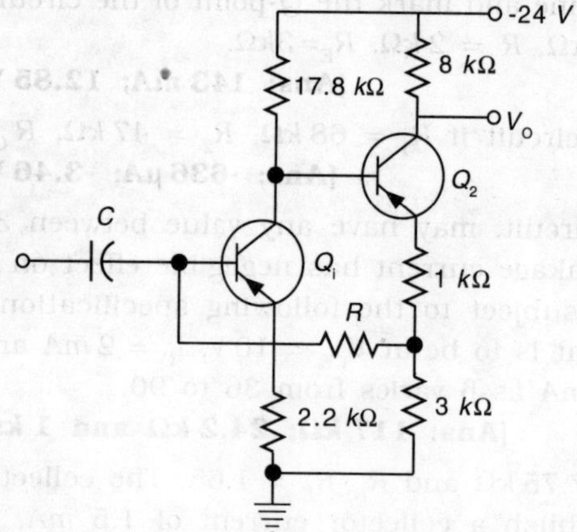


**Fig. 2.17**

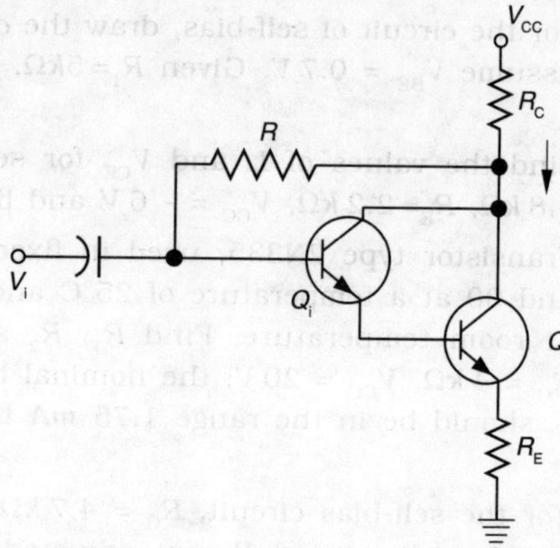


**Fig. 2.18**

# 3 THE TRANSISTOR AT LOW FREQUENCIES

## Chapter Outline

- Introduction
- Graphical Analysis of the *C-E* Configuration
- Two-Port Devices and Hybrid Model
- Transistor Hybrid Model
- The *h* Parameters
- Analysis of a Transistor Amplifier Circuit Using *h* Parameters
- The Emitter Follower
- Miller's Theorem and its Dual

## INTRODUCTION

The biasing of a transistor is purely a *dc* operation. The purpose of biasing is to establish a *Q*-point about which variations in current and voltage can occur in response to an *ac* input signal. In applications where small signal voltages must be amplified — such as from an antenna, a microphone, etc. — variations about the *Q*-point are relatively small. Amplifiers designed to handle these small *ac* signals are known as **small-signal amplifiers**.

In small-signal applications, the device can be replaced by a linear model and the circuit can be analysed as a linear circuit. The model in such a case will be applicable only for a specified operating point, as the model parameters will change with the operating condition. A model is a collection of circuit elements such as *R,C* current source, voltage source, etc., properly chosen, that best approximates the working of the active device.

In practice, a number of stages are used in cascade to amplify a signal from a source. In examining all transistor circuits at low frequencies, the transistor internal capacitances may be neglected.

### AC Quantities

The *dc* quantities are represented by non-italic uppercase (capital) subscripts such as $I_C$, $I_B$, $I_E$, $V_{BE}$ and $I_{CE}$. Instantaneous quantities which vary with time are represented by lowercase letters and subscripts as $i_c$, $i_b$, $i_e$, $v_{be}$ and $v_{ce}$. Maximum, average (*dc*) and effective or root-mean-square (rms) values are represented by the uppercase letters of the proper symbols (*I*, *V*, or *P*). Varying components from some quiescent value are represented by the lowercase subscript of the proper electrode symbol. A single subscript is used if the reference electrode is clearly understood. If there is an ambiguity, the conventional double subscript notation should be used. Figure 3.1 illustrates these quantities for collector and base currents and voltages in C-E transistor configuration. The collector and emitter current and voltage component variations from the corresponding quiescent values are:

$$i_c = i_C - I_C = \Delta i_C \qquad v_c = v_C - V_C = \Delta v_C \qquad \longrightarrow (1)$$

$$i_b = i_B - I_B = \Delta i_B \qquad v_b = v_B - V_B = \Delta v_B$$

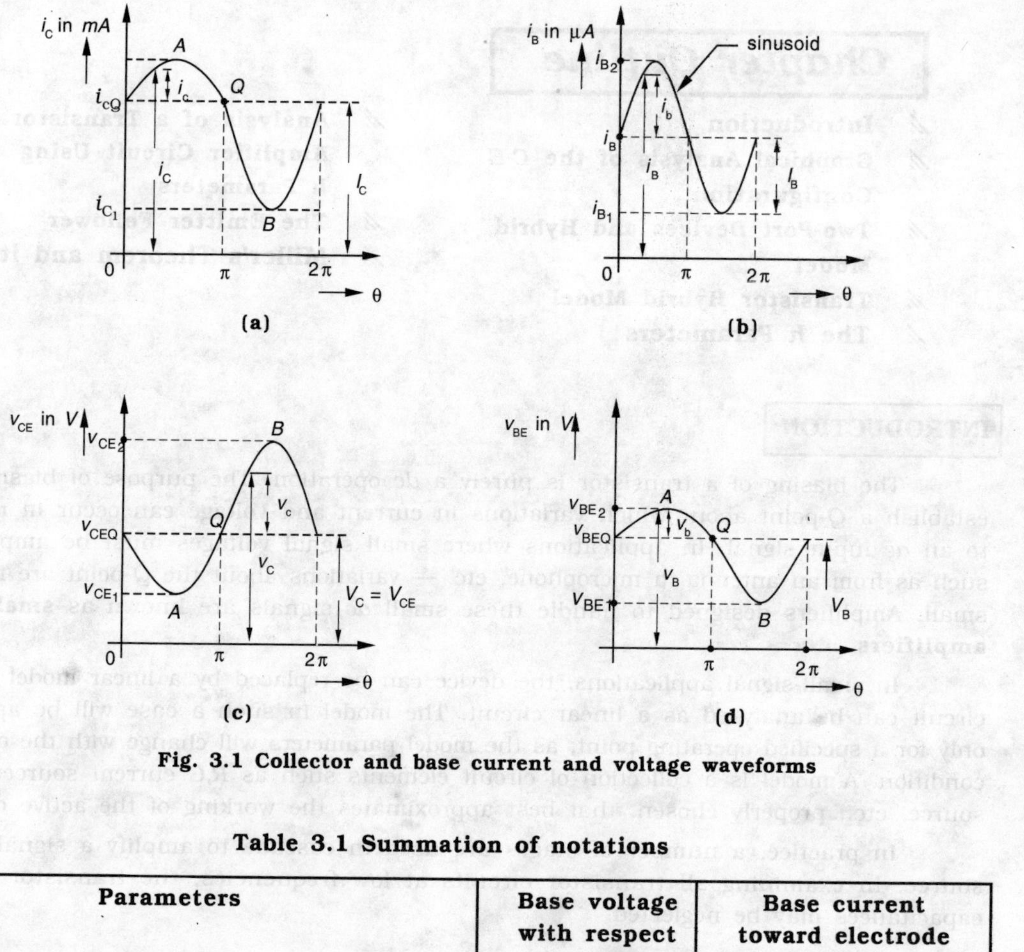


**Fig. 3.1 Collector and base current and voltage waveforms**

**Table 3.1 Summation of notations**

| Parameters | Base voltage with respect to emitter | Base current toward electrode from external circuit |
|---|---|---|
| Instantaneous total value | $v_B$ ($v_C$) | $i_B$ ($i_C$) |
| Quiescent value | $V_B$ ($V_C$) | $I_B$ ($I_C$) |
| Instantaneous value of varying component | $v_b$ ($v_c$) | $i_b$ ($i_c$) |
| Effective value of varying component | $V_b$ ($V_c$) | |
| Supply voltage | $V_{BB}$ ($V_{CC}$) | $I_b$ ($I_c$) |

## GRAPHICAL ANALYSIS OF THE C-E CONFIGURATION

Consider the C-E transistor configuration circuit, shown in Fig. 3.2. Let us understand the graphical operation of the circuit. Figure 3.3 shows the input and output characteristics of a transistor. A load line is drawn in Fig. 3.3(a). The quiescent operating point $Q$ is usually selected at the centre of the load line.

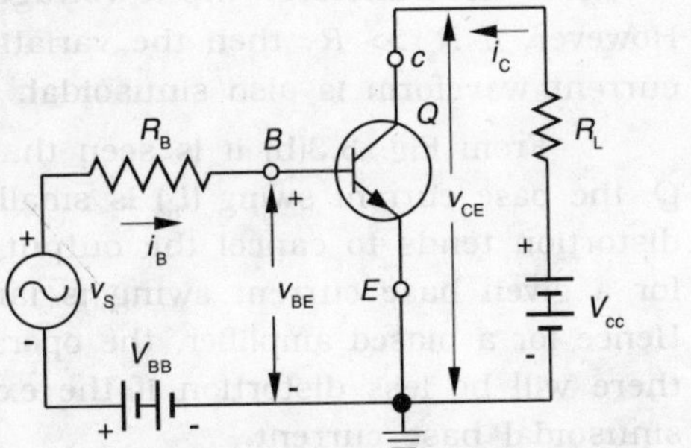


**Fig. 3.2 The C-E transistor configuration**

Assume a 200 $\mu A$ (= $i_b$) peak sinusoidally varying base current signal around the quiescent point $Q$ where $I_B$ = 300 $\mu A$ (= $I_{B_3}$). The extreme points of the base waveform then are A and B, where $i_B$ = $I_{B_5}$ (= 500 $\mu A$) and $i_B$ = $I_{B_1}$ (= 100 $\mu A$) respectively. The corresponding values of $i_C$ and $v_{CE}$ at point A are $i_{C_2}$ and $v_{CE_1}$, and at point B they are $i_{C_1}$ and $v_{CE_2}$. The waveforms of $i_C$ and $v_{CE}$ are plotted in Fig. 3.1(a) and (b). It is seen that the collector current and collector voltage waveforms are not the same as the base current waveform [Fig. 3.1(c)] i.e., not symmetrical because the collector characteristics in the neighbourhood of load line in Fig. 3.3(a) are not parallel lines equally spaced for equal increments in base current. This variation in waveform is known as **output nonlinear distortion**.

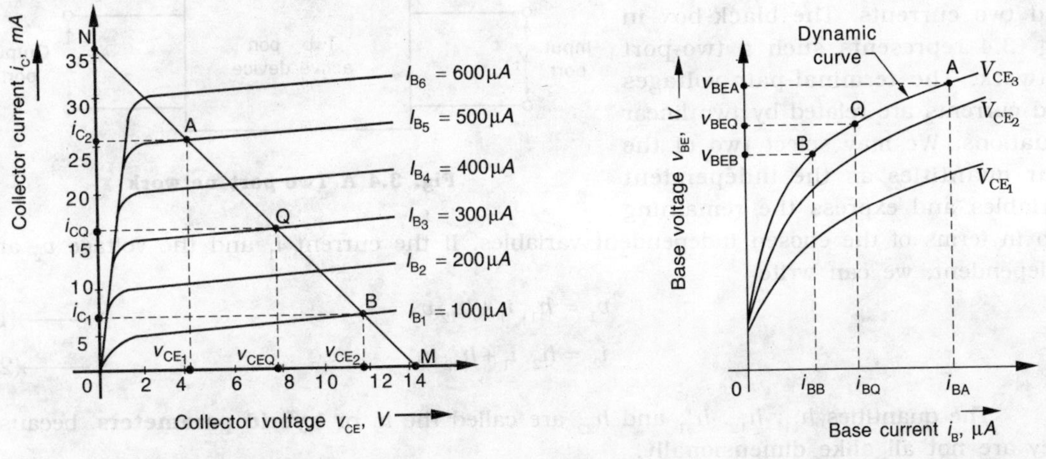


**(a) Output characteristics** **Fig. 3.3** **(b) Input characteristics**

The base-to-emitter voltage $v_{BE}$ for any combination of base current and collector-to-emitter voltage can be obtained from the input characteristic curves [Fig. 3.3(b)]. The dynamic operating curve drawn for the combinations of base current and collector voltage along A-Q-B of the load line of Fig. 3.3(a). The waveform of $v_{BE}$ can be obtained from the dynamic operating curve of Fig. 3.3(b) by knowing the voltage $v_{BE}$ corresponding to a given base current $i_B$. Since the dynamic curve is not a straight line, the waveform of $v_b$ [Fig. 3.1(d)] is not the same as the waveform of $i_b$. This variation in waveform is known as **input nonlinear distortion**. In some cases it is found that $i_b$ will be distorted though

$v_b$ is sinusoidal. The above condition will be true if the sinusoidal voltage source $v_s$ has a small output resistance $R_s$ in comparison with the input resistance $R_i$ of the transistor, so that the transistor input-voltage waveform is the same as the source waveform. However, if $R_s >> R_i$, then the variation in $i_B$ is given by $i_b \approx v_s/R_s$, and hence the base-current waveform is also sinusoidal.

From Fig. 3.3(b) it is seen that for a large sinusoidal base voltage $v_b$ around point $Q$, the base current swing ($i_b$) is smaller to the left of $Q$ than to the right of $Q$. This input distortion tends to cancel the output distortion because, the collector current swing $|i_c|$ for a given base current swing is large over the section BQ than over QA [Fig. 3.3(a)]. Hence for a biased amplifier, the operating point $Q$ is near the centre of $i_C$-$v_{CE}$ plane. Then there will be less distortion if the excitation is a sinusoidal base voltage than if it is a sinusoidal base current.

It is important to note that the dynamic load curve can be approximated by a straight line over a sufficiently small line segment. Hence, if the input signal is small there will be negligible input distortion under any condition of operation.

## TWO-PORT DEVICES AND HYBRID MODEL

Usually an amplifying device has only three independent terminals as in case of VT, BJT and FET. For incremental small signals the device can be characterised as a linear two-port network whose terminal behaviour is specified by two voltages and two currents. The black-box in Fig. 3.4 represents such a two-port network. The terminal-pair voltages and currents are related by two linear equations. We may select two of the four quantities as the independent variables and express the remaining two in terms of the chosen independent variables. If the current $i_1$ and the voltage $v_2$ are independent, we can write

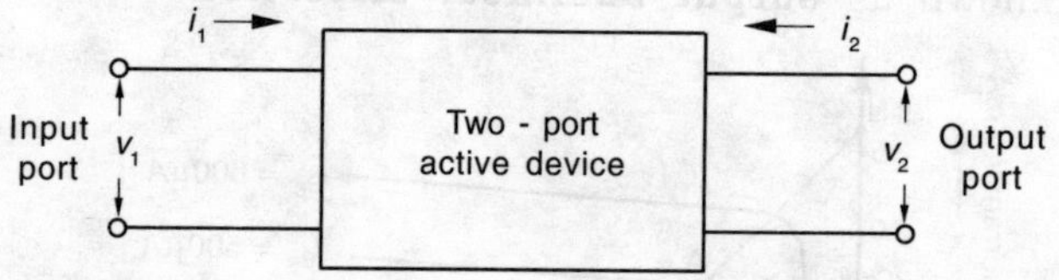


**Fig. 3.4 A Two-port network**

$$v_1 = h_{11}\, i_1 + h_{12}\, v_2 \qquad (1)$$

$$i_2 = h_{21}\, i_1 + h_{22}\, v_2 \qquad (2)$$

The quantities $h_{11}$, $h_{12}$, $h_{21}$ and $h_{22}$ are called the $h$, or **hybrid parameters**, because they are not all alike dimensionally.

**The parameters that completely describe the behaviour of the circuit and are constant for a given circuit are called *h* parameters.**

Assume that there are no reactive elements present in the two-port network. Then, from Eqs. (1) and (2) $h$-parameters are defined as follows:

$$h_{11} = \left.\frac{v_1}{i_1}\right|_{v_2 = 0} = h_i = \text{input resistance with output shorted } (\Omega)$$

$$h_{12} = \left.\frac{v_1}{v_2}\right|_{i_1 = 0} = h_r = \text{reverse open-circuit voltage gain (dimensionless)}$$

$$h_{21} = \left.\frac{i_2}{i_1}\right|_{v_2 = 0} = h_f = \text{short-circuit forward crrent gain (dimensionless)}$$

$$h_{22} = \left.\frac{i_2}{v_2}\right|_{i_1 = 0} = h_o = \text{output admittance with input open (S)}$$

In the case of transistors, another subscript (*b*, *e* or *c*) is added to designate the type of configuration.

The four parameters $h_{11}$, $h_{12}$, $h_{21}$ and $h_{22}$ are real numbers and the voltages and currents $v_1$, $v_2$ and $i_1$, $i_2$ are functions of time.

**The Model:** A circuit model appropriate to the *h* parameters is shown in Fig. 3.5. We can verify that the model of Fig. 3.5 satisfies Eqs. (1) and (2) by writing Kirchhoff's voltage and current laws for the input and output ports, respectively. Note that the input circuit contains a dependent voltage generator whereas the output circuit contains a dependent current source.

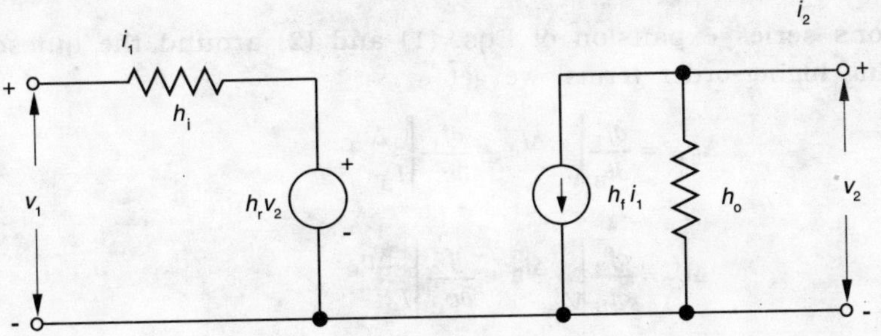


**Fig. 3.5 The hybrid model for the two-port network of Fig. 3.4**

## Advantages of *h*-parameters

1. *h* parameters are real numbers at audio frequencies (20 *Hz* to 20 *kHz*).
2. They can be obtained from the transistor characteristic curves.
3. They are relatively easy to measure.
4. They are convenient to use in circuit analysis and design.
5. They are specified by transistor manufacturers.

The four basic *h* parameters and their descriptions are given in Table 3.2. Each of the four *h* parameters carries a second subscript letter to designate the common-emitter(*e*), common-base(*b*) or common-collector(*c*) amplifier configuration as listed in Table 3.3.

**Table 3.2 Basic *h* parameters**

| ***h* parameter** | **Description** | **Condition** |
|---|---|---|
| $h_i$ | Input resistance | Output shorted |
| $h_r$ | Reverse voltage gain | Input open |
| $h_f$ | Forward current gain | Output shorted |
| $h_o$ | Output admittance | Input open |

**Table 3.3 Subscripts of *h* parameters for the three amplifier configuration**

| Configuration | *h* parameters |
|---|---|
| Common-emitter | $h_{ie}$, $h_{re}$, $h_{fe}$, $h_{oe}$ |
| Common-base | $h_{ib}$, $h_{rb}$, $h_{fb}$, $h_{ob}$ |
| Common-collector | $h_{ic}$, $h_{rc}$, $h_{fc}$, $h_{oc}$ |

## TRANSISTOR HYBRID MODEL

Consider the common-emitter configuration circuit shown in Fig. 3.2. Let $i_B$, $i_C$, $v_{BE} = v_B$ and $v_{CE} = v_C$ represent total instantaneous currents and voltages. Since $v_B$ is a function $f_1$ of $i_B$ and $v_C$ and $i_C$ is another function $f_2$ of $i_B$ and $v_C$, we can write

$$v_B = f_1\ (i_B,\ v_C) \longrightarrow (1)$$

$$i_C = f_2\ (i_B,\ v_C) \longrightarrow (2)$$

Making a Taylor's series expansion of Eqs. (1) and (2) around the quiescent point $I_B$, $V_C$ and neglecting higher-order terms, we get

$$\Delta v_B = \left.\frac{\partial f_1}{\partial i_B}\right|_{V_c} \Delta i_B + \left.\frac{\partial f_1}{\partial v_C}\right|_{I_B} \Delta v_C \longrightarrow (3)$$

$$\Delta i_C = \left.\frac{\partial f_2}{\partial i_B}\right|_{V_c} \Delta i_B + \left.\frac{\partial f_2}{\partial v_C}\right|_{I_B} \Delta v_C \longrightarrow (4)$$

where the quantities $\Delta v_B$, $\Delta v_C$, $\Delta i_B$ and $\Delta i_C$ represent small-signal base and collector voltages and currents, respectively,

Using equations $\quad i_c = i_C - I_C = \Delta i_C,\ v_c = v_C - V_C = \Delta v_C$

$$i_b = i_B - I_B = \Delta i_B \text{ and } v_b = v_B - V_B = \Delta v_B$$

We may write Eqs. (3) and (4) as

$$v_b = h_{ie}\, i_b + h_{re}\, v_c \longrightarrow (5)$$

$$i_c = h_{fe}\, i_b + h_{fe}\, v_c \longrightarrow (6)$$

Where $\quad h_{ie} = \dfrac{\partial f_1}{\partial i_B} = \left.\dfrac{\partial v_B}{\partial i_B}\right|_{V_c},\ h_{re} = \dfrac{\partial f_1}{\partial v_C} = \left.\dfrac{\partial v_B}{\partial v_C}\right|_{I_B} \longrightarrow (7)$

and $\quad h_{fe} = \dfrac{\partial f_2}{\partial i_B} = \left.\dfrac{\partial i_C}{\partial i_B}\right|_{V_c},\ h_{oe} = \dfrac{\partial f_2}{\partial v_C} = \left.\dfrac{\partial i_C}{\partial v_C}\right|_{I_B} \longrightarrow (8)$

The partial derivatives of Eqs. (7) and (8) define the *h* parameters of a transistor for the common-emitter configuration. We find that Eqs. (5) and (6) are exactly the same form as equations $v_1 = h_{11}\, i_1 + h_{12}\, v_2$, and $i_2 = h_{21}\, i_1 + h_{22}\, v_2$. Hence the model of Fig. 3.5 can be used to represent a transistor.

If a parameter is constant, its incremental change is zero. Hence $V_C$ = constant equivalent to $v_c = 0$ and $I_B$ = constant corresponds to $i_b = 0$. With this notation, we can write

$$h_{re} = \frac{v_B}{\partial v_C}\Big|_{I_B} = \frac{v_b}{v_c}\Big|_{i_b=0} \quad \text{or} \quad h_{re} = \frac{V_b}{V_c}\Big|_{I_b=0} \qquad \longrightarrow (9)$$

Figure 3.6 shows the hybrid small-signal models for common-emmitter and common collector configurations. For each configuration, we note from Kirchhoff's current law that

$$i_b + i_c + i_e = 0 \qquad \longrightarrow (10)$$

Note that the circuit models and equations are valid for either an *n-p-n* or a *p-n-p* transistor and are independent of the type of load or the method of biasing.

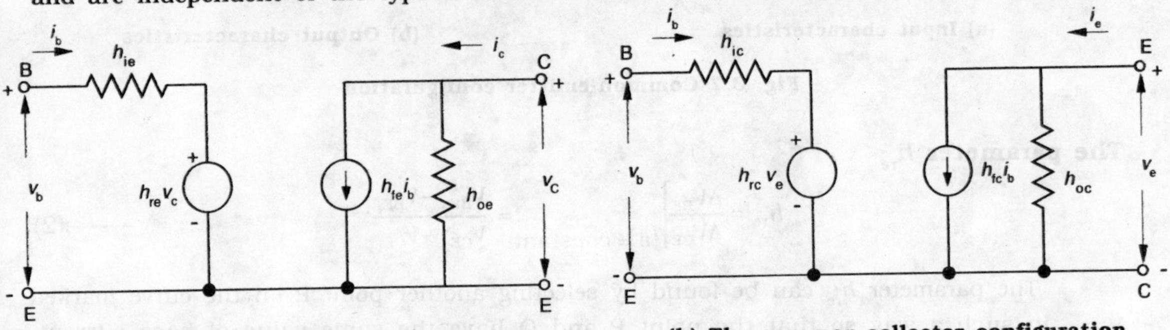


**(a) The common-emitter configuration** **(b) The common-collector configuration**

**Fig. 3.6 Hybrid small-signal models**

## THE *h* PARAMETERS

### Determination of *h* Parameters from the Transistor Characteristics

Consider common-emitter configuration of an *n-p-n* transistor. Two families of curves such as input characteristic and output characteristic curves are usually specified for transistors.

### Determination of $h_{ie}$ and $h_{re}$ from Input characteristics

### The parameter $h_{ie}$

For a common-emitter configuration the input characteristics are shown in Fig. 3.7(a). From the definition of $h_{ie}$, we have

$$h_{ie} = \frac{\Delta V_{BE}}{\Delta I_B}\Big|_{V_{CE}=\text{constant}} = \frac{V_{BE_2} - V_{BE_1}}{I_{B_2} - I_{B_1}} \qquad \longrightarrow (1)$$

Locate a point *Q* on the input characteristics as shown in Fig. 3.7(a). To find the value of $h_{ie}$, draw a tangent to the curve marked $V_{CE_2}$ and passing through the *Q*-point.

The parameter $h_{ie}$ is defined as the ratio of small change in base-to-emitter voltage $\Delta V_{BE}$ to the corresponding change in base current $\Delta I_B$ keeping collector voltage $V_{CE}$ constant at the quiescent point *Q*.

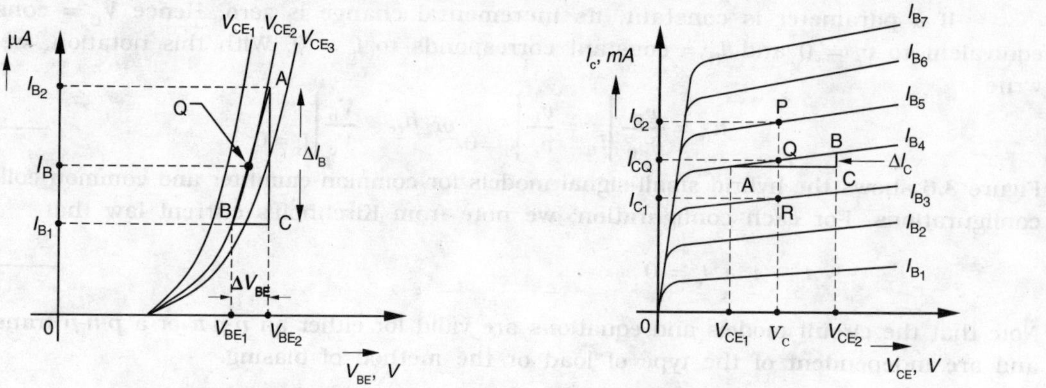


**(a) Input characteristics** **(b) Output characteristics**

**Fig. 3.7 Common-emitter configuration**

### The parameter $h_{re}$

$$h_{re} = \left.\frac{\Delta V_{BE}}{\Delta V_{CE}}\right|_{I_B = \text{constant}} = \frac{V_{BE_2} - V_{BE_1}}{V_{CE_2} - V_{CE_1}} \qquad \longrightarrow (2)$$

The parameter $h_{re}$ can be found by selecting another point P on the curve marked $V_{CE_1}$, in such a way so that the point P and Q have the same value of base current. Then measure the values of changes in collector-to-emitter voltage $\Delta V_{CE}$ i.e., $V_{CE_2} - V_{CE_1}$, and the resulting change in base-to-emitter voltage $\Delta V_{BE}$ i.e., $V_{BE_2} - V_{BE_1}$. Substituting these values in Eq. (2), we get the value of $h_{re}$.

### The parameter $h_{fe}$

The parameter $h_{fe}$ is the most important small-signal parameter of the transistor. The values of $h_{fe}$ and $h_{oe}$ parameters can be determined from the output characteristics shown in Fig. 3.7(b).

$$h_{fe} = \left.\frac{\Delta I_C}{\Delta I_B}\right|_{V_{CE} = \text{constant}} = \frac{I_{C_2} - I_{C_1}}{I_{B_2} - I_{B_1}} \qquad \longrightarrow (3)$$

In order to find the value of $h_{fe}$, select two points P and R (above and below the Q-point) in such a way that these points have the same collector-emitter voltage $V_c$. Then measure the values of selected changes in base current $\Delta I_B$ $(= I_{B_2} - I_{B_1})$ and the corresponding change in collector current $\Delta I_C$ $(= I_{C_2} - I_{C_1})$. Substituting these values in Eq. (3), we get the value of $h_{fe}$.

### The parameter $h_{oe}$

$$h_{oe} = \left.\frac{\Delta I_C}{\Delta V_{CE}}\right|_{I_B = \text{constant}} = \frac{I_{C_2} - I_{C_1}}{V_{CE_2} - V_{CE_1}} \qquad \longrightarrow (4)$$

The parameter $h_{oe}$ can be found by drawing a tangent AB to the output characteristic curve at Q. The slope of the tangent AB gives the value of $h_{oe}$.

## Hybrid-parameter Variations

The $h$ parameters are very effective in characterizing the linear low-frequency behaviour of transistors. Manufacturers often specify $h$ parameters in their data sheets. However, such specification is usually at only one operating point. If the $Q$ point changes, the $h$ parameters will also change. Since the characteristic curves are not in general straight lines, equally spaced for equal changes in $I_B$ or $V_{CE}$, therefore the values of the $h$ parameters depend upon the position of the operating point on the curves. The variation of $h$ parameters for a typical transistor depends upon the following factors:

(i) Collector current, $I_C$
(ii) Collector-to-emitter voltage, $V_{CE}$
(iii) Junction temperature
(iv) Frequency

The most interesting and useful at this stage of the development include the $h$ parameters with junction temperature ($T_j$), collector current ($I_C$) and emitter current ($I_E$).

**Variation of $h$ parameters with $I_C$:** Figure 3.8 (a) shows the effect of the collector current on the $h$ parameters. These curves are plotted with respect to the values of a specific

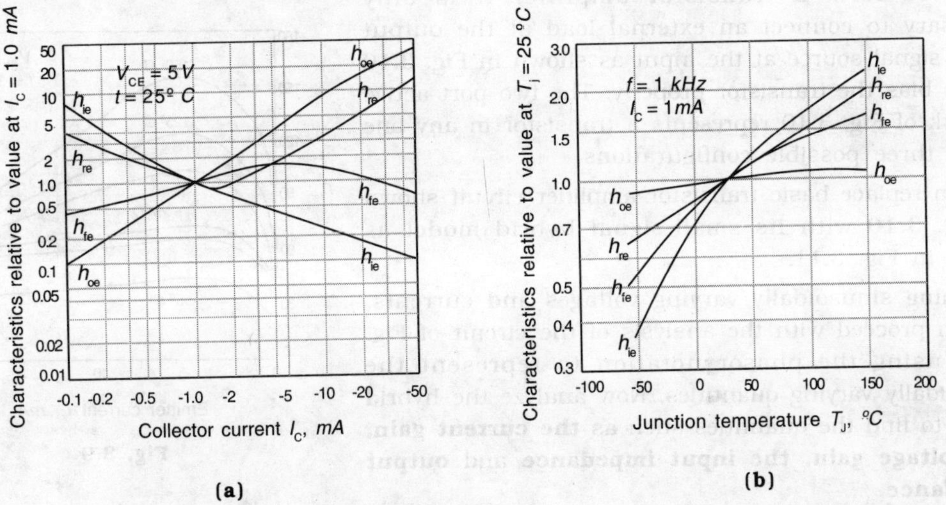


**Fig. 3.8 Variation of common-emitter $h$ parameters (a) with collector current (b) with junction temperature**

operating point. The variation in $h$ parameters as shown in Fig. 3.8(a) is for a constant junction temperature 25 ºC and frequency. Figure 3.8(b) shows the variation of $h$ parameters for changes in junction temperature. The fact that $h_{fe}$ will change from 50% of its normalized value at -50° C to 150% of its normalised value at 150° C indicates clearly that the operating temperature must be carefully considered in the design of transistor circuits.

Manufacturers also provide curves of $h$ parameters versus $V_{CE}$, although this variation with $V_{CE}$ is often not significant. Specifically, $h_{fe}$ is more sensitive to $I_C$ than $V_{CE}$. Most transistors exhibit a well-defined maximum in the value of $h_{fe}$ as a function of collector or emitter current. Such a maximum in the variation of $h_{fe}$ with emitter current and temperature is shown in Fig. 3.9.

Table 3.4 shows values of $h$ parameters for the three different transistor configurations of a typical junction transistor.

**Table 3.4 Typical *h*-parameter values for a transistor (at $I_E$ = 1.3 *mA*)**

| Parameter | C-E | C-B | C-C |
|---|---|---|---|
| $h_{11} = h_i$ | 1,100 Ω | 21.6 Ω | 1,100 Ω |
| $h_{12} = h_r$ | $2.5 \times 10^{-4}$ | $2.9 \times 10^{-4}$ | ~ 1 |
| $h_{21} = h_f$ | 50 | – 0.98 | – 51 |
| $h_{22} = h_o$ | 25 μ*A*/*V* | 0.49 μ*A*/*V* | 25 μ*A*/*V* |
| $1/h_o$ | 40 *k*Ω | 2.04 *M*Ω | 40 *k*Ω |

## ANALYSIS OF A TRANSISTOR AMPLIFIER CIRCUIT USING *h* PARAMETERS

To form a transistor amplifier it is only necessary to connect an external load at the output and a signal source at the input as shown in Fig. 3.10 and to bias the transistor properly. The two port active network of Fig. 3.10 represents a transistor in any one of the three possible configurations.

We can replace basic transistor amplifier circuit shown in Fig. 3.10 with its small-signal hybrid model as shown in Fig. 3.11.

Assuming sinusoidally varying voltages and currents, we can proceed with the analysis of the circuit of Fig. 3.11, using the phasor notation to represent the sinusoidally varying quantities. Now analyze the hybrid model to find the quantities such as **the current gain, the voltage gain, the input impedance** and **output impedance.**

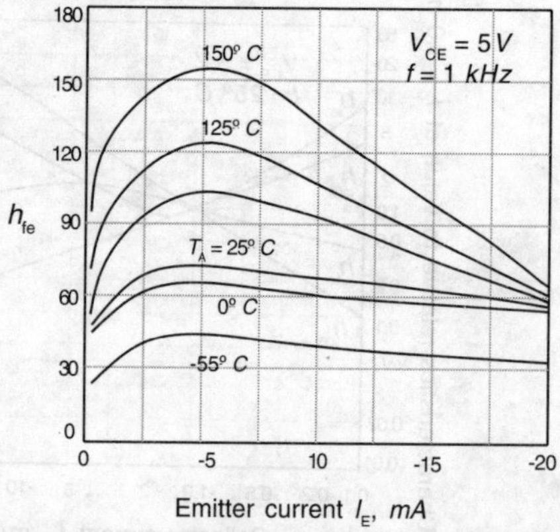


**Fig. 3.9**

**Current Gain ($A_I$):** For transistor amplifier, current gain is defined as the ratio of output current to input current. It is given by

$$A_I = \frac{I_L}{I_1} = \frac{-I_2}{I_1} \quad \text{(because } I_L = -I_2) \longrightarrow (1)$$

From the circuit of Fig. 3.11, we obtain

$$I_2 = h_f I_1 + h_o V_2 \longrightarrow (2)$$

Substituting $V_2 = -I_2 Z_L$ in Eq. (2), we get

$$I_2 = h_f I_1 + h_o (-I_2 Z_L)$$

or

$$I_2 (1 + h_o z_L) = h_f I_1$$

or

$$A_I = \frac{-I_2}{I_1} = -\frac{h_f}{1 + h_o Z_L} \longrightarrow (3)$$

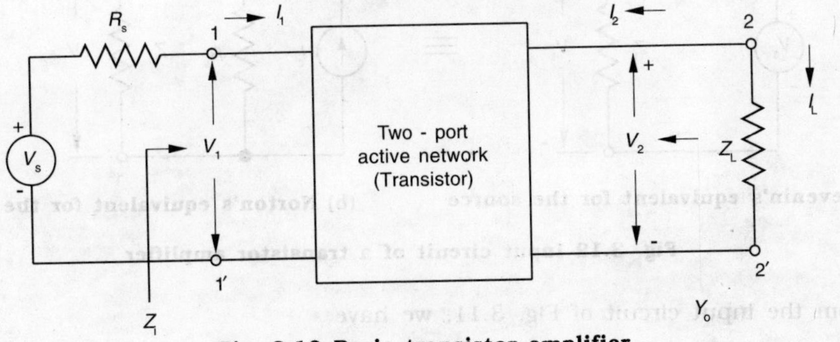


Fig. 3.10 Basic transistor amplifier

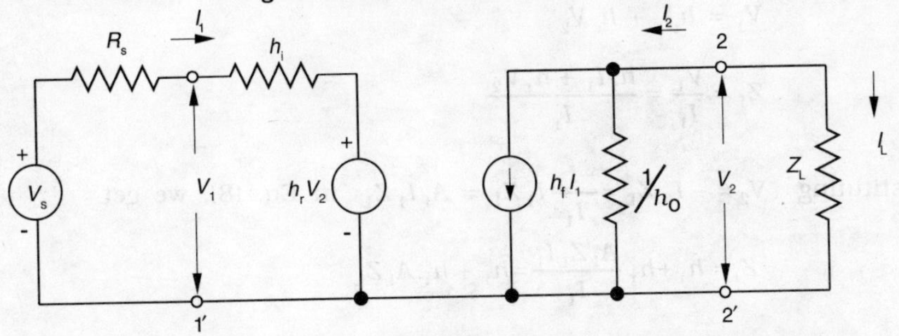


Fig. 3.11 Transistor amplifier in its *h*-parameter model

## Current Gain ($A_{IS}$), Taking into Account the Source Resistance $R_s$

If the hybrid model is driven by a current generator $I_s$ instead of a voltage generator (Fig. 3.12), then the **overall current gain** is defined as

$$A_{I_s} = \frac{-I_2}{I_s} = \frac{-I_2}{I_1} \cdot \frac{I_1}{I_s}$$

Since $A_I = \dfrac{-I_2}{I_1}$,

$$A_{I_s} = A_I \cdot \frac{I_1}{I_s} \longrightarrow (4)$$

From Fig. 3.12(b), we have $I_1 = \dfrac{I_s R_s}{Z_i + R_s}$

and hence

$$A_{I_s} = \frac{A_I R_s}{Z_i + R_s} \longrightarrow (5)$$

If $R_s = \infty$, then $A_{I_s} = A_I$. Hence **$A_I$ is current gain for an ideal current source.**

**Input Impedance ($Z_i$):** As shown in Fig. 3.11, the impedance looking into the amplifier input terminals $(1, 1')$ is the **amplifier input impedance** $Z_i$. It is given by

$$Z_i = \frac{V_1}{I_1} \longrightarrow (6)$$

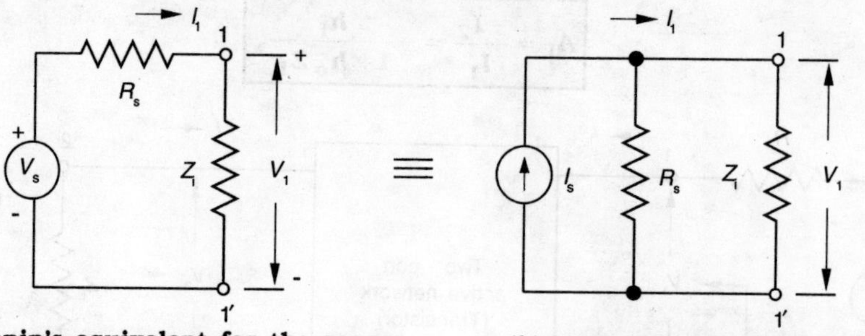


**(a) Thevenin's equivalent for the source** **(b) Norton's equivalent for the source**

**Fig. 3.12 Input circuit of a transistor amplifier**

From the input circuit of Fig. 3.11, we have

$$V_1 = h_i I_1 + h_r V_2 \qquad \longrightarrow (7)$$

Hence

$$Z_i = \frac{V_1}{I_1} = \frac{h_i I_1 + h_r V_2}{I_1} \qquad \longrightarrow (8)$$

Substituting $V_2 = -I_2 Z_L = \frac{-I_2}{I_1} I_1 Z_L = A_I I_1 Z_L$ in Eq. (8), we get

$$Z_i = h_i + h_r \frac{A_I Z_L I_1}{I_1} = h_i + h_r A_I Z_L$$

From Eq. (3)

$$Z_i = h_i - \frac{h_f h_r Z_L}{1 + h_o Z_L} = h_i - \frac{h_f h_r}{(1/Z_L) + h_o} \qquad \left(\text{because } Y_L = \frac{1}{Z_L}\right)$$

$$\boldsymbol{Z_i = h_i - \frac{h_f h_r}{Y_L + h_o}} \qquad \longrightarrow (9)$$

From Eq. (9), it is seen that **input impedance is a function of the load impedance.**

**Voltage Gain ($A_V$):** It is the ratio of output voltage $V_2$ to input voltage $V_1$ and is given by

$$A_V = \frac{V_2}{V_1} \qquad \longrightarrow (10)$$

Substituting $V_2 = A_I I_1 Z_L$ in Eq. (10), we get

$$A_V = \frac{A_I I_1 Z_L}{V_1}$$

Since $Z_i = \frac{V_1}{I_1}$,

$$\boldsymbol{A_V = \frac{A_I Z_L}{Z_i}} \qquad \longrightarrow (11)$$

**Voltage Gain $A_{V_s}$, taking into account the source Resistance ($R_s$)**

The overall voltage gain $A_{V_s}$ is given by

$$A_{V_s} = \frac{V_2}{V_s} = \frac{V_2}{V_1}\frac{V_1}{V_s}$$

$$A_{V_s} = A_V \cdot \frac{V_1}{V_s} \longrightarrow (12)$$

From the equivalent input circuit of the amplifier, shown in Fig. 3.12(a),

$$V_1 = \frac{V_s Z_i}{Z_i + R_s}$$

Then

$$A_{V_s} = \frac{A_V Z_i}{Z_i + R_s} \longrightarrow (13)$$

From Eq. (11), $A_V Z_i = A_I Z_L$. Then

$$A_{V_s} = \frac{A_I Z_L}{Z_i + R_s} \longrightarrow (14)$$

Combining,

$$\boldsymbol{A_{V_s} = \frac{A_V Z_i}{Z_i + R_s} = \frac{A_I Z_L}{Z_i + R_s}} \longrightarrow (15)$$

If $R_S = 0$, then $A_{V_s} = A_V$. Hence $A_V$ is the **voltage gain of an ideal voltage source.**

If $Z_i$ is resistive and equal in magnitude to $R_S$, then

$$A_{V_s} = \frac{1}{2} A_V \longrightarrow (16)$$

**Output Admittance ($Y_o$):** It is the ratio of output current $I_2$ to the output voltage $V_2$. The reciprocal of $Z_O$ is output admittance. The output impedance is obtained by setting the source voltage $V_S$ to zero and the load impedance $Z_L$ to infinity and by driving the output terminals from a generator $V_2$.

i.e.

$$Y_o = \frac{I_2}{V_2} \text{ with } V_S = 0 \text{ and } R_S = \infty \longrightarrow (17)$$

Substituting the value of $I_2$ from Eq. (2) in Eq. (17), we get

$$Y_o = \frac{h_f I_1 + h_o V_2}{V_2} = h_f \frac{I_1}{V_2} + h_o \longrightarrow (18)$$

From Fig. 3.11, with $V_S = 0$, we can write

$$R_S I_1 + h_i I_1 + h_r V_2 = 0 \longrightarrow (19)$$

$$I_1 (h_i + R_S) = -h_r V_2$$

or

$$\frac{I_1}{V_2} = \frac{-h_r}{h_i + R_s} \longrightarrow (20)$$

Substituting the value of $I_1/V_2$ in Eq. (18), we get

$$Y_o = h_f \left( \frac{-h_r}{h_i + R_s} \right) + h_o$$

or

$$\boldsymbol{Y_o = h_o - \frac{h_f h_r}{h_i + R_s}} \longrightarrow (21)$$

Hence **output admittance $Y_o$ is a function of the source resistance $R_s$.**

**Power Gain ($A_p$):** It is the ratio of average power delivered to the load $Z_L$ to the input power.

Output power, $P_2 = V_2 I_L = -V_2 I_2$ ⟶(22)

Input power, $P_1 = V_1 I_1$ ⟶(23)

$$\therefore \quad A_p = \frac{-V_2 I_2}{V_1 I_1} = A_V A_I$$

Since $A_V = \frac{A_I Z_L}{Z_i}$,

$$A_p = \frac{A_I^2 Z_L}{Z_i} \quad \longrightarrow(24)$$

**Relation Between $A_{V_s}$ and $A_{I_s}$**

From Eqs. (5) and (15), $A_{I_s} = \frac{A_I R_s}{Z_i + R_s}$ and $A_{V_s} = \frac{A_I Z_L}{Z_i + R_s}$

or $\frac{A_{Vs}}{A_{Is}} = \frac{Z_L}{R_s}$

or

$$A_{V_s} = A_{I_s} \frac{Z_L}{R_s} \quad \longrightarrow(25)$$

The important formulae are summarised in Table 3.5.

**Table 3.5 Small-signal analysis of a transistor amplifier**

| Sl. No. | Formula | Sl. No. | Formula |
|---|---|---|---|
| 1. | $A_I = \frac{-h_f}{1 + h_o Z_L}$ | 5. | $A_{V_s} = \frac{A_I Z_L}{Z_i + R_s}$ |
| 2. | $A_{I_s} = \frac{A_I R_s}{Z_i + R_s}$ | 6. | $Y_o = h_o - \frac{h_f h_r}{h_i + R_s}$ |
| 3. | $Z_i = h_i - \frac{h_f h_r}{Y_L + h_o}$ | 7. | $A_p = \frac{A_I^2 Z_L}{Z_i}$ |
| 4. | $A_V = \frac{A_I Z_L}{Z_i}$ | 8. | $A_{V_s} = A_{I_s} \frac{Z_L}{R_s}$ |

## THE EMITTER FOLLOWER

The common-collector transistor amplifier is usually referred to as an **emitter follower** and its circuit diagram is shown in Fig. 3.13. It has a voltage gain that is close to unity because a change in base voltage appears as an equal change across the load without any phase inversion at the emitter. Since there is no phase inversion and because the voltage gain is close to unity, the output voltage closely **follows** the input voltage in both phase and amplitude; thus the term **emitter follower.**

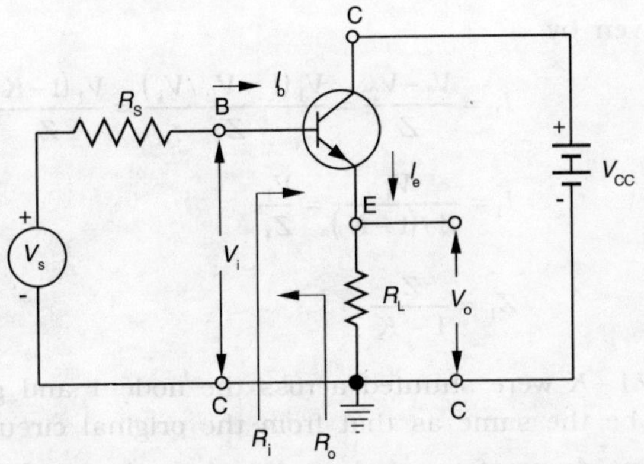


**Fig. 3.13 A common-collector transistor amplifier**

The emitter-follower configuration is frequently used for impedance matching purposes. Its input resistance ($R_i$) is very high and the output resistance ($R_o$) is very low. In addition the emitter follower increases the power level of the signal.

Current gain, $$A_I = \frac{-I_e}{I_b} = \frac{-h_{fc}}{1 + h_{oc} R_L} = \frac{1 + h_{fe}}{1 + h_{oe} R_L} \longrightarrow (1)$$

Input resistance, $$R_i = \frac{V_i}{I_b} = h_{ic} + h_{rc} A_I R_L = h_{ie} + A_I R_L \longrightarrow (2)$$

Voltage gain, $$A_V = \frac{V_o}{V_i} = \frac{A_I R_L}{R_i} = \frac{R_i - h_{ie}}{R_i} = 1 - \frac{h_{ie}}{R_i} \longrightarrow (3)$$

Output admittance, $$Y_o = h_{oc} - \frac{h_{fc} h_{rc}}{h_{ic} + R_s} = h_{oc} + \frac{1 + h_{fe}}{1 + h_{ie} R_s} \longrightarrow (4)$$

Overall voltage gain, $$A_{V_s} = \frac{A_V R_i}{R_i + R_s} \longrightarrow (5)$$

## MILLER'S THEOREM AND ITS DUAL

The transistor circuits can be analyzed more conveniently by using Miller's theorem. Miller's theorem is used for converting any circuit having configuration of Fig. 3.14(a) to the equivalent circuit of Fig. 3.14(b).

Consider an arbitrary circuit configuration with $N$ distinct nodes 1, 2, 3 ---, $N$, as shown in Fig. 3.14(a). Let the node voltages be $V_1$, $V_2$, $V_3$, ----, $V_N$, where, $V_N = 0$. $N$ is the reference, or ground node. The impedance $Z$ is connected between the nodes 1 and 2. The impedance $Z$ has been split into two impedances $Z_1$ and $Z_2$, where $Z_1$ is connected between node and ground and $Z_2$ is connected between node 2 and ground. We postulate that we know the ratio $V_2/V_1$ $(= K)$. The current $I_1$ drawn from node 1 through $Z$ can be obtained by disconnecting terminal 1 from $Z$ and by bridging an impedance $Z/(1-K)$ from node 1 to ground as shown in Fig. 3.14(b).

The current $I_1$ is given by

$$I_1 = \frac{V_1 - V_2}{Z} = \frac{V_1(1 - V_2/V_1)}{Z} = \frac{V_1(1-K)}{Z}$$

$$I_1 = \frac{V_1}{Z/(1-K)} = \frac{V_1}{Z_1} \longrightarrow (1)$$

where
$$Z_1 = \frac{Z}{1-K} \longrightarrow (2)$$

Therefore, if $Z_1 = Z/1-K$ were shunted across the node 1 and ground, the current drawn from node 1 would be the same as that from the original circuit.

Similarly, the current drawn from node 2 may be calculated by removing $Z$ and connecting between node 2 and ground.

$$I_2 = \frac{V_2 - V_1}{Z} = \frac{V_2}{V_2}\left(\frac{V_2 - V_1}{Z}\right)$$

$$= \frac{V_2}{Z}\left(\frac{V_2 - V_1}{V_2}\right) = \frac{V_2}{Z}\left(1 - \frac{V_1}{V_2}\right) = \frac{V_2\,(1 - 1/K)}{Z}$$

$$I_2 = \frac{V_2}{Z/(1-1/K)} = \frac{V_2}{Z_2} \longrightarrow (3)$$

where
$$Z_2 = \frac{Z}{1-1/K} = \frac{ZK}{K-1} \longrightarrow (4)$$

Since identical nodal equations are obtained from the configurations of Fig. 3.14(a) and (b), these two networks are equivalent.

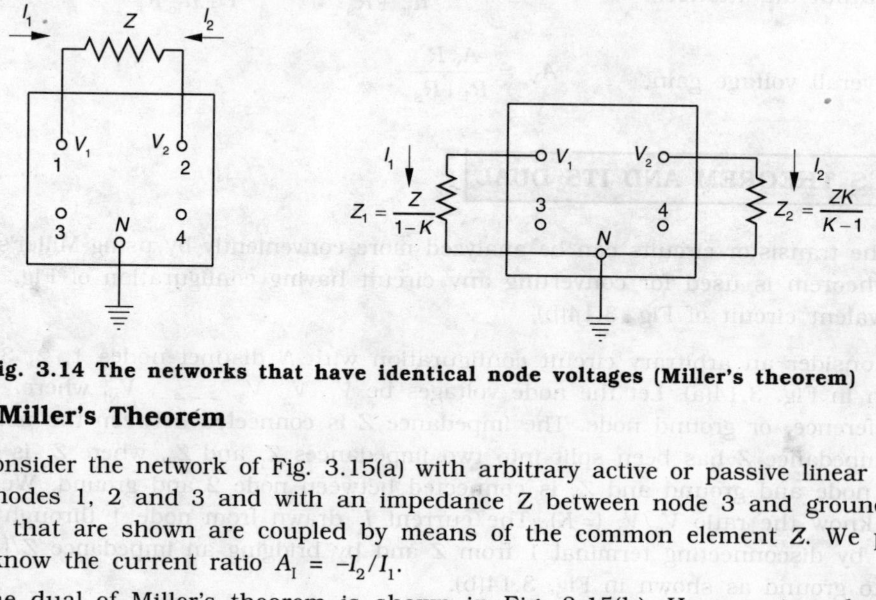


**Fig. 3.14 The networks that have identical node voltages (Miller's theorem)**

## Dual of Miller's Theorem

Consider the network of Fig. 3.15(a) with arbitrary active or passive linear elements between nodes 1, 2 and 3 and with an impedance $Z$ between node 3 and ground $N$. The two loops that are shown are coupled by means of the common element $Z$. We postulate that we know the current ratio $A_I = -I_2/I_1$.

The dual of Miller's theorem is shown in Fig. 3.15(b). Here an impedance $Z_1$ is connected in mesh 1 and $Z_2$ in mesh 2.

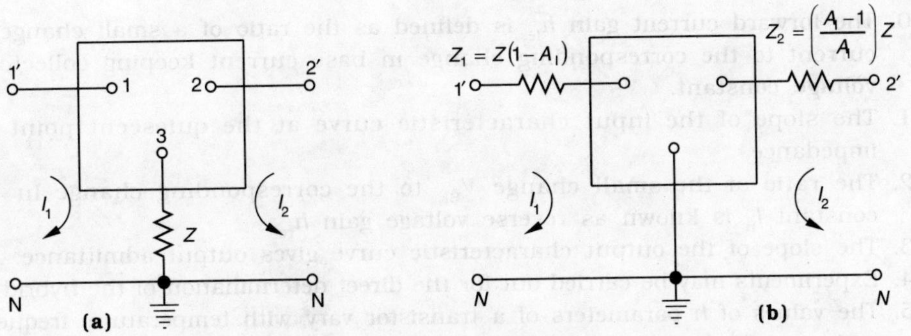


**Fig. 3.15 The networks that have the same currents**

It is seen that $V_{1'N} = (I_1 + I_2)\ Z = I_1 Z_1$

$$(I_1 + I_2)Z = I_1\left(1 + \frac{I_2}{I_1}\right)Z$$

$$= I_1\ (1 - A_I)\ Z$$

$$= I_1\ Z_1 \qquad \longrightarrow (5)$$

where $Z_1 = Z\ (1 - A_I)$ $\longrightarrow$(6)

Hence $V_{1'N}$ is the same in the two circuits in Fig. 3.15 (a) and (b).

Similarly, $V_{2'N}$ has the same value in the two circuits if $Z_2 = \left(\frac{A_I - 1}{A_I}\right)Z$. The two networks are therefore identical in the sense that if the same voltages $V_{1'N}$ and $V_{2'N}$ are impressed on both, the same current $I_1$ will flow in mesh 1 and $I_2$ in mesh 2.

## IMPORTANT POINTS TO REMEMBER

1. The term small signal refers to the use of signals that take up a relatively small percentage of an amplifier's operational range.
2. For small signals the transistor operates linearly in the active region.
3. The collector characteristics in the neighbourhood of the load line are not parallel lines equally spaced for equal increments in base current. These give rise to a distortion known as output nonlinear distortion.
4. Since the small-signal low-frequency response of a transistor is linear, it can be obtained analytically rather than graphically.
5. For incremental small signals the amplifying device *viz.* VT, BJT and FET can be characterised as a linear two-port network whose terminal behaviour is specified by two voltages and two currents.
6. The parameters that completely describe the behaviour of the transistor circuit and are constant for a given circuit are called *h*-parameters.
7. The quantities $h_{11}$, $h_{12}$, $h_{21}$ and $h_{22}$ are called the *h* or hybrid parameters.
8. The four hybrid parameters $h_{11}$, $h_{12}$, $h_{21}$ and $h_{22}$ are real numbers and the voltages and currents $V_1$, $V_2$ and $I_1$, $I_2$ are functions of time.
9. Two families of curves such as output characteristic curves and input characteristic curves are specified for transistors.

10. The forward current gain $h_{fe}$ is defined as the ratio of a small change in collector current to the corresponding change in base current keeping collector to emitter voltage constant.
11. The slope of the input characteristic curve at the quiescent point gives input impedance.
12. The ratio of the small change $V_{BE}$ to the corresponding change in $V_{CE}$ keeping constant $I_B$ is known as reverse voltage gain $h_{re}$.
13. The slope of the output characteristic curve gives output admittance.
14. Experiments may be carried out for the direct determination of the hybrid parameters.
15. The values of $h$ parameters of a transistor vary with temperature, frequency, voltage and current.
16. Hybrid parameters of a transistor are functions of frequency.
17. When the output is taken from the emitter terminal of the transistor, the network is referred to as emitter follower.
18. The emitter follower is frequently used for impedance matching.
19. Miller theorem is used to analyse certain configurations in more simple way.

## KEY FORMULAE

1. Input voltage of the amplifier, $v_1 = h_{11}\, i_1 + h_{12}\, v_2$
2. Output current of the amplifier, $i_2 = h_{21}\, i_1 + h_{22}\, v_2$
3. $h_{11} = h_i = \left.\dfrac{v_1}{i_1}\right|_{v_2 = 0}$ = input resistance with output short circuit
4. $h_{12} = h_r = \left.\dfrac{v_1}{v_2}\right|_{i_1 = 0}$ = reverse voltage gain with input open
5. $h_{21} = h_f = \left.\dfrac{i_2}{i_1}\right|_{v_2 = 0}$ = forward current gain with output short-circuit
6. $h_{22} = h_o = \left.\dfrac{i_2}{v_2}\right|_{i_1 = 0}$ = output admittance with input open
7. $v_{be} = h_{ie}\, i_b + h_{re}\, v_{ce}$
8. $i_C = h_{fe}\, i_b + h_{oe}\, v_{ce}$
9. Current gain, $A_I = \dfrac{-h_f}{1 + h_o Z_L}$
10. Overall current gain, $A_{I_s} = \dfrac{A_I R_s}{Z_i + R_s}$
11. Voltage gain, $A_v = \dfrac{A_I Z_L}{Z_i}$

(Formulae 9–11: For C-E configuration transistor)

12. Overall voltage gain, $A_{V_s} = \dfrac{A_I Z_L}{Z_i + R_s}$

13. Input impedance, $Z_i = h_i - \dfrac{h_f h_r}{Y_L + h_o}$

14. Output admittance, $Y_o = h_o - \dfrac{h_f h_r}{h_i + R_s}$

15. Power gain, $A_p = \dfrac{A_I^2 Z_L}{Z_i}$

(12–15: For C-E configuration transistor)

16. Current gain, $A_I = \dfrac{-h_{fc}}{1 + h_{oc} R_L}$

17. Input resistance, $R_i = h_{ic} + h_{rc} A_I R_L$

18. Voltage gain, $A_V = \dfrac{A_I R_L}{R_i}$

19. Output admittance, $Y_o = h_{oc} - \dfrac{h_{fc} h_{rc}}{h_{ic} + R_s}$

(16–19: For C-C configuration transistor)

## SOLVED PROBLEMS

**1. A C-E amplifier uses $R_L$ = 200 Ω. The *h*-parameters are $h_{ie}$ = 1100 Ω, $h_{re}$ = 2.5 × $10^{-4}$, $h_{fe}$ = 50 and $h_{oe}$ = 22 μA/V. Calculate: (i) current gain (ii) input impedance and (iii) voltage gain.** **[Feb. 2000, VTU]**

**Data:** $R_L$ = 200 Ω, $h_{ie}$ = 1100 Ω, $h_{re}$ = 2.5 × $10^{-4}$, $h_{fe}$ = 50 and $h_{oe}$ = 22 μA/V

**To find:** $A_I$, $Z_i$ and $A_V$

**Solution:** (i) Current gain, $A_I = \dfrac{h_{fe}}{1 + h_{oe} R_L}$

$$= \frac{50}{1 + 22 \times 10^{-6} \times 200} = \mathbf{49.78}$$

(ii) Output impedance, $Z_i = h_{ie} + h_{re} A_I R_L$

$$= 1100 + 2.5 \times 10^{-4} \times 49.78 \times 200 = \mathbf{1102.5\ \Omega}$$

(iii) Voltage gain, $A_V = \dfrac{A_I R_L}{Z_i}$

$$= \frac{49.78 \times 200}{1102.5} = \mathbf{9.03}$$

**2. The transistor circuit shown in Fig. 3.16 has $h_{ie}$ = 1.1 kΩ, $h_{re}$ = 2.5 × $10^{-4}$, $h_{fe}$ = 50, $h_{oe}$ = 25 μA/V if $R_L$ = $R_s$ = 1 kΩ. Find voltage gain, current gain, input impedance and output admittance.** **[Sept. 2000, VTU]**

**Data:** $h_{ie}$ = 1.1 kΩ, $h_{re}$ = 2.5 × $10^{-4}$, $h_{fe}$ = 50, $h_{oe}$ = 25 μA/V, and $R_L = R_s$ = 1 kΩ

**To Find:** $A_V$, $A_I$, $Z_i$ and $Y_o$

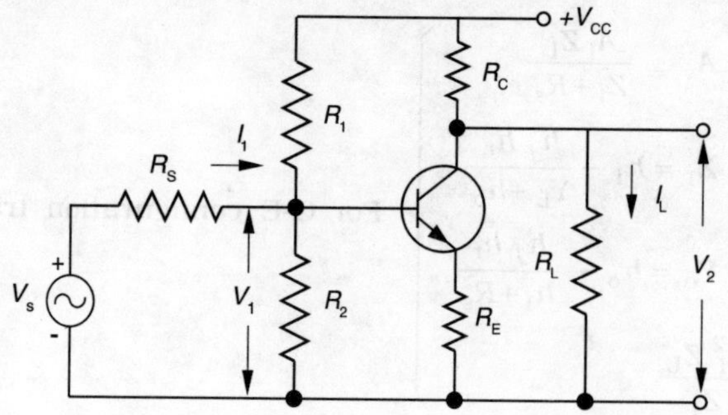


**Fig. 3.16**

**Solution:** (i) Current gain,
$$A_I = \frac{h_{fe}}{1+h_{oe}R_L}$$

$$= \frac{50}{1+25\times10^{-6}\times10^3} = \mathbf{48.78}$$

(ii) Input impedance,
$$Z_i = h_{ie} + h_{re}\,A_I R_L$$

$$= 1100 + 2.5\times10^{-4}\times48.78\times1000$$

$$= \mathbf{1112.2\Omega}$$

(iii) Voltage gain,
$$A_V = \frac{A_I R_L}{Z_i}$$

$$= \frac{48.78\times1000}{1112.2} = \mathbf{43.86}$$

(iv) Output admittance,
$$Y_o = h_{oe} - \frac{h_{fe}+h_{re}}{h_{ie}+R_s}$$

$$= 25\times10^{-6} - \frac{50\times2.5\times10^{-4}}{1100+1000}$$

$$= 25\times10^{-6} - 5.952\times10^{-6} = \mathbf{19.048\,\mu S}$$

Output impedance,
$$Z_o = \frac{1}{Y_o}$$

$$= \frac{1}{19.048\times10^{-6}} = \mathbf{52.632\,k\Omega}$$

Output impedance considering load resistance is given by

$$Z_o' = Z_o \,||\, R_L$$

$$= 52.632 \,||\, 1 = \mathbf{981.35\Omega}$$

**3. Show that the exact expression for $h_{fe}$ in terms of the C-B hybrid parameters is**

$$h_{fe} = \frac{h_{fb}(1-h_{rb})+h_{ib}h_{ob}}{(1+h_{fb})(1-h_{rb})+h_{ob}h_{ib}}$$

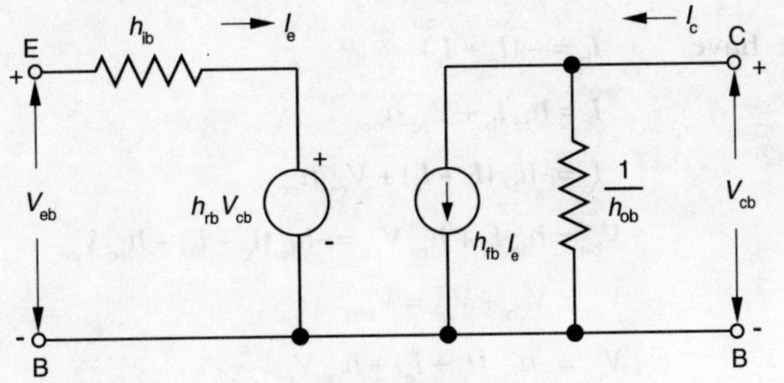


**Fig. 3.17**

**Solution:** By definition $$h_{fe} = \left.\frac{I_c}{I_b}\right|_{V_{ce}=0} \longrightarrow (1)$$

From Fig. 3.17, we can write

$$I_c = h_{fb}\,I_e - I = h_{fb}\,I_e - V_{bc}\,h_{ob} \longrightarrow (2)$$

With $V_{ce} = 0$, $$V_{be} = V_{bc} = -V_{cb}$$

And $$I_e = -\frac{V_{be} + h_{rb}V_{cb}}{h_{ib}} = -\frac{V_{cb}(h_{rb}-1)}{h_{ib}} \longrightarrow (3)$$

Substituting $I_e$ in Eq. (2), we get

$$I_c = h_{fb}\left[\frac{V_{cb}(1-h_{rb})}{h_{ib}}\right] - V_{bc}\,h_{ob} \longrightarrow (4)$$

We know that $$I_b = -(I_c + I_e)$$

$$= -\left[\frac{h_{fb}(1-h_{rb})}{h_{ib}} + h_{ob}\right]V_{cb} - \frac{(1-h_{rb})V_{cb}}{h_{ib}}$$

$$= -\left[\frac{h_{fb} - h_{fb}h_{rb} + h_{ob}h_{ib} + 1 - h_{rb}}{h_{ib}}\right]V_{cb}$$

$$I_b = -\left[\frac{(1+h_{fb})(1-h_{rb})}{h_{ib}} + h_{ob}\right]V_{cb} \longrightarrow (5)$$

$$h_{fe} = \frac{I_c}{I_b} = -\left[\frac{h_{fb}}{h_{ib}}(1-h_{rb}) + h_{ob}\right] \Big/ \left[\frac{(1+h_{fb})(1-h_{rb})}{h_{ib}} + h_{ob}\right]$$

$$h_{fe} = -\frac{h_{fb}(1-h_{rb}) + h_{ib}h_{ob}}{(1+h_{fb})(1-h_{rb}) + h_{ib} + h_{ob}}$$

**4. Show that the exact expression for $h_{fb}$ in terms of the C-E hybrid parameters is**

$$\boldsymbol{h_{fb} = -\frac{h_{fe}(1-h_{re}) + h_{ie}h_{oe}}{(1+h_{fe})(1-h_{re}) + h_{oe}\,h_{ie}}}.$$

**Solution:** By definition $$h_{fb} = \left.\frac{I_c}{I_b}\right|_{V_{cb}=0}$$

From Fig. 3.6(a), we have $I_b = -(I_c + I_e)$

$$I_c = h_{fe} I_b + V_{ce} h_{oe}$$

$$I_c = -h_{fe}(I_c + I_e) + V_{ce} h_{oe} \quad \longrightarrow (1)$$

And $V_{be} = h_{ie} I_b + h_{re} V_{ce} = -h_{ie}(I_c + I_e) + h_{re} V_{ce} \quad \longrightarrow (2)$

For $V_{cb} = 0$, Then $V_{ce} = V_{cb} + V_{be} = V_{be}$

Eq. (2) becomes, $V_{ce} = -h_{ie}(I_c + I_e) + h_{re} V_{ce}$

or $V_{ce}(1 - h_{re}) = -h_{ie}(I_c + I_e)$

or $$V_{ce} = -\frac{h_{ie}}{1 - h_{re}}(I_c + I_e) \quad \longrightarrow (3)$$

Substituting the value of $V_{ce}$ in Eq. (1), we get

$$I_c = -h_{fe}(I_c + I_e) - \frac{h_{ie}}{1 - h_{re}}(I_c + I_e)h_{oe}$$

or $$I_c\left[1 + h_{fe} + \frac{h_{ie} h_{oe}}{1 - h_{re}}\right] = -\left[h_{fe} + \frac{h_{ie} h_{oe}}{1 - h_{re}}\right] I_e$$

Hence, $$h_{fb} = \frac{I_c}{I_e} = -\left(h_{fe} + \frac{h_{ie} h_{oe}}{1 - h_{re}}\right) \Bigg/ \left[1 + h_{fe} + \frac{h_{ie} h_{oe}}{1 - h_{re}}\right]$$

$$h_{fb} = -\frac{h_{fe}(1 - h_{re}) + h_{ie} h_{oe}}{(1 - h_{re})(1 + h_{fe}) + h_{ie} h_{oe}}$$

**5. Given a single-stage transistor amplifier with $h$ parameters specified in Table 3.2, calculate $A_I$, $A_V$, $A_{VS}$, $R_i$ and $R_o$ for C-C transistor configuration with $R_S = R_L = 10\ k\Omega$.**

**Data:** $h_{ic} = 1100\Omega$, $h_{rc} = 1$, $h_{fc} = -51$, $h_{oc} = 25\ \mu A/V$, $1/h_o = 40\ k\Omega$, and $R_s = R_L = 10\ k\Omega$

**To find:** $A_I$, $A_V$, $A_{Vs}$, $R_i$ and $R_o$

**Solution:** (i) Current gain, $$A_I = \frac{h_{fc}}{1 + h_{oc} R_L} = \frac{51}{1 + 25 \times 10^{-6} \times 10^4} = \mathbf{40.8}$$

(ii) Input resistance, $$R_i = h_{ic} + h_{rc} A_I R_L$$

$$= 1.1 + 1 \times 40.8 \times 10 = \mathbf{409.1\ k\Omega}$$

(iii) Voltage gain, $$A_V = \frac{A_I R_L}{R_i} = \frac{40.8 \times 10}{409.1} = \mathbf{0.998}$$

(iv) Output admittance, $$Y_o = h_{oc} - \frac{h_{fc} h_{rc}}{h_{ic} + R_s}$$

$$= 25 \times 10^{-6} + \frac{51 \times 10^{-3}}{1.1 + 10} = \mathbf{4.625 \times 10^{-3}}$$

$$R_o = \frac{1}{Y_o} = \frac{1}{4.625 \times 10^{-3}} = \mathbf{217\,\Omega}$$

(v) Overall voltage gain, $A_{V_s} = \dfrac{A_V R_i}{R_i + R_s}$

$$= \frac{0.998 \times 409.1}{409.1 + 10} = \mathbf{0.974}$$

**6. For any single-transistor amplifier prove that** $\boldsymbol{R_i = \dfrac{h_i}{1 - h_r A_V}}$**.**

**Solution:** We know $R_i = h_i + h_r A_I R_L$ and $R_L = \dfrac{A_V R_i}{A_I}$

Substituting the value of $R_L$, we get

$$R_i = h_i + h_r \frac{A_I A_V R_i}{A_I} = h_i + h_r A_V R_i$$

or $$R_i(1 - h_r A_V) = h_i$$

$$\boxed{R_i = \frac{h_i}{1 - h_r A_V}}$$

**7. (a) For a C-E configuration, what is the maximum value of $R_L$ for which $R_i$ differs by no more than 10 percent of its value at $R_L = 0$? Use the transistor parameters given in Table 3.2.**

**(b) What is the maximum value of $R_s$ for which $R_o$ differs by no more than 10 percent of its value for $R_s = 0$?**

**Data:** $h_{ie} = 1100\,\Omega$, $h_{re} = 2.5 \times 10^{-4}$, $h_{fe} = 50$, $h_{oe} = 25\,\mu A/V$

**To find:** Maximum value of $R_L$ for which $R_i$ differs by more than 10% of its value at $R_L = 0$.

**Solution:** (a) We know $R_i = h_{ie} - \dfrac{h_{fe} h_{re}}{h_{oe} + 1/R_L}$

If $R_L = 0$, $R_i = h_{ie} - 0 = h_{ie}$

The value of $R_L$ for which $R_i = 0.9\,h_{ie}$ is found from

$$0.9 h_{ie} = h_{ie} - \frac{h_{fe} h_{re}}{h_{oe} + 1/R_L}$$

or $$R_L = \frac{0.1 h_{ie}}{h_{fe} h_{re} - 0.1 h_{ie} h_{oe}}$$

$$= \frac{0.1 \times 1100}{50 \times 2.5 \times 10^{-4} - 0.1 \times 1100 \times 24 \times 10^{-6}} = \mathbf{11.3\,k\Omega}$$

(b) For a C-E Configuration, if $R_s$ increases, then $R_o$ decreases slightly. If $R_o$ decreases by 10%, $Y_o = \dfrac{1}{R_o}$ increases by 11%. Then $Y_o\big|_{R_s = 0} = h_{oe} - \dfrac{h_{fe} h_{re}}{h_{ie}}$

$$= 25 \times 10^{-6} - \frac{50 \times 2.5 \times 10^{-4}}{1100}$$

or $$Y_o\Big|_{R_s=0} = 13.63\times10^{-6}$$

Thus, $$1.11\times13.63\times10^{-6} = 25\times10^{-6} - \frac{125\times10^{-4}}{(1.1+R_s)10^3} = 15.15\times10^{-6}$$

or $$9.85\times10^{-6} = \frac{1.25\times10^{-6}}{(1.1+R_s)}$$

or $$1.1+R_s = \frac{12.5}{9.85} = 1.27$$

∴ $$R_s = 0.17k\Omega = \mathbf{170\ \Omega}$$

**8. For the emitter follower with $R_s$ = 0.5 kΩ and $R_L$ = 5 kΩ, calculate $A_I$, $R_i$, $A_V$, $A_{Vs}$ and $R_o$. Assume $h_{fe}$ = 50, $h_{ie}$ = 1 kΩ, $h_{oe}$ = 25 μA/V.**

**Data:** $R_s$ = 0.5 kΩ, $R_L$ = 5 kΩ, $h_{fe}$ = 50, $h_{ie}$ = 1 kΩ, and $h_{oe}$ = 25 μA/V

**To find:** $A_I$, $R_i$, $A_V$, $A_{Vs}$ and $R_o$

**Solution:** We know $$A_I = \frac{1+h_{fe}}{1+h_{oe}R_L} = \frac{1+50}{1+25\times10^{-6}\times5\times10^3} = \mathbf{45.3}$$

Input resistance, $$R_i = h_{ie} + A_I R_L = 1 + 45.3\times5 = \mathbf{227\ k\Omega}.$$

Voltage gain, $$A_V = 1 - \frac{1}{227} = \mathbf{0.9956}$$

Overall voltage gain, $$A_{V_s} = A_V\frac{R_i}{R_i+R_s} = \frac{0.9956\times227}{227+0.5} = \mathbf{0.9934}$$

Output admittance, $$Y_o = 25 + \frac{51}{1.5} = 34.025\times10^{-3}\ \mathbf{A/V}$$

Output resistance, $$R_o = \frac{1}{Y_o} = \frac{1}{34.025}\times10^3 = \mathbf{29.4\ \Omega}$$

**9. (a) Design an emitter follower having $R_i$ = 500 kΩ and $R_o$ = 20 Ω. Assume $h_{fe}$ = 50, $h_{ie}$ = 1kΩ, $h_{oe}$ = 25 μA/V.**

**(b) Find $A_I$ and $A_V$ for the emitter follower of part (a).**

**(c) Find $R_i$ and the necessary $R_L$ so that $A_V$ = 0.999.**

**Data:** $R_i$ = 500 kΩ, $R_o$ = 20 Ω, $h_{fe}$ = 50, $h_{ie}$ = 1kΩ, $h_{oe}$ = 25 μA/V

**To find:** $A_I$, $A_V$, $R_L$, $Y_o$ and $R_s$

**Solution:** (a) (i) Input resistance, $$R_i = h_{ie} + A_I R_L$$

$$500 = 1 + A_I R_L$$

or $$A_I R_L = 499$$

(ii) Current gain, $$A_I = \frac{1+h_{fe}}{1+h_{oe}R_L}$$

$$A_I = \frac{1+50}{1+25\times10^{-6}\times R_L}$$

$$51 = A_I + 25 \times 10^{-6} A_I R_L$$

$$51 = A_I + 25 \times 10^{-6} \times 499 \times 10^3$$

$$A_I = 38.5$$

or $$R_L = \frac{499}{A_I} = \frac{499}{38.5} = \mathbf{13\ k\Omega}$$

(iii) Output admittance, $$Y_o = h_{oe} + \frac{1+h_{fe}}{h_{ie}+R_s}$$

$$0.05 = 25\times10^{-6} + \frac{51}{1000+R_s}$$

or $$R_s = \mathbf{20\ \Omega}$$

(b) $A_I = 38.5$, $$A_V = 1 - \frac{h_{ie}}{R_i}$$

$$= 1 - \frac{1000\ \Omega}{500\ k\Omega} = \mathbf{0.998}$$

(c) Given $A_V = 0.999$ $$A_V = 1 - \frac{h_{ie}}{R_i}$$

$$0.999 = 1 - \frac{1000}{R_i}$$

or $$R_i = \mathbf{1\ M\Omega}$$

$$R_i = h_{ie} + A_I R_L = 10^6\ \Omega$$

For $h_{ie} = 10^3\ \Omega$, $$A_I R_L = 10^6 = \frac{(1+h_{fe})}{1+h_{oe}R_L} R_L$$

$$10^6 + 25\ R_L = 51\ R_L$$

or $$R_L = \frac{10^6}{26} = \mathbf{38.4\ k\Omega}$$

**10. (a) In the circuit shown in Fig. 3.18, find the input impedance $Z_i$ in terms of the CE $h$ parameters, $R_L$ and $R_E$.**

**(b) If $R_L = R_E = 1\ k\Omega$ and the $h$ parameters are given in Table 3.2, what is the value of $R_i$?**

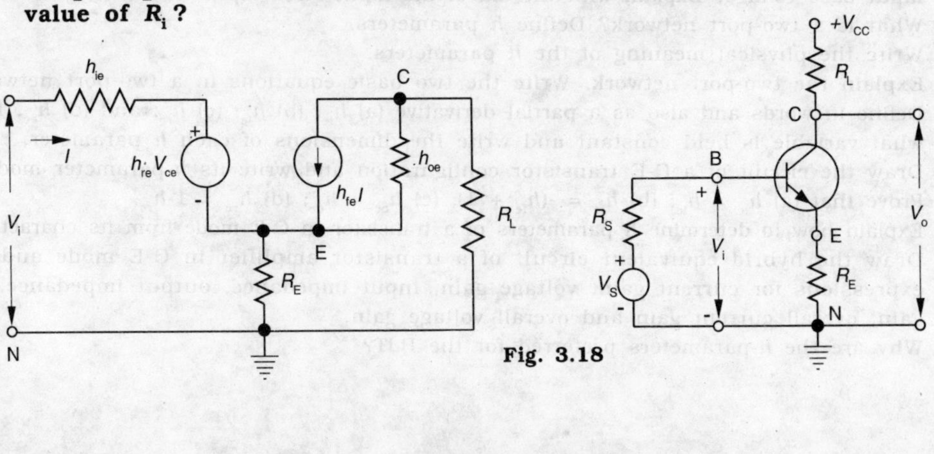


**Fig. 3.18**

**Data:** $h_{ie}$ = 1100 $k\Omega$, $h_{re}$ = $2.5 \times 10^{-4}$, $h_{fe}$ = 50, $h_{re}$ = 25 $\mu A/V$, $R_L$ = 1 $k\Omega$

**To find:** Input impedance $Z_i$

**Solution:** (a) Input impedance, $$Z_i = \frac{V_i}{I}$$

where $$V_i = Ih_{ie} + h_{re}V_{ce} + V_{en}$$

$$V_{en} = (I + h_{fe}I + V_{ce}h_{oe})R_E$$

$$V_{ce} = -V_{en} - R_L(h_{fe}I + V_{ce}h_{oe})$$

$$= -[I + h_{fe}I + V_{ce}h_{oe}]R_E - R_L[h_{fe}I + V_{ce}h_{oe}]$$

or $$V_{ce}[1 + h_{oe}R_E + h_{oe}R_L] = -[(1 + h_{fe})R_E + h_{fe}R_L]$$

Thus $$V_i = I[h_{ie} + (1 + h_{fe})R_E] + V_{ce}[h_{re} + h_{oe}R_E]$$

$$V_i = \left\{ I[h_{ie} + (1 + h_{fe})R_E] - (h_{re} + h_{oe}R_E)\frac{[(1 + h_{fe})R_E + h_{fe}R_L]}{1 + h_{oe}R_E + h_{oe}R_L} \right\}$$

Hence $$Z_i = \mathbf{h_{ie} + (1 + h_{fe})R_E - (h_{re} + h_{oe}R_E)\frac{(1 + h_{fe})R_E + h_{fe}R_L}{1 + h_{oe}(R_E + R_L)}}$$

(b) If $R_L$ = $R_E$ = 1 k$\Omega$, then $$Z_i = 1.1 + 51 - \frac{(2.5 \times 10^{-4} + 0.025)(51 + 50)}{1 + 25 \times 10^{-6}(2 \times 10^3)}$$

$$= \left[52.1 - \frac{0.02525 \times 101}{1.05}\right] k\Omega$$

$$= 52.1 - 2.4 = \mathbf{49.7 k\Omega}$$

## QUESTIONS

1. What is meant by small-signal linear model?
2. Distinguish between small-signal model and large-signal model of a transistor.
3. Is nonlinear distortion greater for a sinusoidal-input base current or for a sinusoidal-input base voltage? Explain with the aid of the input and output transistor characteristics.
4. What is a two-port network? Define $h$ parameters.
5. Write the physical meaning of the $h$ parameters.
6. Explain the two-port network. Write the two basic equations in a two-port network.
7. Define in words and also as a partial derivative (a) $h_{ie}$; (b) $h_{fe}$; (c) $h_{re}$; and (d) $h_{oe}$. Indicate what variable is held constant and write the dimensions of each $h$ parameter.
8. Draw the circuit of a C-E transistor configuration and write its $h$-parameter model.
9. Prove that (a) $h_{ic}$ = $h_{ie}$; (b) $h_{fc}$ = $-(h_{fe} + 1)$; (c) $h_{oc}$ = $h_{oe}$; (d) $h_{rc}$ = $1-h_{re}$.
10. Explain how to determine $h$ parameters of a transistor in C-E mode from its characteristics.
11. Draw the hybrid equivalent circuit of a transistor amplifier in C-E mode and derive expressions for current gain, voltage gain, input impedance, output impedance, power gain, overall current gain and overall voltage gain.
12. Why are the $h$-parameters preferred for the BJT?

13. Derive the expression $A_V$ in terms of $A_I$.
14. Explain briefly the general guidelines for analysis of a transistor circuit.
15. Explain with a circuit diagram the operation of an emitter follower. Write its main application.
16. Write three most important characteristics of an emitter follower.
17. State Miller's theorem with the aid of a circuit diagram.
18. Write a short note on dual of Miller's theorem.
19. Explain with relevant graphs the variations of hybrid parameters.

## EXERCISES

1. A BJT has the following C-B parameters: $h_{ib}$ = 20 Ω, $h_{rb}$ = $2 \times 10^{-4}$, $h_{fb}$ = -0.98 and $h_{ob}$ = $10^{-6}$ S. Estimate its $h$-parameters for the C-E and C-C configurations.
**[Ans: $h_{ie}$ = 2 kΩ, $h_{re}$ = $8 \times 10^{-4}$, $h_{fe}$ = 49, $h_{oe}$ = 0.5 mS, $h_{ic}$ = 1 kΩ, $h_{rc}$ = 0.9992, $h_{fc}$ = -50, $h_{oc}$ = 0.5 mS]**

2. The C-E $h$ parameters of a BJT are as follows: $h_{ie}$ = 1 kΩ, $h_{re}$ = $2 \times 10^{-4}$, $h_{fe}$ = 200, $h_{oe}$ = $10^{-4}$ S. Find the $h$-parameters for C-C and C-B configurations.
**[Ans: $h_{ic}$ = 1 kΩ, $h_{rc}$ = 0.9998, $h_{fc}$ = -201, $h_{oc}$ = $10^{-4}$ S, $h_{ib}$ = 4.97Ω, $h_{rb}$ = $2.97 \times 10^{-4}$, $h_{fb}$ = -0.995, $h_{ob}$ = $0.49 \times 10^{-6}$ S]**

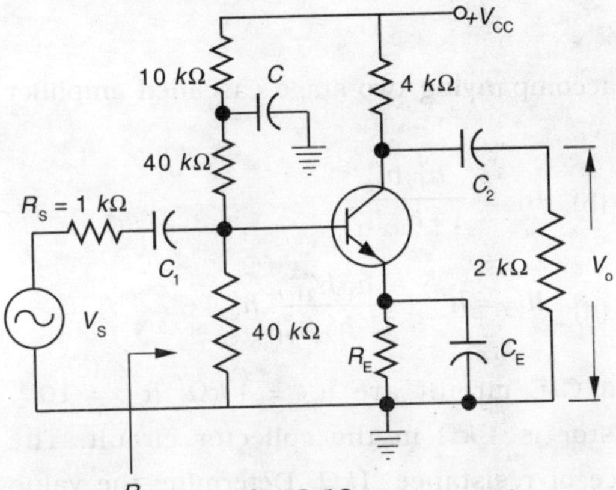

Fig. 3.19

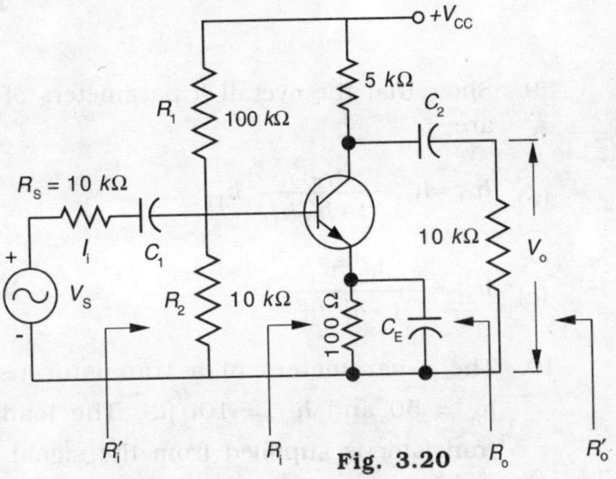

Fig. 3.20

3. A transistor with $h_{ie}$ = 1.1 kΩ, $h_{fe}$ = 50, $h_{re}$ = $205 \times 10^{-4}$, $h_{oe}$ = 25 μS is connected in C-E configuration as shown in Fig. 3.19. Calculate: $A_I$, $A_{Is}$, $A_V$, $A_{Vs}$ and $R_o$.
**[Ans: $A_I$ = –31.595, $A_V$ = -40.43, $A_{Vs}$ = – 20.63, $A_{Is}$ = – 15.47, $R_o$ = 1.33 kΩ]**

4. The amplifier shown in Fig. 3.20 uses a transistor with the following parameters: $h_{ie}$ = 1.1 kΩ, $h_{fe}$ = 50, $h_{re}$ = $25 \times 10^{-6}$ S. Calculate $A_I$, $A_V$, $A_{V_s}$, $R_o$, $R_o'$, $R_i$ and $R_i'$.
**[Ans: $R_i$ = 1.1 kΩ, $R_i'$ = 981.27Ω, $A_I$ = -50, $A_{I_s}$ = – 14.87, $A_V$ = -151.52, $A_{V_s}$ = – 13.54, $R_o$ = ∞, $R_o'$ = 3.33 kΩ]**

5. Prove that $Y_o = h_o \left( \dfrac{R_s + R_{i\infty}}{R_s + R_{io}} \right)$ where $R_{i\infty}$ = $R_i$ for $R_L$ = ∞ and $R_{io}$ = $R_i$ for $R_L$ = 0.

6. The C-B hybrid parameters of a BJT are as follows: $h_{ib}$ = 15 Ω, $h_{rb}$ = 10 × $10^{-4}$, $h_{fb}$ = − 0.99 and $h_{ob}$ = 1.5 μS. Estimate the C-E and C-C h-parameters.
**[Ans: $h_{ie}$ = 1.5 kΩ, $h_{re}$ = 12.5 × $10^{-4}$, $h_{fe}$ = 99, $h_{oe}$ = 1.5 × $10^{-4}$ S, and $h_{ic}$ = 1.5 kΩ, $h_{rc}$ = 1, $h_{fc}$ = − 100 and $h_{oc}$ = 1.5 × $10^{-4}$ S]**

7. Find $h_{re}$ in terms of the C-B h-parameters. **[Ans: $h_{re} \approx \dfrac{h_{ib}h_{ob}}{1+h_{fb}} - h_{rb}$ ]**

8. The transistor of Fig. 3.11 is connected as a common-emitter amplifier and the h parameters are those given in Table 3.2. If $R_L$ = 10 kΩ and $R_s$ = 1kΩ, find the various gains and the input and output impedances.
**[Ans: $A_I$ = − 40, $R_i$ = 1 kΩ, $A_V$ = − 400, $A_{V_s}$ = − 200, $A_{I_s}$ = − 20, $Y_o$ = 19 μS and $Z_o$ = 52.6 kΩ]**

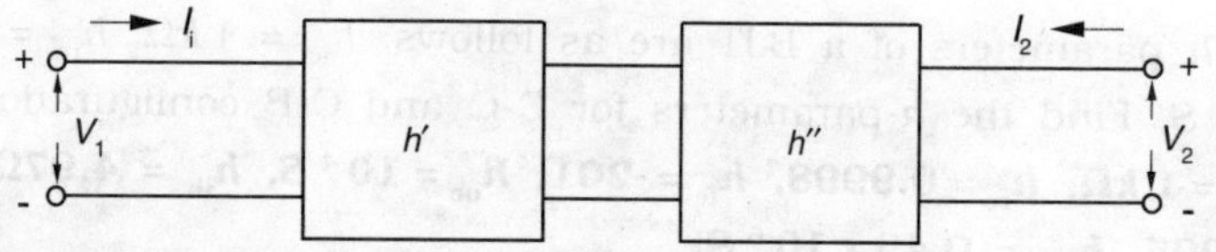

**Fig. 3.21**

9. Show that the overall h parameters of the accompanying two-stage cascaded amplifier are

(a) $h_{11} = h'_{11} - \dfrac{h'_{12}h'_{21}}{1+h'_{22}h''_{11}} . h''_{11}$  (b) $h_{12} = \dfrac{h'_{12}h''_{12}}{1+h'_{22}h''_{11}}$

(c) $h_{21} = -\dfrac{h'_{21}h''_{21}}{1+h'_{22}h''_{11}}$  (d) $h_{22} = h''_{22} - \dfrac{h''_{12}h''_{21}}{1+h'_{22}h''_{11}} . h'_{22}$

10. The h-parameters of a transistor used in C-E circuit are $h_{ie}$ = 1 kΩ, $h_{re}$ = $10^{-4}$, $h_{fe}$ = 50 and $h_{oe}$ = 100 μS. The load resistor is 1 kΩ in the collector circuit. The transistor is supplied from the signal source of resistance 1kΩ. Determine the value of impedance, voltage and current gains. **[Ans: $A_I$ = − 45.5, $Z_i$ = 995 Ω, $A_V$ = -45.7]**

# 4 THE TRANSISTOR AT HIGH FREQUENCIES

## Chapter Outline

- Introduction
- Hybrid-π conductances
- Hybrid capacitances

## INTRODUCTION

The high-frequency response of a transistor amplifier, unlike the low-frequency response, depends primarily on the capability of the transistor itself. At low frequencies, it is assumed that the transistor responds instantly to changes of input voltage or current. In order to understand how the transistor behaves at high frequencies, it is important to examine the diffusion mechanism of charge carriers from emitter to collector. When high-frequency applications are considered, the low-frequency models have to be modified and it is found that all the $h$ parameters become complex. They are also functions of the frequency. On the other hand, the model specially developed by **Giacolleto** for high frequency range known as the **hybrid π model** gives a reasonable compromise between accuracy and simplicity. The circuit elements in hybrid π model have frequency independent parameters. Note that if the capacitances of the hybrid π model are neglected it becomes the low-frequency model.

### THE HYBRID-PI (π) COMMON-EMITTER TRANSISTOR MODEL

The common-emitter is the most important practical transistor configuration. A common-emitter hybrid-π model which is valid at high frequencies is shown in Fig. 4.1. The resistive components, in this circuit can be obtained from the low-frequency $h$ parameters. The parameters resistances and capacitances are assumed to be independent of frequency. They may vary with the quiescent operating point. But they are reasonably constant for small signal swings under given bias conditions.

The resistance $r_{bb'}$ is the base spreading resistance, which represents the bulk resistance of the base. Resistance $r_{b'e}$ is that portion of the base and emitter which may be thought of as being "in series with" the collector junction. This establishes a virtual base B' for the junction capacitances to be connected instead of B. The input resistance from base-to-emitter with the output shorted is the same as $h_{ie}$. Thus

$$h_{ie} = r_{bb'} + r_{b'e} \longrightarrow (1)$$

Resistance $r_{b'c}$ that shunts $C_{b'c}$ is very large ($\cong 4\ M\Omega$).

Resistance $r_{ce}$ is the output resistance and since $r_{ce} >> R_L$, we can neglect $r_{ce}$ when $R_L$ is connected from collector to emitter.

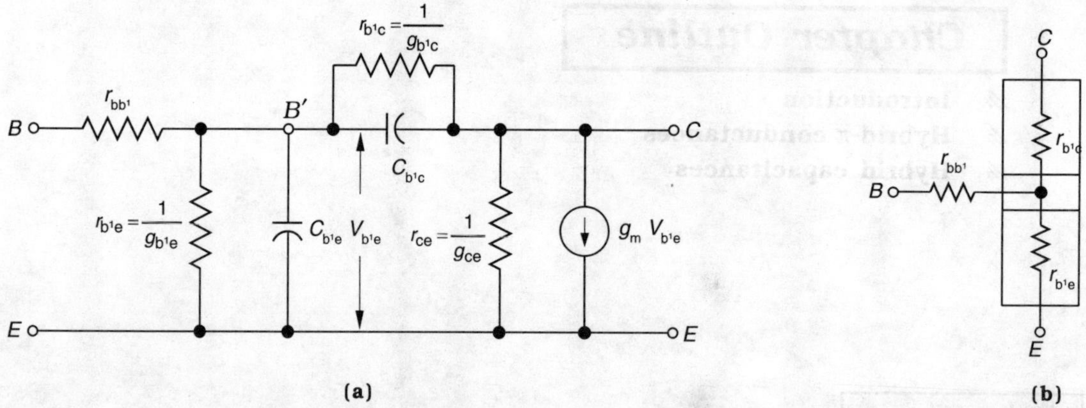


**Fig. 4.1 The hybrid-π model for a transistor in the C-E configuration**

For small changes in the voltage $V_{b'e}$ across the emitter junction, the excess minority carrier concentration injected into the base is proportional to $V_{b'e}$. The resulting small-signal collector current, with the collector shorted to the emitter is therefore proportional to $V_{b'e}$. This effect accounts for the current generator $g_m V_{b'e}$.

The increase in minority carriers in the base results in increased recombination base current. This effect is taken into account by inserting a conductance $g_{b'e}$ between $B'$ and $E$. The capacitance $C_{b'e}$ represents the capacitance of the forward-biased emitter-base junction.

The varying voltage across the collector-to-emitter junction results in **base-width modulation**. A change in the effective base width causes the emitter (and hence collector) current to change. This feedback effect between input and output is taken into account by connecting $g_{b'c}$ between $B'$ and $C$. The conductance between $C$ and $E$ is $g_{ce}$. The capacitance $C_{b'c}$ represents the capacitance of the reverse-biased collector-base junction.

## HYBRID-π CONDUCTANCES

Figure 4.2 shows a *p-n-p* transistor in the *C-E* configuration with the collector shorted to the emitter for time-varying signals. Let us understand how all the resistive components in the hybrid-π model can be obtained from the *h* parameters.

**1. Transconductance ($g_m$)**

In the active region the collector current is given by

$$I_C = I_{CO} - \propto_o I_E \longrightarrow (1)$$

Since the short-circuit current is $g_m V_{b'e}$, the transconductance is defined by

$$g_m = \left.\frac{\partial I_C}{\partial V_{B'E}}\right|_{V_{CE}} = -\propto_o \frac{\partial I_E}{\partial I_E} = \propto_o \frac{\partial I_E}{\partial V_E} \longrightarrow (2)$$

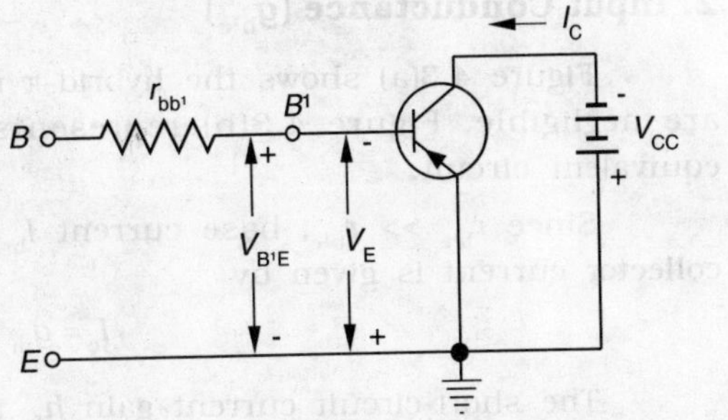

**Fig. 4.2 A *p-n-p* transistor in *C-E* configuration**

For a *p-n-p* transistor $V_E = -V_{B'E}$. If the emitter diode resistance is $r_e$, then

$$r_e = \frac{\partial V_E}{\partial I_E} \longrightarrow (3)$$

Hence $$g_m = \frac{\propto_o}{r_e} \longrightarrow (4)$$

Since the dynamic resistance of a forward-biased diode is $r_e = V_T / I_E = \bar{k}T / qI_E$ where $V_T = \bar{k}T/q$, then

$$g_m = \frac{\propto_o I_E}{V_T} \longrightarrow (5)$$

From Eq. (1), $\propto_o I_E = I_{CO} - I_C$

$$\therefore \quad g_m = \frac{I_{CO} - I_C}{V_T} \longrightarrow (6)$$

For a *p-n-p* transistor $I_C$ is negative. For an *n-p-n* transistor $I_C$ is positive.

Hence for an *n-p-n* transistor $\quad g_m = \dfrac{I_c - I_{co}}{V_T}$

Since $|I_c| >> |I_{co}|$, then $g_m$ is given by

$$\boldsymbol{g_m \cong \frac{|I_c|}{V_T}} \longrightarrow (7)$$

where $V_T = T/11{,}600$. It is clear that $g_m$ is directly proportional to current and inversely proportional to temperature. At room temperature

$$\boldsymbol{g_m \cong \frac{|I_c|\ mA}{26}} \longrightarrow (8)$$

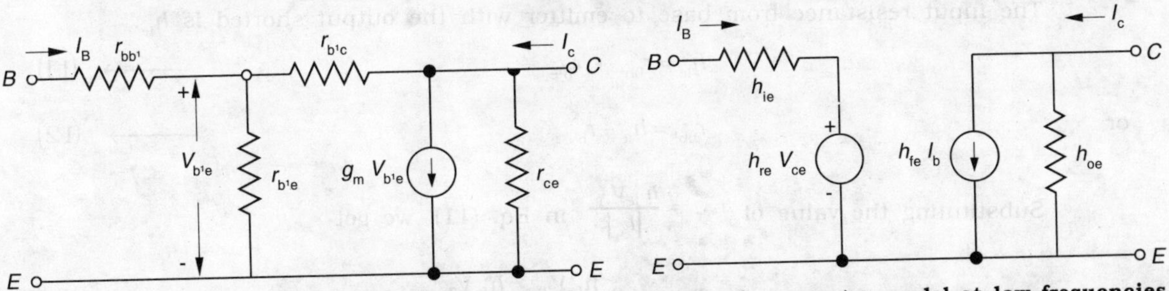

**(a) The hybrid-π model at low frequencies** **(b) The *h*-parameter model at low frequencies**

**Fig. 4.3**

**2. Input Conductance ($g_{b'e}$)**

Figure 4.3(a) shows the hybrid-π model at low frequencies, where all capacitances are negligible. Figure 4.3(b) represents the same transistor, using the *h*-parameter equivalent circuit.

Since $r_{b'c} >> r_{b'e}$, base current $I_b$ flows into $r_{b'e}$ and $V_{b'e} = I_b\, r_{b'e}$. The short-circuit collector current is given by

$$I_c = g_m V_{b'e} \approx g_m I_b\, r_{b'e}$$

The short-circuit current gain $h_{fe}$ is defined by

$$h_{fe} = \left.\frac{I_c}{I_b}\right|_{V_{CE}} = g_m r_{b'e}$$

or $$r_{b'e} = \frac{h_{fe}}{g_m} = \frac{h_{fe} V_T}{|I_c|}$$

or $$g_{b'e} = \frac{g_m}{h_{fe}} \quad \longrightarrow (8)$$

**It is seen that $r_{b'e}$ is directly proportional to temperature and inversely proportional to current.**

In Fig. 4.3(a) we find that at both very low and very high currents, $h_{fe}$ decreases.

**3. Feedback Conductance ($g_{b'c}$)**

With the input open-circuited, $h_{re}$ is defined as the reverse voltage gain with $I_b = 0$.

∴ $$h_{re} = \left.\frac{V_{b'e}}{V_{ce}}\right|_{I_b=0} = \frac{r_{b'e}}{r_{b'e} + r_{b'c}} \quad \longrightarrow (9)$$

or $$r_{b'e}(1 - h_{re}) = h_{re}\, r_{b'c}$$

Since $h_{re} << (10^{-4})$, $r_{b'e} = h_{re}\, r_{b'c}$, $$\frac{1}{g_{b'e}} = \frac{h_{re}}{g_{b'c}}$$

or $$g_{b'c} = h_{re}\, g_{b'e} \quad \longrightarrow (10)$$

Since $h_{re}$ is very small, $r_{b'c} >> r_{b'e}$.

**4. Base-spreading Resistance ($r_{bb'}$)**

The input resistance from base to emitter with the output shorted is $h_{ie}$.

$$h_{ie} = r_{bb'} + r_{b'e} \quad \longrightarrow (11)$$

or $$r_{bb'} = h_{ie} - r_{b'e} \quad \longrightarrow (12)$$

Substituting the value of $r_{b'e} = \dfrac{h_{fe} V_T}{|I_c|}$ in Eq. (11), we get

$$h_{ie} = r_{bb'} + \frac{h_{fe} V_T}{|I_c|} \approx \frac{h_{fe} V_T}{|I_c|} \quad \longrightarrow (13)$$

From the above equation, it is seen that the short-circuit input impedance $h_{ie}$ varies with current and temperature.

**5. Output Conductance ($g_{ce}$)**

With the input open-circuited, the output conductance is defined as $g_{ce}$. For $I_b = 0$, we have

$$I_c = \frac{V_{ce}}{r_{ce}} + \frac{V_{ce}}{r_{b'c} + r_{b'e}} + g_m V_{b'e} \longrightarrow (14)$$

With $I_b = 0$, we have, from Eq. (9), $V_{b'e} = h_{re} V_{ce}$ and from Eq. (13), we find

$$h_{oe} = \frac{I_c}{V_{ce}} = \frac{1}{r_{ce}} + \frac{1}{r_{b'c}} + g_m h_{re} \qquad (r_{b'c} >> r_{b'e}) \longrightarrow (15)$$

Substituting Eqs. (8) and (10), we have

$$h_{oe} = g_{ce} + g_{b'c} + g_{b'e} h_{fe} \frac{g_{b'c}}{g_{b'e}}$$

or

$$\mathbf{g_{ce} = h_{oe} - (1 + h_{fe})\, g_{b'c}} \longrightarrow (16)$$

Since $h_{fe} >> 1$,

$$g_{ce} = h_{oe} - h_{fe}\, g_{b'c} \longrightarrow (17)$$

**Summary:** If the *C-E* *h* parameters at low frequencies are known at a given collector current $I_c$, the conductances or resistances in the hybrid-$\pi$ circuit can be obtained from the following equations:

(i) $g_m = \dfrac{|I_c|}{V_T}$

(ii) $r_{b'e} = \dfrac{h_{fe}}{g_m} = \dfrac{h_{fe} V_T}{|I_c|}$ or $g_{b'e} = \dfrac{g_m}{h_{fe}}$

(iii) $r_{bb'} = h_{ie} - r_{b'e}$

(iv) $r_{b'c} = \dfrac{r_{b'e}}{h_{re}}$ or $g_{b'c} = \dfrac{h_{re}}{r_{b'e}}$

(v) $g_{ce} = h_{oe} - (1 + h_{fe}) g_{b'c} = \dfrac{1}{r_{ce}}$

## HYBRID CAPACITANCES

The hybrid-$\pi$ model for a transistor as shown in Fig. 4.1 consists of two capacitances. The collector-junction capacitance $C_{b'c}$ is the measured output capacitance with input open ($I_E = 0$). Since $C_{b'c}$ is the transition capacitance and varies as $V_{CB^{-n}}$, where $n = \frac{1}{2}$ or $\frac{1}{3}$ for an abrupt or graded junction, respectively.

$C_{b'e}$, the capacitance of the forward-biased emitter-base junction is the sum of the emitter diffusion capacitance $C_{De}$ and the emitter junction capacitance $C_{Te}$.

$$C_{b'e} = C_{De} + C_{Te}$$

Since $C_{De} >> C_{Te}$, $\quad C_{b'e} = C_{De}$ $\longrightarrow$ (1)

## Diffusion Capacitance

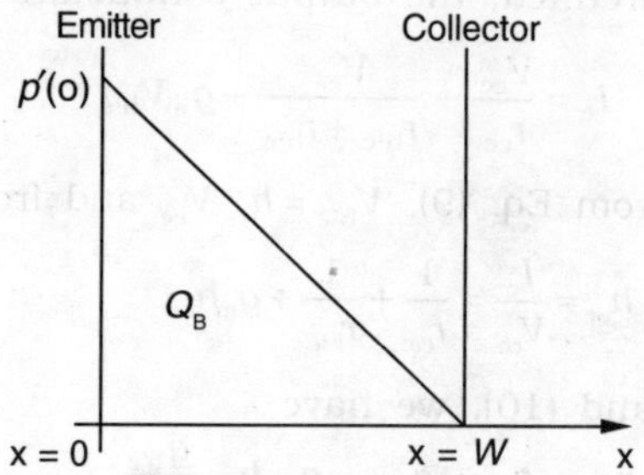


**Fig. 4.4 Minority carrier charge distribution in the base region.**

The injected hole concentration versus distance in the base region of *p-n-p* transistor is shown in Fig. 4.4. The base width *W* is assumed to be small compared with the diffusion length $L_B$ of the minority carriers. Since the collector junction is reverse-biased, the injected charge concentration $p'$ at the collector junction is essentially zero. If $W << L_B$, then $p'$ varies linearly from the value $p'(o)$ at the emitter to zero at the collector as shown in Fig. 4.4. The stored base charge $Q_B$ is the average concentration $p'(o)/2$ times the volume of the base *WA* (where *A* is the base cross-sectional area) times the electron charge *q*; i.e.,

$$Q_B = \frac{1}{2}p'(o)AWq \longrightarrow (2)$$

Diffusion current *I* is given by

$$I = -AqD_B\frac{dp'}{dx} = AqD_B\frac{p'(o)}{W} \longrightarrow (3)$$

where $D_B$ is the diffusion constant for minority carriers in the base. Combining Eqs. (1) and (3), we have

$$Q_B = \frac{IW^2}{2D_B} \longrightarrow (4)$$

The static emitter diffusion capacitance $C_{De}$ is defined as the rate of change of $Q_B$ with respect to emitter voltage *V*.

$$C_{De} = \frac{dQ_B}{dV} = \frac{d}{dV}\left(\frac{IW^2}{2D_B}\right) = \frac{W^2}{2D_B}\frac{dI}{dV}$$

$$C_{De} = \frac{W^2}{2D_B}\frac{1}{r_e} \longrightarrow (5)$$

where $r_e = \frac{dV}{dI} = \frac{V_T}{I_E}$, $\quad C_{De} = \frac{W^2 I_E}{2D_B V_T} = g_m\frac{W^2}{2D_B}$ $\longrightarrow$ (6)

The above equation shows that **the diffusion capacitance is proportional to the emitter-bias current.** From the Einstein relationship $D = \mu V_T$, it is clear that the diffusion constant

$D$ varies at $T^{-m}$ over a temperature range of 100 to 400 $K$. For silicon, $m$ = 1.5 (1.7) for electrons (holes) and for germanium, $m$ = 0.66 (1.33) for electrons (holes). Thus $C_{DE}$ varies as $T^{+n}$, where for silicon $n$ = + 0.5 (+ 0.7) for electrons (holes) and for germanium $n$ = - 0.34 (+ 0.33) for electrons (holes).

It can be shown that for a transistor the capacitance for a sinusoidal input equals two-thirds the static capacitance.

Capacitance $C_{be}$ can be determined experimentally from a measurement of $f_T$, the frequency at which the C-E short-circuit current gain drops to unity. That is

$$C_{b'e} \approx \frac{g_m}{2\pi f_T}$$

**Note:** Maximum attainable bandwidth for the current gain of a CE amplifier is given by

$$f_\beta = \frac{g_{b'e}}{2\pi(C_{b'e} + C_{b'c})}$$

## IMPORTANT POINTS TO REMEMBER

1. The hybrid-π model of a transistor at high frequencies includes the capacitive effects of the *p-n* junctions and involves a base-spreading resistance that generates a virtual base.
2. Giacoletto developed the hybrid-π model.
3. The circuit elements in hybrid-π model have frequency independent parameters.
4. The resistance $r_{b'e}$ and $C_{b'e}$ represent the resistance and capacitance of the forward-biased emitter-base junction.
5. The resistance $r_{b'c}$ and $C_{b'c}$ represent the resistance and capacitance of the reverse-biased collector-base junction.
6. The transconductance $g_m$ is directly proportional to current and inversely proportional to temperature.
7. The input conductance $g_{b'e}$ is directly proportional to current and inversely proportional to temperature.
8. The base-spreading resistance varies with current and temperature.
9. The capacitance $C_{b'e}$ is the sum of the emitter diffusion capacitance and emitter junction capacitance.
10. The diffusion capacitance is proportional to the emitter bias current.

## KEY FORMULAE

1. Input resistance, $h_{ie} = r_{bb'} + r_{b'e}$
2. Transconductance, $g_m = \dfrac{I_C - I_{CO}}{V_T} \approx \dfrac{|I_C|}{V_T}$
3. Input conductance, $g_{b'e} = \dfrac{g_m}{h_{fe}}$
4. Feedback conductance, $g_{b'c} = h_{re}\, g_{b'e} = 1/r_{b'c}$
5. Base-spreading resistance, $r_{bb'} = h_{ie} - r_{b'e}$

6. Output conductance, $g_{ce} = h_{oe} - (1 + h_{fe})g_{b'c}$

7. Capacitance, $C_{b'e} = C_{De} + C_{Te} = \dfrac{g_m}{2\pi f_T} - C_{b'c}$

8. Diffusion capacitance, $C_{De} = \dfrac{W^2 I_E}{2D_B V_T} = g_m \dfrac{W^2}{2D_B}$

## SOLVED PROBLEMS

**1. A junction transistor is to be operated at the following operating point: $|I_C| = 2\ mA, V_{CEQ} = 20\,V, I_{BQ} = 20\,\mu A$. The following values for this operating point are listed: $\beta_o = 100$, $f_T = 50\ MHz$, $C_{ob} = 3\ pF$ and $h_{ie} = 1400\ \Omega$ (low frequency). Determine the parameters of the hybrid-π model. Assume that the operating temperature is 300 K.**

**Data:** $|I_C| = 2\ mA, V_{CEQ} = 20V, I_{BQ} = 20\,\mu A, \beta_o = 100, f_T = 50\,MHz, C_{ob} = 3\,pF$ and $h_{ie} = 1400\,\Omega$, $T = 300\,K$

**To find:** Parameters of the hybrid-π model

**Solution:** (i) Transconductance $g_m = \dfrac{|I_C|}{V_T} = \dfrac{11{,}600}{T}|I_C|$

$$g_m = \frac{11{,}600}{T} \times 2 \times 10^{-3} = \mathbf{77{,}333\ \mu S}$$

(ii) Input resistance $r_{b'e} = \dfrac{h_{fe}}{g_m} = \dfrac{100}{77{,}333 \times 10^{-6}} = \mathbf{1293\,\Omega}$

(iii) Capacitance, $C_{b'c} = C_{ob} = \mathbf{3\ pF}$

(iv) Capacitance, $C_{b'e} = \dfrac{g_m}{\omega_T} - C_{b'c}$

$$= \frac{77{,}333 \times 10^{-6}}{2\pi \times 50 \times 10^6} - 3 \times 10^{-12} = \mathbf{243\ pF}$$

(v) Base spreading resistance $r_{bb'} = h_{ie} - r_{b'e}$

$$= 1400 - 1293 = \mathbf{107\,\Omega}$$

**2. The following low-frequency parameters are known for a given transistor at $I_C = 10\ mA$, $V_{CE} = 10\ V$ and at room temperature, $h_{ie} = 500\ \Omega$, $h_{oe} = 4 \times 10^{-5}\ S$, $h_{fe} = 100$, $h_{re} = 10^{-4}$.**

**At the same operating point, $f_T = 50\ MHz$ and $C_{ob} = 3\ pF$, compute the values of all the hybrid-π parameters.**

**Data:** $h_{ie} = 500$, $h_{fe} = 100$, $h_{oe} = 4 \times 10^{-5}\ S$, $h_{re} = 10^{-4}$, $f_T = 50\,MHz$, $C_{ob} = 3\ pF$

**To find:** Hybrid π parameters

**Solution:** (i) Transconductance, $g_m = \frac{|I_C|}{V_T} = \frac{11{,}600}{T}|I_C|$

$$= \frac{10\,mA}{26\,mV} = \mathbf{385\,mA/V}$$

(ii) Input resistance, $r_{b'e} = \frac{h_{fe}}{g_m} = \frac{100}{0.385} = 260\,\Omega$

(iii) Base spreading resistance, $r_{b'b} = h_{ie} - r_{b'e}$

$$= 500 - 260 = \mathbf{240\,\Omega}$$

(iv) Feedback conductance, $g_{b'c} = h_{re}\, g_{b'e} = \frac{1}{260 \times 10^{+6}} = \mathbf{0.38 \times 10^{-6}}$

(v) Output conductance, $g_{ce} = h_{oe} - (1 + h_{fe})g_{b'c}$

$$= 4 \times 10^{-5} - \frac{101}{2.6 \times 10^{6}} = \mathbf{0.12 \times 10^{-5}}$$

(vi) Capacitance, $C_{b'e} = \frac{g_m}{2\pi f_T} = \frac{385 \times 10^{-3}}{2\pi \times 50 \times 10^{6}} = \mathbf{1224\,pF}$

(vii) Capacitance $C_C = 3\,pF$

**3. Given the following transistor measurements made at $I_C$ = 5 *mA*, $V_{CE}$ = 10 *V*, and at room temperature: $h_{fe}$ = 100, $h_{ie}$ = 600 Ω, [$A_{ie}$] = 10 at 10 *MHz*, $C_{b'c}$ = 3 *pF*. Find $f_\beta$, $f_T$, $C_{b'e}$, $\tau_{b'e}$ and $r_{bb'}$**

**Data:** $I_C$ = 5 *mA*, $V_{CE}$ = 10 *V*, $h_{fe}$ = 100, $h_{ie}$ = 600 Ω, [$A_{ie}$] = 10 at 10 *MHz* and $C_{b'c} = 3\,pF$

**To find:** $f_\beta$, $f_T$, $C_{b'e}$, $\tau_{b'e}$ and $r_{bb'}$

**Solution:** Current gain $|A_{ie}| = \left|\frac{-h_{fe}}{1 + jf/f_B}\right| = \frac{h_{fe}}{\sqrt{1 + (f/f_B)^2}}$

$$10 = \frac{100}{\sqrt{1 + (f/f_B)^2}} \Rightarrow 1 + (f/f_B)^2 = 100$$

$$(f/f_B)^2 = 99 \quad \text{or} \quad f/f_B = 9.95$$

$$f_B = \frac{f}{9.95} = \frac{10\,MHz}{9.95} = \mathbf{1.005\,MHz}$$

We know $f_T = h_{fe} f_\beta = 100 \times 1.005 = \mathbf{100.5\,MHz}$

Capacitance, $C_{b'e} = \frac{g_m}{2\pi f_T} = \frac{|I_C|}{V_T}\frac{1}{2\pi f_T}$

$$= \frac{5 \times 10^{-3}}{26 \times 10^{-3}} \frac{1}{2\pi \times 100.5 \times 10^{6}} = \mathbf{304\,pF}$$

Resistance, $$r_{b'e} = \frac{h_{fe}}{g_m} = \frac{100}{5/26} = \mathbf{520\,\Omega}$$

Base spreading resistance, $r_{bb'} = h_{ie} - r_{b'e} = 600 - 520 = \mathbf{80\,\Omega}$

**4. A silicon *p-n-p* transistor has an $f_T$ = 400 *MHz*. What is the base thickness?**

**Data:** $f_T$ = 400 *MHz*

**To find:** Base thickness, $W$

**Solution:** We know that $$f_T = h_{fe} f_\beta = \frac{g_m}{2\pi C_{b'e}} \longrightarrow (1)$$

$$C_{b'e} = C_{De} = g_m \frac{W^2}{2D_B} \longrightarrow (2)$$

Substituting the value of $C_{b'e}$ in Eq. (1), we get

$$f_T = \frac{g_m}{2\pi\left(g_m W^2/2D_B\right)} = \frac{D_B}{\pi W^2}$$

or $$W^2 = \frac{D_B}{\pi f_T}$$

For a *p-n-p* transistor $D_B = D_p = 13\ cm^2/sec$

$$\therefore \quad W^2 = \frac{13\times10^{-4}}{3.14\times400\times10^6} = \mathbf{1.03\,\mu m}$$

**5. Given a germanium *p-n-p* transistor whose base width is $10^{-4}$ *cm*. At room temperature and for a *dc* emitter current of 2 *mA*, find (a) the emitter diffusion capacitance and (b) $f_T$.**

**Data:** $W = 10^{-6}\ m$, $I_E = 2\ mA$

**To find:** Emitter diffusion capacitance, $C_{b'e}$ and $f_T$

**Solution:** (a) Emitter diffusion capacitance, $C_{b'e} = C_{De}$

We know that $$C_{b'e} = g_m \frac{W^2}{2D_B} \longrightarrow (1)$$

Since $g_m = \frac{|I_C|}{V_T} = \frac{|I_E|}{V_T}$, $$C_{b'e} = \frac{|I_E|}{V_T}\frac{W^2}{2D_B}$$

$$= \frac{2\times10^{-3}}{26\times10^{-3}}\frac{\left(10^{-6}\right)^2}{2\times47\times10^{-4}} \qquad \left(\because D_B = 47\times10^{-4}\right)$$

$$= \mathbf{8.2\ pF}$$

(b) We also know that $$f_T = \frac{g_m}{2\pi C_{b'e}} = \frac{|I_C|}{V_T 2\pi C_{b'e}}$$

$$f_T = \frac{2\times10^{-3}}{26\times10^{-3}\times2\times3.14\times8.2\times10^{-12}} = \mathbf{1500\ MHz}$$

## QUESTIONS

1. Explain what is meant by the term "virtual base".
2. Explain with a circuit diagram the operation of small-signal high-frequency C-E model of a transistor.
3. Determine the hybrid-π parameters of a transistor experimentally.
4. How does the transistor conductance vary with temperature? Why?
5. (a) What is the physical origin of the two capacitors in the hybrid-π model? (b) What is the order of magnitude of each capacitance?
6. How does transistor conductance vary with (a) $|I_C|$; (b) $|V_{CE}|$; and (c) $T$?
7. Prove that $g_m = |I_C|/V_T$
8. Prove that $g_{b'e} = g_m/h_{fe}$ or $h_{fe} = g_m r_{b'e}$.
9. (a) Prove that $h_{ie} = r_{bb'} + r_{b'e}$. (b) Assuming $r_{bb'} << r_{b'e}$, how does $h_{ie}$ vary with $|I_C|$?
10. How does capacitance $C_{b'c}$ vary with $|I_C|$ and $|V_{CE}|$?
11. How does the capacitance $C_{b'e}$ vary with $|I_C|$ and $|V_{CE}|$?
12. Write a short note on hybrid-π capacitances.

## EXERCISES

1. Show that at low frequencies the hybrid-π model with $r_{b'c}$ and $r_{ce}$ taken as infinite reduces to the approximate C-E $h$-parameters model.
2. Consider the hybrid-π circuit at low frequencies, so that $C_{b'e}$ and $C_{b'c}$ may be neglected. Omit none of the other elements in the circuit. If the load resistance is $R_L = 1/g_L$, prove that

$$K = \frac{V_{ce}}{V_{b'e}} = \frac{-g_m + g_{b'c}}{g_{b'c} + g_{ce} + g_L}$$

3. Using Miller's theorem, draw the equivalent circuit between B and E of a transistor. Prove that the current gain under load is

$$A_I = \frac{g_L}{(g_{b'c} + g_{b'e})/k - g_{b'c}}$$

4. A 40397 silicon transistor used in video applications is connected in the C-E configuration with its collector-to-emitter voltage held constant at 1 $V$. A 70 $mV$ change in the base-emitter voltage causes a change in collector current from 1 $mA$ to 10 $mA$.
   (a) What is the approximate value of the transistor's $g_m$?
   (b) Calculate $g_m$ at $I_C$ = 1 $mA$, 4 $mA$ and 6 $mA$, assuming room temperature and $g_m = |I_C|/26\,mV$.
   (c) If the temperature is increased, what factors would have to be taken into consideration to determine the new values of $g_m$?

   **[Ans: (a) 130 $mS$; (b) 38.5 $mS$ at $I_C$ = 1 $mA$, 154 $mS$ at $I_C$ = 4 $mA$ and 231 $mS$ at $I_C$ = 6 $mA$; (c) An increase in temperature would increase $I_C$, but also would increase $V_T$, so $g_m$ would be fairly constant]**

# 5 MULTISTAGE AMPLIFIERS

## Chapter Outline

- Introduction
- Classification of Amplifiers
- Distortion in Amplifiers
- Frequency Response
- The RC-Coupled Amplifier

## INTRODUCTION

In most applications, an amplifier in a single stage cannot supply enough signal output. Hence, two or more amplifier stages are cascaded to provide the greater signal. This is achieved by coupling (cascading) number of amplifier stages, known as **multistage amplifier**. The basic purpose of a multistage arrangement is to increase the overall voltage gain. But cascading results in new problems to be faced in the multistage design. For example, problems such as the interaction between stages due to input impedance effects, the consequences of applying feedback in multistages like instability, hum and noise, etc. have to be considered and tackled.

### Multistage Voltage Gain

The overall voltage gain $A_{ov}$ of **cascaded** amplifiers, as shown in Fig. 5.1, is the product of the individual voltage gains.

$$A_{ov} = A_{v1} \times A_{v2} \times A_{v3} \times \ldots \times A_{vn}$$

where $n$ is the number of **stages.**

For $n$ identical stages each having gain $A_v$, the overall voltage gain is

$$A_{ov} = (A_v)^n$$

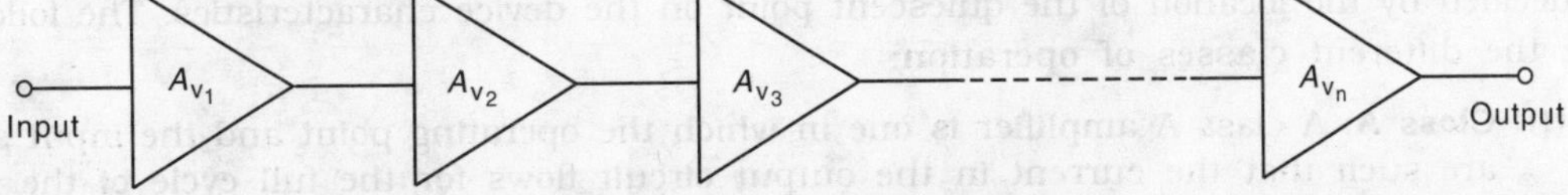


**Fig. 5.1 Cascaded amplifiers**

## CLASSIFICATION OF AMPLIFIERS

Amplifiers are classified in many ways, according to their frequency range, the method of operation, the ultimate use, the type of load, the method of interstage coupling, etc.

### 1. In terms of Frequency

(a) *DC* amplifiers (from zero frequency)
(b) Audio amplifiers (20 *Hz* to 20 *kHz*)
(c) Video or pulse amplifiers (upto a few *MHz*)
(d) Radio frequency amplifiers (a few *kHz* to hundreds of *MHz*)
(e) Ultra-high frequency amplifiers (hundreds or thousands of *MHz*)

### 2. In terms of Signal Strength

(a) Small-signal amplifiers
(b) Large-signal amplifiers

### 3. In terms Coupling

(a) Resistance-Capacitance coupling
(b) Transformer coupling
(c) Direct coupling

### 4. In terms Voltage, Current and Power

(a) Voltage amplifiers
(b) Current amplifiers
(c) Power amplifiers

### 5. In terms of Biasing condition and Amplitude of the input signal (class of operation)

(a) Class A amplifier
(b) Class B amplifier
(c) Class AB amplifier
(d) Class C amplifier

### 6. For Radio Frequency Applications

(a) Single-tuned voltage amplifiers
(b) Double-tuned voltage amplifiers
(c) Stagger-tuned voltage amplifiers

Large signal amplifiers can be classified according to the class of operation which is decided by the location of the quiescent point on the device characteristics. The following are the different classes of operation:

(i) **Class A:** A class A amplifier is one in which the operating point and the input signal are such that the current in the output circuit flows for the full cycle of the input signal. It operates over a linear portion of transistor characteristics.

(ii) **Class B:** A class B amplifier is one in which the operating point is at an extreme end of its characteristic. The current in the output circuit flows for only half the cycle of the input signal. The current will be zero for the other half-cycle. Because of the small power dissipation, efficiency is high.

(iii) **Class AB:** A class AB amplifier is one operating between the two extremes defined for class A and class B. In class AB operation, the $Q$-point is so located that the output current flows for more than half the cycle and less than the full cycle of the input signal.

(iv) **Class C:** A class C amplifier is one in which the operating point is located beyond the cutoff. The output current flows for less than half the input signal cycle. The current flows in the form of pulses. Class C provides very high efficiency. Most of the transmitter amplifiers are class C type.

## DISTORTION IN AMPLIFIERS

An ideal amplifier should magnify the input without loss of fidelity. This means that the output should follow the input in all respects, except amplitude. In practice, the output waveform is not an exact replica of the input signal waveform because of the presence of distortion. The distortions may arise either from the inherent nonlinearity in transistor characteristics or from the influence of the associated circuit.

The distortions are classified as:

1. Nonlinear or amplitude distortion
2. Frequency distortion
3. Phase distortion
4. Inter-modulation distortion

**The change of output wave shape from the input wave shape of an amplifier is known as distortion.**

### Nonlinear Distortion

Nonlinear distortion is produced by operating the amplifier over the nonlinear part of the transfer characteristic of the amplifier. Since the amplifier amplifies different parts of the input signal differently, new frequencies called harmonics are generated in the output. The percentage harmonic distortion for the $n^{th}$ harmonic is given by

$$D_n = \frac{A_n \left(\text{Amplitude of the } n^{th} \text{ harmonic}\right)}{A_1 \left(\text{Amplitude of the fundamental}\right)} \times 100\,\% \qquad \longrightarrow (1)$$

The component of frequency at the output same as the input signal, is called the fundamental frequency. The total harmonic distortion is given by

$$D_T = \sqrt{D_2^{\,2} + D_3^{\,2} + \ldots + D_n^{\,2}} \qquad \longrightarrow (2)$$

where $D_2$, $D_3$, .......... are harmonic components, second, third, ....... and so on.

The total distortion is measured by a distortion-factor meter. The spectrum or wave analyser can be used to measure the amplitude of each harmonic.

### Frequency Distortion

Frequency distortion results when the signal components of different frequencies are amplified differently. This distortion is due to the various frequency-dependent reactances (both capacitive and inductive) associated with the circuit or the active device (BJT or FET) itself. In the case of audio signals, the frequency distortion leads to a change in the quality of sound, because all the different frequencies have different amplitudes. Hence in the design of untuned or wideband amplifiers, the amplifier should provide the same gain for all the frequencies so that the frequencies will have the same relative amplitudes in the output. If the frequency-response characteristic is not flat over the range of frequencies under consideration, the circuit is said to have frequency distortion.

### Phase Distortion

Phase distortion results from unequal phase shifts of signals of different frequencies. The changes in phase angles are also due to frequency-dependent capacitive and inductive reactances associated with the circuit and the active device of the amplifier.

This type of distortion which is due to the non-uniform phase shift of different frequency components, is difficult to eliminate. Note that if all the frequency components in a signal are shifted in phase by an integral multiple of $\pi$, the resultant output waveform is not changed, although the polarity of the wave may be altered but changes in polarity do not constitute distortion.

### Inter-modulation Distortion

The harmonics introduced in the amplifier can combine with each other or with the original frequencies to produce new frequencies that are not harmonics of the fundamental. This effect is called **inter-modulation distortion,** because the frequencies combine with each other. This distortion results in unpleasant hearing.

## FREQUENCY RESPONSE OF AN AMPLIFIER

Amplifiers can be classified according to the frequency range of the signals to be handled. A criterion that may be used to compare one amplifier with another with respect to fidelity of reproduction of the input signal is suggested by the following considerations:

Any arbitrary wave may be resolved into a Fourier Spectrum. If the waveform is periodic, the spectrum then consists of a series of sines and cosines whose frequencies are all integral multiples of a fundamental frequency. The fundamental frequency is the reciprocal of the time that must elapse before the waveform repeats itself. If the waveform is not periodic, the fundamental period extends from a time $-\infty$ to a time $+\infty$. Then the fundamental frequency is infinitesimally small. Hence the frequencies of successive terms in the Fourier series differ by an infinitesimal small amount and thereby Fourier series becomes instead a Fourier integral. In either case the spectrum includes terms whose frequencies extend from zero frequency to infinity.

### Fidelity Considerations

Consider a sinusoidal signal of angular frequency $\omega$ represented by $v = V_m \sin(\omega t + \phi)$. If $A$ is the voltage gain of the amplifier and if the signal suffers a phase change (lead angle) $\theta$, then the output will be given by

$$v_o = AV_m \sin(\omega t + \phi + \theta) = A\,V_m \sin\left[\omega\left(t + \frac{\theta}{\omega}\right) + \phi\right] \longrightarrow (1)$$

Hence, **if the amplification $A$ is independent of frequency and if the phase shift $\theta$ is proportional to frequency (or is zero), then the amplifier will preserve the form of input signal, although the signal will be shifted in time by an amount $D = \frac{\theta}{\omega}$.**

The above discussion suggests that the extent to which an amplifier's amplitude response is not uniform and its time delay is not constant with frequency, may serve as a measure of the lack of fidelity anticipated. In practice, both amplitude and time delay response are related and one having been specified, the other is uniquely determined. Hence it is not necessary to specify both amplitude and time delay response. However, the time-delay or amplitude response is the more sensitive indicator of frequency distortion.

In practical amplifier circuits, capacitors are often used to block *dc* and couple only the *ac* signal to the load or the input of another stage of the amplifier. Figure 5.4 shows the typical frequency response of an amplifier.

For an amplifier stage the frequency characteristics may be divided into three regions: (i) There is a range, called **mid-frequency range**, over which the amplification $A_o$ is constant and over which the delay is also constant. This middle range is where the amplifier is supposed to operate. (ii) In the **low-frequency** region below midband, an amplifier stage behaves like the high-pass circuit as shown in Fig. 5.2(a). At **high frequencies**, the reactance of $C_1$ will be small and hence it acts as a short without any signal voltage loss across it. As the frequency is decreased, the output voltage falls. At zero frequency (*dc*), the output will be zero. Thus the series capacitor makes the amplifier behave as a high-pass filter.

(iii) In the **high-frequency** region above midband, the circuit often behaves like the low-pass network as shown in Fig. 5.2(b). At low frequencies, capacitor $C$ has high reactance. Hence it will behave like an open circuit. As the frequency is increased, the reactance of $C$ decreases. Therefore more voltage is dropped in $R_s$ and less is available at the output. Thus the voltage gain of the amplifier decreases at high frequencies. Thus the shunt capacitor makes the amplifier behave as a low-pass filter. The total frequency characteristic for all three regions is shown in Fig. 5.4.

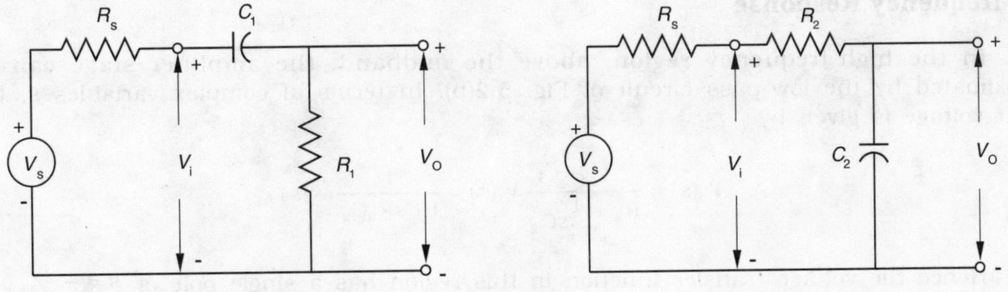


(a) High-pass circuit **Fig. 5.2** (b) Low-pass circuit

## Low-frequency Response

In the low-frequency region, below the midband, the amplifier stage can be approximated by the high-pass circuit of Fig. 5.2(a). In terms of complex variable $s$, output voltage is given by

$$V_o(s) = \frac{V_i(s) R_1}{R_1 + 1/sC_1} = V_i(s)\frac{s}{s + 1/R_1C_1} \qquad \longrightarrow (1)$$

Thus the voltage transfer function at low frequencies, $A_L(s) = \dfrac{V_o(s)}{V_i(s)}$, has one zero at $s = 0$ and one pole at $s = -1/R_1C_1$. For real frequencies ($s = j\omega = j2\pi f$), Eq. (1) becomes

$$A_L(jf) = \frac{1}{1 - j(f_L/f)} \longrightarrow (2)$$

where

$$f_L = \frac{1}{2\pi R_1 C_1} \longrightarrow (3)$$

The magnitude ($A_L$) of voltage gain is given by

$$|A_L(jf)| = \frac{1}{\sqrt{1 + (f_L/f)^2}} \longrightarrow (4)$$

The phase lead $\theta_L$ of the gain is given by

$$\theta_L = \arctan\left(\frac{f_L}{f}\right) \longrightarrow (5)$$

At $f = f_L$, $A_L = \dfrac{1}{\sqrt{2}} = 0.707$, whereas in the midband region ($f >> f_L$), $A_L$ tends to unity. Hence the magnitude of the voltage gain falls off to 70.7% of the midband value at $f = f_L$. Such a frequency is called the **lower cut-off** or **lower 3 *dB* frequency.** From Eq. (3) we see that $f_L$ is that frequency for which the resistance $R_1$ equals the capacitive reactance, $X_1 = \dfrac{1}{2\pi f_L C_1}$.

## High-frequency Response

In the high-frequency region, above the midband, the amplifier stage can be approximated by the low-pass circuit of Fig. 5.2(b). In terms of complex variables $s$, the output voltage is given by

$$V_o(s) = \frac{1/sC_2}{R_2 + 1/sC_2} V_i(s) = \frac{1}{1 + sR_2C_2} V_i(s) \longrightarrow (6)$$

Hence the voltage transfer function in this region has a single pole of $s = -1/R_2C_2$. For real frequencies ($s = j\omega = j2\pi f$), Eq. (2) becomes

$$|A_H(jf)| = \left|V_o(s)/V_i(s)\right| s = j2\pi f$$

Magnitude $|A_H|$ of voltage gain is given by

$$|A_H(jf)| = \frac{1}{\sqrt{1+\left(\frac{f}{f_H}\right)^2}} \longrightarrow (7)$$

where
$$f_H = \frac{1}{2\pi R_2 C_2} \longrightarrow (8)$$

The phase $\theta_H$ of the gain is given by

$$\theta_H = -\arc\tan\left(\frac{f}{f_H}\right) \longrightarrow (9)$$

At $f = f_H$, $A_H = \left(\frac{1}{\sqrt{2}}\right)A = 0.707A$, then $f_H$ is called the **upper cut-off** or **upper 3-*dB* frequency**. It also represents the frequency at which the resistance $R_2$ = capacitive reactance of $C_2$ $\left(= \frac{1}{2\pi f_H C_2}\right)$.

From the above discussion we find that at frequencies $f_L$ and $f_H$ the voltage gain falls to $\frac{1}{\sqrt{2}}$ of the midband voltage gain. Hence the power gain falls to half the value

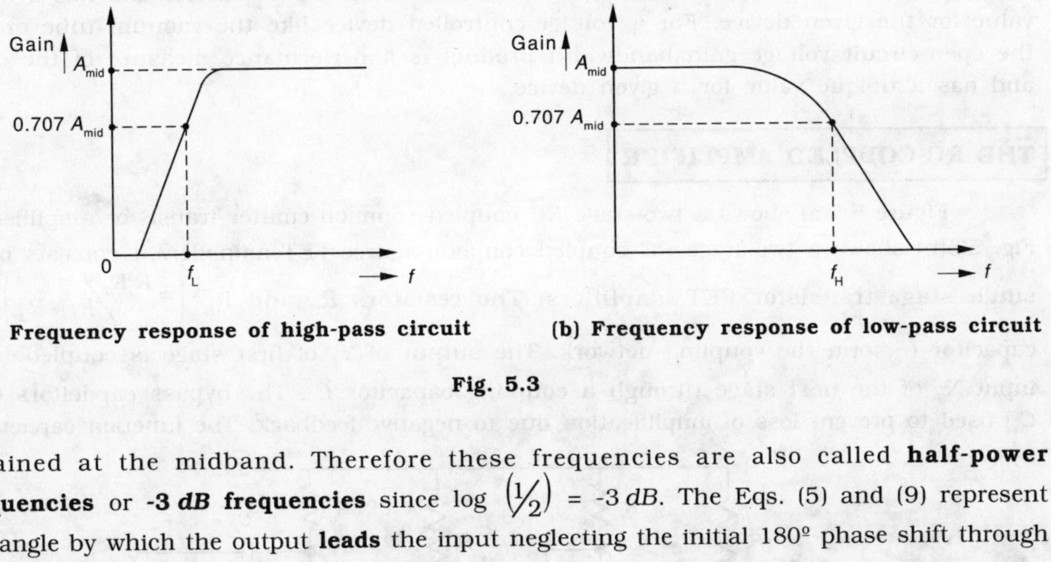


**(a) Frequency response of high-pass circuit** **(b) Frequency response of low-pass circuit**

**Fig. 5.3**

obtained at the midband. Therefore these frequencies are also called **half-power frequencies** or **-3 *dB* frequencies** since log $\left(\frac{1}{2}\right)$ = -3 *dB*. The Eqs. (5) and (9) represent the angle by which the output **leads** the input neglecting the initial 180º phase shift through the amplifier.

## Frequency-response Plots

The gain and phase plots versus frequency can be approximately sketched by using straight-line segments called asymptotes. Such plots are called **Bode plots**. They are very convenient for a quick evaluation of the frequency response of cascaded amplifiers.

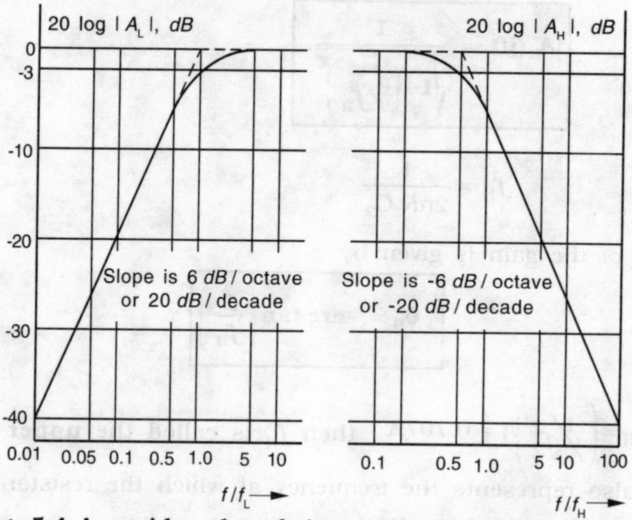


**Fig. 5.4 A semi-log plot of the amplitude frequency-response (Bode) characteristic of an *RC* coupled amplifier**

**Bandwidth:** The frequency range from $f_L$ to $f_H$ is called the bandwidth of the amplifier stage. The product of the mid-frequency gain and the 3 *dB* bandwidth of an amplifier is called the **gain-bandwidth product**. For a current-controlled device like the BJT, the short-circuit current gain-bandwidth product is a performance measure for the device and has a unique value for the given device. For a voltage-controlled device like the vacuum tube or FET, the open-circuit voltage gain bandwidth product is a performance measure of the device and has a unique value for a given device.

## THE RC-COUPLED AMPLIFIER

Figure 5.5(a) shows a two-stage *RC* coupled common-emitter transistor amplifier and Fig. 5.5(b) shows a two-stage *RC* coupled common-source FET amplifier. It consists of two single-stage transistor/FET amplifiers. The resistors $R_C$ and $R_B$ $\left(= R_1R_2 \big/ R_1 + R_2\right)$ and capacitor $C_c$ form the coupling network. The output of $Y_1$ of first stage is coupled to the input $X_2$ of the next stage through a coupling capacitor $C_c$. The bypass capacitors $C_E$ (= $C_s$) used to prevent loss of amplification due to negative feedback. The junction capacitance

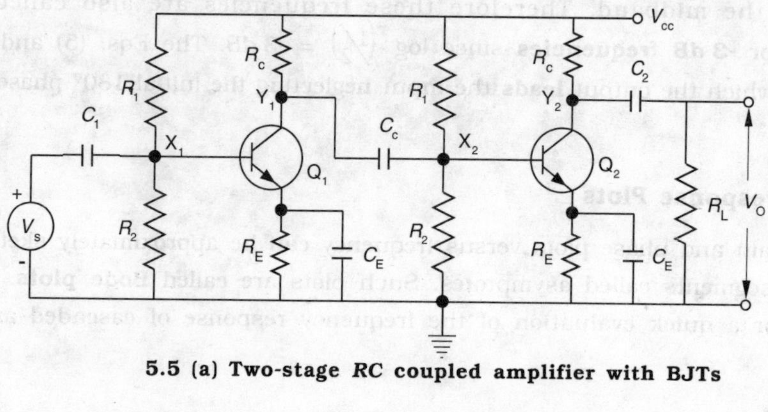


**5.5 (a) Two-stage *RC* coupled amplifier with BJTs**

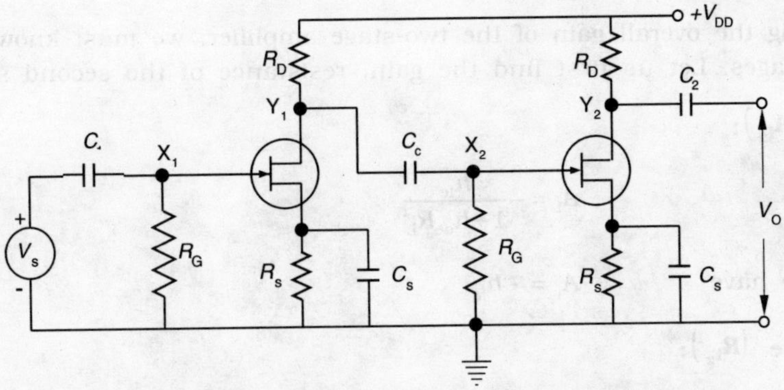


**5.5 (b) Two-stage *RC* coupled amplifier with FETs**

should be taken into account when high-frequency response is considered.

The operation of the circuit may be understood from the example that when an *ac* signal is applied to the input of the first stage, it is amplified by a transistor/FET and appears across the collector resistor $R_c$/drain resistor $R_D$. This signal is given to the input of the second stage through a coupling capacitor $C_c$. The second stage does further amplification of the signal. In this way, the cascaded stages amplify the signal and the overall gain is equal to the product of the individual stage gains.

## Analysis of Two-Stage *RC*-Coupled Amplifier

Assuming all capacitances are arbitrarily large and act as *ac* short circuits across $R_E$, then *h*-parameter equivalent circuit for cascade amplifier obtained is shown in Fig. 5.6.

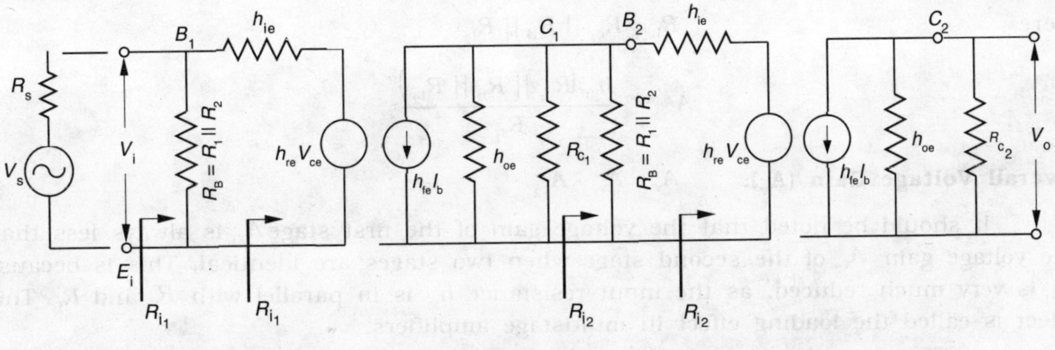


**Fig. 5.6 *h*-parameter equivalent circuit for *RC* coupled amplifier**

The *dc* power supply is also replaced by a short-circuit. The transistors are replaced by their *h*-parameter approximate models.

The parallel combination of resistors $R_1$ and $R_2$ is replaced by a single resistor $R_B$ i.e.,

$$R_B = R_1 \,||\, R_2 = \frac{R_1 R_2}{R_1 + R_2}$$

For finding the overall gain of the two-stage amplifier, we must know the gains of the individual stages. Let us first find the gain, resistance of the second stage.

**Current Gain $(A_{i_2})$:**

We know that $$A_i = -\frac{h_{fe}}{1+h_{oe}R_L}$$

Neglecting $h_{oe}$ we have $$A_i = -h_{fe}$$

**Input Resistance $(R_{i_2})$:**

We know that $$R_i = h_{ie} + h_{re} A_i R_L$$

$\therefore$ $$R_i = h_{fe} \quad \text{and} \quad R_{i_2} = h_{ie}$$

**Voltage Gain $(A_{v_2})$:**

We know that $$A_v = \frac{A_i R_L}{R_i}$$

$\therefore$ $$A_{v_2} = -\frac{h_{fe} R_{c_2}}{R_{i_2}}$$

**Current Gain $A_{i_1}$:** $$A_{i_1} = -h_{fe}$$

**Input Resistance $(R_{i_1})$:** $$R_{i_1} = h_{ie}$$

**Voltage Gain $(A_{v_1})$:** $$A_v = \frac{A_i R_L}{R_{i_1}}$$

Here $$R_L = R_{c_1} \,||\, R_B \,||\, R_{i_2}$$

$\therefore$ $$A_{v_1} = -\frac{h_{fe}\left(R_{c_1} \,||\, R_B \,||\, R_{i_2}\right)}{R_{i_1}}$$

**Overall Voltage Gain $(A_v)$:** $$A_v = A_{v_1} \times A_{v_2}$$

It should be noted that the voltage gain of the first stage $A_{v_1}$ is always less than the voltage gain $A_{v_2}$ of the second stage when two stages are identical. This is because $R_{i_1}$ is very much reduced, as the input resistance $h_{ie}$ is in parallel with $R_{c_1}$ and $R_B$. This effect is called the loading effect in multistage amplifiers.

## IMPORTANT POINTS TO REMEMBER

1. Single-stage amplifiers can be connected in sequence with various coupling methods to form multistage amplifiers.
2. The total gain of a multistage amplifier is the product of the individual gains (sum of *dB* gains).
3. Amplifiers can be classified according to their frequency range, the method of

operation, the ultimate use, the type of load, the method of interstage coupling, etc.

4. Large-signal amplifiers are power amplifiers.
5. Power amplifiers (class *C*) are normally used as the final stage of a communication receiver or transmitter to provide signal power to speakers or to a transmitting antenna.
6. A class *A* power amplifier operates entirely in the linear region of the transistor's characteristic curves. The transistor conducts during the full 360° of the input cycle.
7. A class *B* amplifier operates in the linear region for half of the input cycle (180°) and it is cutoff for the other half.
8. A class *AB* amplifier is biased slightly above cutoff and operates in the linear region for slightly more than 180° of the input cycle.
9. A class *C* amplifier operates in the linear region for only a small part of the input cycle.
10. The efficiency of class *C* amplifier is very high i.e., it can approach 100 percent.
11. Distortion is the change of output wave shape from the input wave shape of an amplifier.
12. The various distortions that arise in amplifiers are (i) nonlinear or amplitude distortion (ii) frequency distortion (iii) phase distortion and (iv) inter-modulation distortion.
13. Non-linear distortion is produced by operating the amplifier over the non-linear part of the transfer characteristic of the amplifier.
14. Frequency distortion results when the gain of the amplifier varies with frequency.
15. Phase distortion results from unequal phase shifts of signals of different frequencies.
16. When an amplifier is deliberately overdriven to obtain harmonics at the output, we get inter-modulation distortion.
17. The signals handled by amplifiers are quite often nonsinusoidal.
18. By the use of the Fourier-series concept, we can consider the nonsinusoidal signals to be made up of a series of sinusoidal components of different frequencies.
19. For an amplifier stage, the frequency characteristics may be divided into three regions *viz* (i) low-frequency region (ii) mid-frequency region and (iii) high-frequency region.
20. In most amplifiers, voltage gain falls off at low as well as high frequencies.
21. In low-frequency region, the magnitude of the voltage gain falls off to 70.7% of the midband value at $f = f_L$.
22. In high-frequency region, the magnitude of the voltage gain falls to $1/\sqrt{2}$ of the midband voltage gain.
23. The amplifier behaves as a bandpass amplifier amplifying the frequencies between the two cutoff frequencies lower cutoff frequency ($f_L$) and upper cutoff frequency ($f_H$).
24. Since the human ear is not very sensitive to power gain variations of about 3 *dB*, the effective passband called the bandwidth of the amplifier is taken as its 3 *dB* bandwidth.
25. The gain and phase plots versus frequency can be approximately sketched by using straight-line segments called asymptotes. Such plots are called Bode plots.
26. The frequency range from lower cutoff frequency $f_L$ to upper cutoff frequency $f_H$ is called the bandwidth of the amplifier stage.

## KEY FORMULAE

1. Overall voltage gain, $A_v = A_{v_1} \times A_{v_2} \times A_{v_3} \times .............. \times A_{v_n}$

2. Percentage harmonic distortion, $D_n = \dfrac{A_n\left(\text{Amplitude of the } n^{\text{th}} \text{ harmonic}\right) \times 100\%}{A_1\left(\text{Amplitude of the fundamental}\right)}$.

3. Magnitude $|A_L|$ of voltage gain in low-frequency region is $|A_L(jf)| = \dfrac{1}{\sqrt{1+\left(f_L/f\right)^2}}$

4. In low-frequency region, the phase lead of the gain is $\theta_L = \text{arc tan}\left(f_L/f\right)$

5. In high-frequency region, the magnitude of voltage gain is $|A_H(jf)| = \dfrac{1}{\sqrt{1+\left(f/f_H\right)^2}}$

6. In high-frequency region, the phase of the gain is $\theta_H = -\text{arc tan}\left(f/f_H\right)$

7. Bandwidth of an amplifier, $\text{BW} = f_H - f_L$

8. Gain-bandwidth product = $A \times 3\,dB$ bandwidth

## SOLVED PROBLEMS

**1. A given amplifier arrangement has the following voltage gains: $A_{v_1} = 10$, $A_{v_2} = 20$ and $A_{v_3} = 40$. What is the overall voltage gain?. Also express each gain in *dB* and determine the total *dB* voltage gain.**

**Data:** $A_{v_1} = 10$, $A_{v_2} = 20$ and $A_{v_3} = 40$

**To find:** Overall voltage gain $A_v$ and each gain and total gain in *dB*.

**Solution:** Overall voltage gain $A_v = A_{v_1} \times A_{v_2} \times A_{v_3}$

$= 10 \times 20 \times 40 = \mathbf{8000}$

Total *dB* voltage gain, $G_v = G_{v_1} + G_{v_2} + G_{v_3}$

$G_{v_1} = 20 \ln A_{v_1} = 20 \ln 10 = \mathbf{20\,dB}$

$G_{v_2} = 20 \ln A_{v_2} = 20 \ln 20 = \mathbf{26\,dB}$

$G_{v_3} = 20 \ln A_{v_3} = 20 \ln 40 = \mathbf{32\,dB}$

$G_v = 20 + 26 + 32 = \mathbf{78\,dB}$

**2. An amplifier has an input power of $5\,\mu W$. The power gain of the amplifier is $40\,dB$. Find the output power of the amplifier.**

**Data:** $P_i = 5\,\mu W$, $A_p = 40\,dB$

**To find:** Output power, $P_o$

**Solution:** Power gain in $dB$ $= 10 \log_{10}\left(P_o / P_i\right) = 40$

$$\frac{P_o}{P_i} = \text{antilog}_{10} 4 = 10^4$$

$\therefore$ Output power, $P_o = P_i \times 10^4 = 5 \times 10^{-6} \times 10^4$

$$= \mathbf{50\ mW}$$

**3. The bandwidth of an amplifier extends from 20 *Hz* to 20 *kHz*. Find the frequency range over which the gain is down less than 1 *dB* from its midband value. Assume that the low- and high-frequency response is given by Eqs. $V_o(s) = V_i(s)\dfrac{s}{s + 1/R_1C_1}$ and $V_o(s) = \dfrac{1}{1 + sR_2C_2}.V_i(s)$, multiplied by a constant $A_{v_o}$.**

**Data:** $f_L = 20\ Hz$, $f_H = 20\ kHz$

**To find:** Frequency range over which $A$ is down less than $1dB$ from $A_m$.

**Solution:** At low frequencies,

$$A_V = \frac{V_o(s)}{V_i(s)} = \frac{A_{V_o} s}{s + 1/R_1C_1}$$

or

$$\frac{A_V}{A_{V_o}} = \frac{s}{s + 1/R_1C_1}$$

We know that

$$\frac{A_V}{A_{V_o}} = \frac{1}{1 - jf_L/f} = \frac{1}{\sqrt{1 + \left(f_L/f\right)^2}}$$

Since $\dfrac{A_V}{A_{V_o}}$ is $-1dB$,

$$20 \log \left(1 + \left(f_L/f_1\right)^2\right)^{1/2} = 1$$

$$\log \left(1 + \left(f_L/f_1\right)^2\right) = 0.1$$

or

$$1 + \left(f_L/f_1\right)^2 = 1.26$$

Solving,

$$1 + \left(20/f_1\right)^2 = 1.26$$

$$\mathbf{f_1 = 39.3\ Hz}$$

Similarly, at high frequencies, $20 \log \left(1 + \left(f_2/f_H\right)^2\right)^{1/2} = 1$ or $1 + \left(f_2/f_H\right)^2 = 1.26$

or

$$f_2 = \mathbf{10.2\ kHz}$$

**4. Prove that over the range of frequencies from $10 f_L$ to $0.1 f_H$ the voltage amplification is constant to within 0.5 percent and the phase shift to within ±0.1 rad.**

**Solution:** At low frequencies, we have

$$\frac{A_L}{A_o} = \frac{1}{\sqrt{1+\left(f_L/f\right)^2}} = \frac{1}{\sqrt{1+0.1^2}} = \frac{1}{\sqrt{1.01}}$$

$$= \frac{1}{1.005} = 0.995$$

At high frequencies,
$$\frac{A_H}{A_o} = \frac{1}{\sqrt{1+\left(f/f_H\right)^2}} = \frac{1}{\sqrt{1+0.1^2}} = 0.995$$

Hence, the gain is constant within 0.5% in the frequency range $10\,f_L \le f \le 0.1\,f_H$.

We know
$$\theta_1 = -\tan^{-1}\left(\frac{f_L}{f}\right) = -\tan^{-1}(0) \approx -0.1 \text{ rad} \qquad (\because \tan\theta \approx \theta \text{ as } \theta \to 0)$$

Similarly,
$$\theta_2 = \tan^{-1}\left(\frac{f}{f_H}\right) = \tan^{-1}(0.1) \approx -0.1 \text{ rad}$$

Hence, the phase shift is constant to within 0.1 rad = 5.73° for $10 f_L \le f \le 0.1 f_H$.

## QUESTIONS

1. What is meant by a multistage amplifier? Explain it briefly.
2. Explain multistage amplifiers and analyse their operation.
3. Explain the need of cascading amplifiers.
4. What is a cascade amplifier? State the various methods of cascading transistor amplifiers.
5. Define the following modes of operation of an amplifier: (a) class A; (b) class B; (c) class AB and (d) class C.
6. What are the types of distortion in an amplifier?
6. Explain briefly the following types of distortion: (a) nonlinear; (b) frequency; (c) phase distortion and (d) inter-modulation distortion.
7. What are the causes for the various types of distortion in an amplifier?
8. Under what conditions does an amplifier preserve the form of the input signal?
9. Explain the frequency-response magnitude characteristic of an amplifier.
10. Draw a typical frequency response curve. Indicate the low and high 3-*dB* frequencies.
11. Why does the gain of an amplifier fall off at low and high frequencies?
12. Explain capacitive coupling in multistage amplifiers.
13. What is the effect of cascading on frequency response and bandwidth?
14. Why the intermediate transistor should be connected in C-E intermediate stages to meet the following requirements:

    (i) To match a low impedance source (ii) No loading to source (iii) To match an impedance load.

15. Write a short note on typical frequency response of an amplifier.
16. Explain the effect of coupling, bypass and spurious shunting capacitors on frequency response of small signal and cascaded amplifiers.
17. What is meant by lower 3 *dB* and upper 3 *dB* frequency?
18. Define bandwidth and gain bandwidth product of an amplifier?
19. Explain briefly and derive expressions for voltage gain and phase in low-frequency and high frequency regions of a multistage amplifier.
20. What are Bode plots?

## EXERCISES

1. A certain cascaded amplifier arrangement has the following voltage gains: $A_{v_1} = 10$, $A_{v_2} = 15$ and $A_{v_3} = 20$. What is the overall voltage gain? Also express each gain in *dB* and determine the total voltage gain in *dB*.

   **[Ans: 3000; 20 *dB*, 23.5 *dB*, 26 *dB* and 69.5 *dB*)**

2. In a certain multistage amplifier, the individual stages have the following voltage gains: $A_{v_1} = 25$, $A_{v_2} = 5$ and $A_{v_3} = 12$. What is the overall gain? Express each gain in *dB* and determine the total voltage gain in *dB*.

   **[Ans: 1500; 27.96 *dB*; 13.98 *dB*; 21.58 *dB* and 63.52 *dB*]**

# 6 FEEDBACK AMPLIFIERS

## Chapter Outline

- Introduction
- Classification of Amplifiers
- The Feedback Concept
- Transfer gain with Feedback
- Characteristics of negative Feedback Amplifiers

## INTRODUCTION

Feedback is one of the fundamental aspects inherent in nature. Also it plays an important role in almost all electronic circuits. **Feedback is the process whereby a portion of the output signal of the amplifier is fed back to the input of the amplifier.** The feedback signal can be either a voltage or a current, being applied in series or shunt respectively with the input signal. The path over which the feedback is applied is the **feedback loop**. There are two types of feedback used in electronic circuits. (i) If the feedback voltage or current is in phase with the input signal and adds to its magnitude, the feedback is called **positive** or **regenerative feedback**. (ii) If the feedback voltage or current is opposite in phase to the input signal and opposes it, the feedback is called **negative** or **degenerative feedback.**

In this chapter we introduce the concept of feedback and show how to modify the characteristics of an amplifier by combining a portion of the output signal with the input signal.

## CLASSIFICATION OF AMPLIFIERS

Before proceeding with the concept of feedback, it is useful to classify amplifiers based on the magnitudes of the input and output impedances of an amplifier relative to the source and load impedances respectively as (i) **voltage** (ii) **current** (iii) **transconductance** and (iv) **transresistance amplifiers.**

### 1. Voltage Amplifier

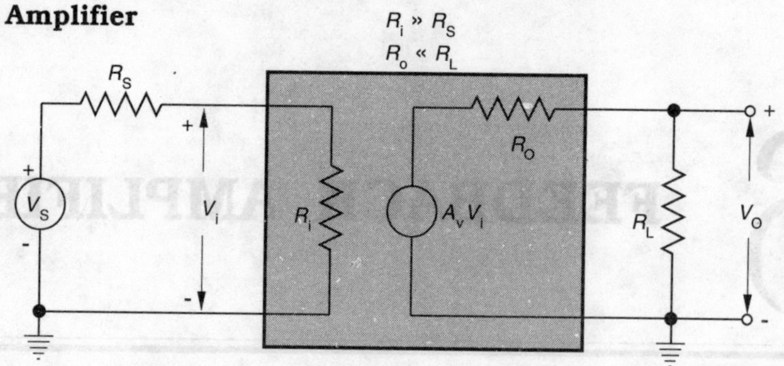


**Fig. 6.1 Thevenin's equivalent circuit of a voltage amplifier**

Figure 6.1 shows a Thevenin's equivalent circuit of an amplifier. If the amplifier input resistance $R_i$ is large compared with the source resistance $R_s$, then $V_i = V_s$. If the external load $R_L$ is large compared with the output resistance $R_o$ of the amplifier, then $V_o = A_vV_s$. This type of amplifier provides a voltage output proportional to the voltage input and **the proportionality factor does not depend on the magnitudes of the source and load resistances**. Hence, this amplifier is known as voltage amplifier.

An ideal voltage amplifier must have infinite input resistance $R_i$ and zero output resistance. In Fig. 6.1, $A_v$ represents $V_o/V_i$ with $R_L = \infty$ and hence called the open-circuit voltage gain. A practical voltage amplifier must have $R_i >> R_s$ and $R_L >> R_o$.

### 2. Current Amplifier

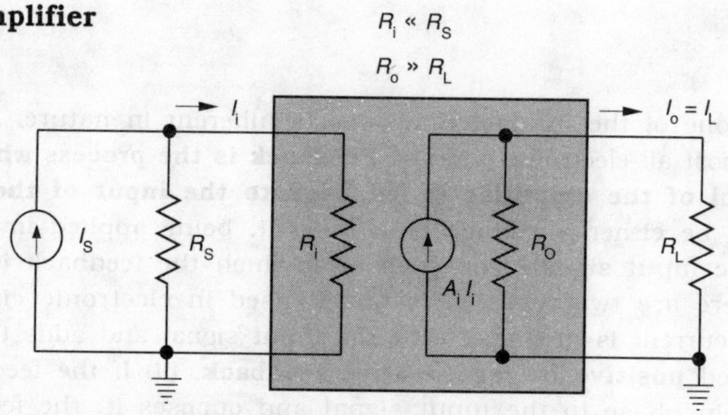


**Fig. 6.2 Norton's equivalent circuit of a current amplifier**

Figure 6.2 shows a Norton's equivalent circuit of a current amplifier. If the amplifier input resistance $R_i$ is very low compared with source resistance $R_s$, then $I_i = I_s$. If the amplifier output resistance $R_o$ is very large compared to external load $R_L$, then $I_L = A_iI_i = A_iI_s$. This type of amplifier provides an output current proportional to the signal current and **the proportionality factor is independent of source and load resistances.** Hence, this amplifier is called a **current amplifier**.

An ideal current amplifier must have zero input resistance $R_i$ and infinite output resistance $R_o$. In Fig. 6.2, $A_i$ represents $I_L/I_i$ with $R_L = 0$ and hence called **short-circuit current gain**. A practical current amplifier must have $R_i << R_s$ and $R_o >> R_L$.

**3. Transconductance Amplifier**

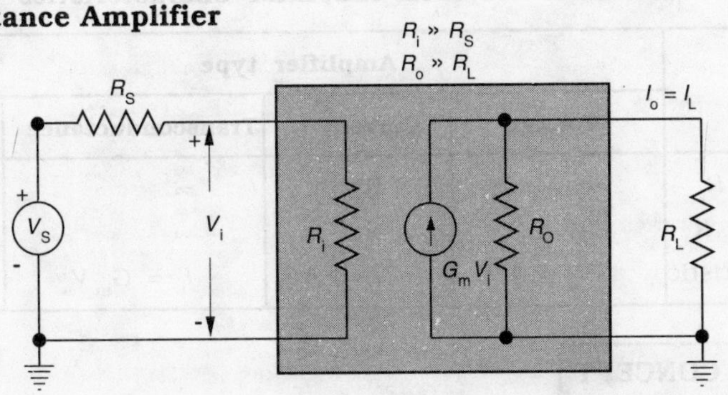


**Fig. 6.3 Equivalent circuit of a transconductance amplifier**

Figure 6.3 shows the equivalent circuit of transconductance amplifier. In this amplifier, the output current $I_o$ is proportional to the signal voltage $V_s$ and the proportionality factor is independent of the magnitudes of source and load resistances.

An ideal transconductance amplifier must have an infinite input resistance $R_i$ and infinite output resistance $R_o$. A practical transconductance amplifier must have $R_i >> R_s$ and $R_o >> R_L$.

**4. Transresistance Amplifier**

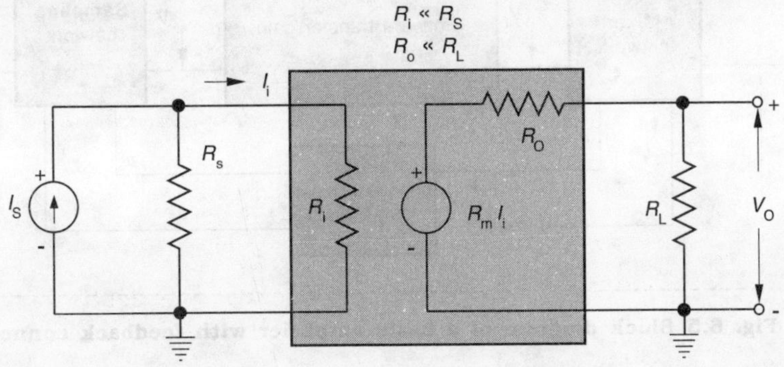


**Fig. 6.4 Equivalent circuit of a transresistance amplifier**

Figure 6.4 shows the equivalent circuit of transresistance amplifier. In this amplifier the output voltage $V_o$ is proportional to the signal current $I_s$ and the proportionality factor is independent of magnitudes of source and load resistances. If $R_s >> R_i$, then $I_i = I_s$ and if $R_o << R_L$, then $V_o = R_m I_i = R_m I_s$. Note $R_m = V_o / I_i$, with $R_L = \infty$, called the open-circuit transresistance.

An ideal transresistance amplifier must have zero input resistance and zero output resistance. A practical transresistance amplifier must have $R_i << R_s$ and $R_o << R_L$.

The characteristics of the four ideal amplifier types are summarised in Table 6.1.

Table 6.1 Ideal Amplifier Characteristics

| Parameter | Amplifier type | | | |
|---|---|---|---|---|
| | Voltage | Current | Transconductance | Transresistance |
| Input resistance, $R_i$ | $\infty$ | 0 | $\infty$ | 0 |
| Output resistance $R_o$ | 0 | $\infty$ | $\infty$ | 0 |
| Transfer characteristic | $V_o = A_v V_s$ | $I_L = A_i I_s$ | $I_L = G_m V_s$ | $V_o = R_m I_s$ |

## THE FEEDBACK CONCEPT

In each of the amplifier circuit we can sample the output voltage or current by means of a suitable sampling network and apply this signal to the input through a feedback two-port network, as shown in Fig. 6.5. At the input the feedback signal is combined with the source signal through a network called **mixer** and is finally fed into the amplifier.

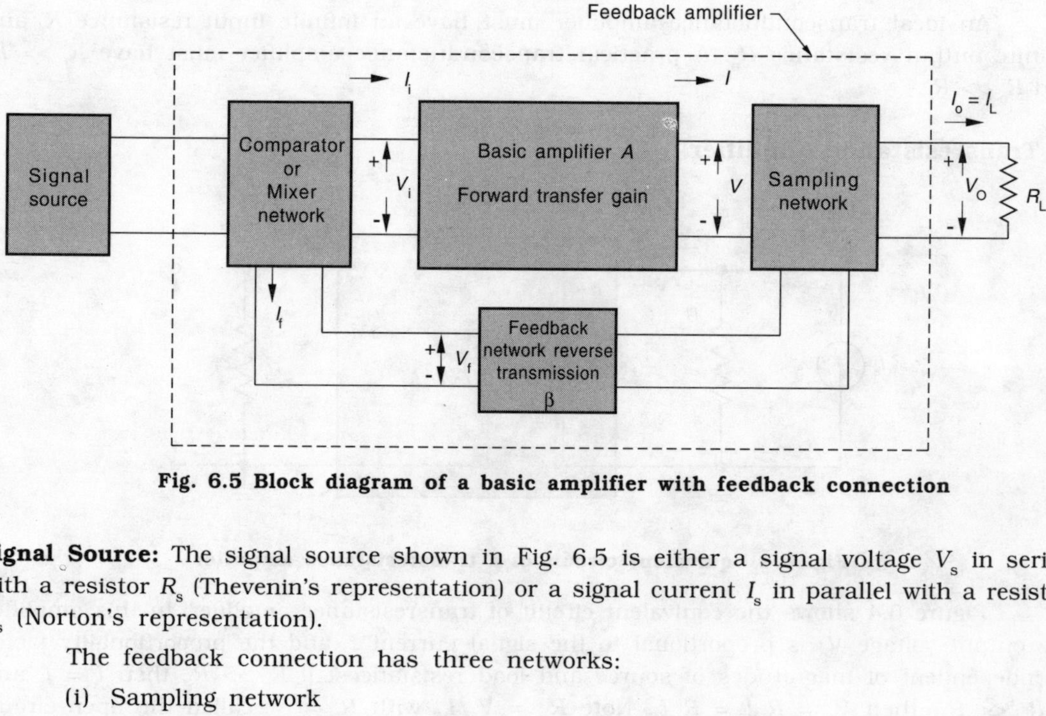


**Fig. 6.5 Block diagram of a basic amplifier with feedback connection**

**Signal Source:** The signal source shown in Fig. 6.5 is either a signal voltage $V_s$ in series with a resistor $R_s$ (Thevenin's representation) or a signal current $I_s$ in parallel with a resistor $R_s$ (Norton's representation).

The feedback connection has three networks:

(i) Sampling network

(ii) Feedback network

(iii) Mixer network

### (i) Sampling Network

There are two ways to sample the output, depending on the sampling parameter either voltage or current. The output voltage is sampled by connecting the feedback network

**in shunt** across the output, as shown in Fig. 6.6(a). This type of connection is known as **voltage** or **node sampling**. If the output current is sampled by connecting the feedback network **in series** with the output [Fig. 6.6(b)], the type of connection is called **current** or **loop sampling**.

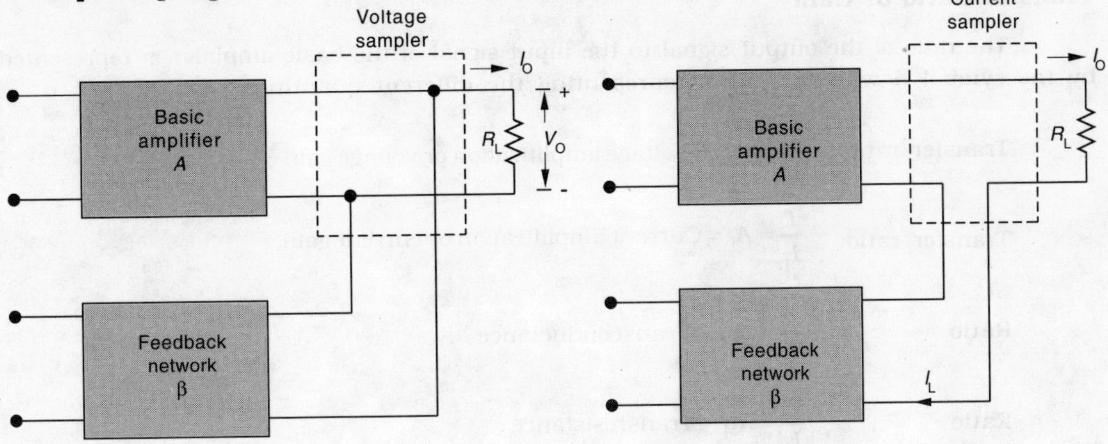


**(a) Voltage or node sampling** **Fig. 6.6** **(b) Current or loop sampling**

## (ii) Feedback Network

The feedback network is usually a passive two-port network consisting of resistors, capacitors and inductors. It provides a fraction of the output voltage as feedback signal $V_f$ to the input mixer network. The feedback voltage is given by

$$V_f = \beta V_o$$

where β is a **feedback ratio** or **feedback factor** or **reverse transmission factor**. It lies between 0 and 1.

## (iii) Mixer Network

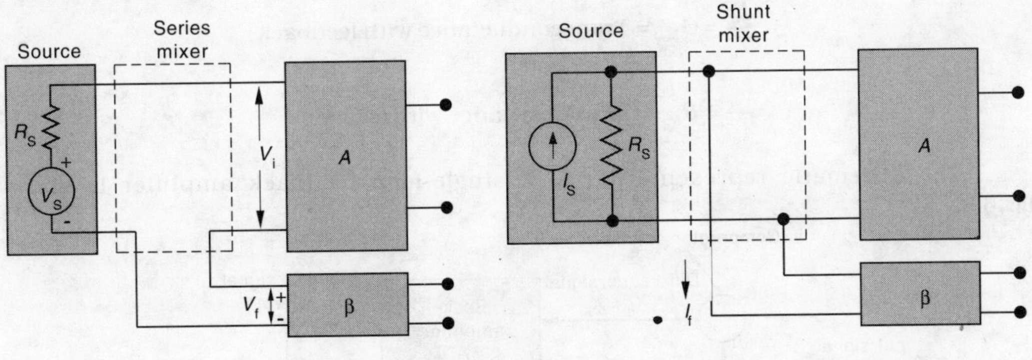


**(a) Series mixing** **(b) Shunt mixing**

**Fig. 6.7 Feedback connections of the input of a basic amplifier**

There are two ways of mixing feedback signal with the input signal as shown in Fig. 6.7. When a fraction of the output voltage is applied in **series (loop)** with the input voltage through the feedback network as shown in Fig. 6.7(a) it is called **series mixing**. When a fraction of the output voltage is applied in parallel to the input of the amplifier as shown in Fig. 6.7(b), it is called **shunt (node)** feedback.

## Transfer Ratio or Gain

The ratio of the output signal to the input signal of the basic amplifier is represented by the symbol $A$ with the suffix representing the different quantities.

Transfer ratio $\dfrac{V}{V_i} = A_V$ = Voltage amplification or voltage gain

Transfer ratio $\dfrac{I}{I_i} = A_I$ = Current amplification or current gain

Ratio $\dfrac{I}{V_i} = G_M$ = Transconductance

Ratio $\dfrac{V}{I_i} = R_M$ = Transresistance

Actually $G_M$ and $R_M$ do not represent amplification. Nevertheless, it is convenient to refer to each of the four quantities $A_V$, $A_I$, $G_M$ and $R_M$ as **transfer gain of the basic amplifier without feedback**.

The symbol $A_f$ is used to define the ratio of the output signal to the input signal of the amplifier with feedback as shown in Fig. 6.5, and is called the **transfer gain of the amplifier with feedback**. Hence $A_f$ is used to represent any one of the following ratios:

$\dfrac{V_o}{V_s} = A_{V_f}$ = Voltage gain with feedback

$\dfrac{I_o}{I_s} = A_{I_f}$ = Current gain with feedback

$\dfrac{I_o}{V_s} = G_{M_f}$ = Transconductance with feedback

$\dfrac{V_o}{I_s} = R_{M_f}$ = Transresistance with feedback

The schematic representation of a single-loop feedback amplifier is shown in Fig. 6.8.

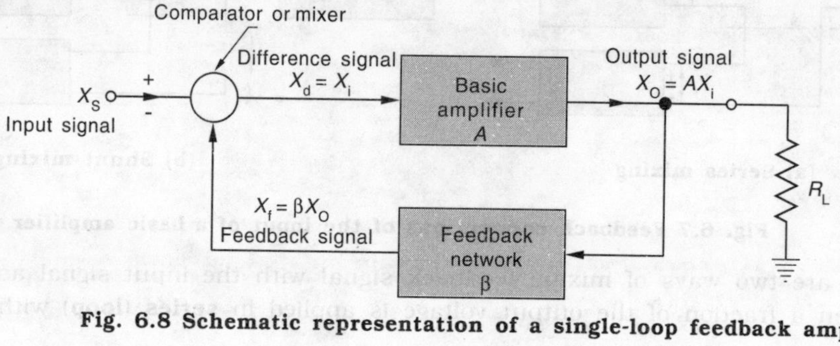


**Fig. 6.8 Schematic representation of a single-loop feedback amplifier**

## Advantages of Negative Feedback

When a fraction of the output signal is subtracted from the input signal the feedback is negative. The negative feedback reduces the gain of the amplifier. It also provides other advantages.

The salient features of negative feedback in amplifiers are:

1. High input resistance of a voltage amplifier can be made higher.
2. Low output resistance of a voltage amplifier can be lowered.
3. The transfer gain $A_f$ of the amplifier can be stabilised against the variations of parameters of transistors.
4. The frequency response of the amplifier can be improved.

Note that all the advantages mentioned above are obtained at the expense of the gain $A_f$ with feedback, which is lowered in comparison with the transfer gain $A$ of an amplifier without feedback.

The analysis of the feedback amplifier can be carried out by replacing each active element (BJT, FET, or VT) by its small-signal model and by writing Kirchhoff's loop, or nodal equations. Consider the schematic representation of a single-loop feedback amplifier, shown in Fig. 6.8. The basic amplifier of Fig. 6.8 may be a voltage, transconductance, current or transresistance amplifier connected in a feedback configuration, as shown in Fig. 6.9. The four basic types of feedback are:

(i) Voltage-series feedback

(ii) Current-series feedback

(iii) Current-shunt feedback

(iv) Voltage-shunt feedback

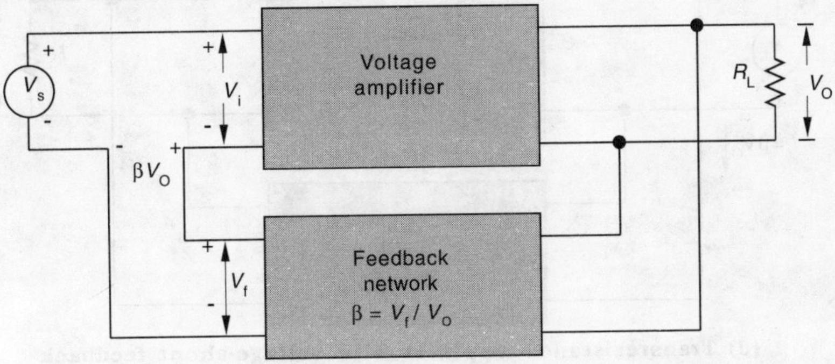


**(a) Voltage amplifier with voltage-series feedback**
**Fig. 6.9 Types of amplifiers with negative feedback**

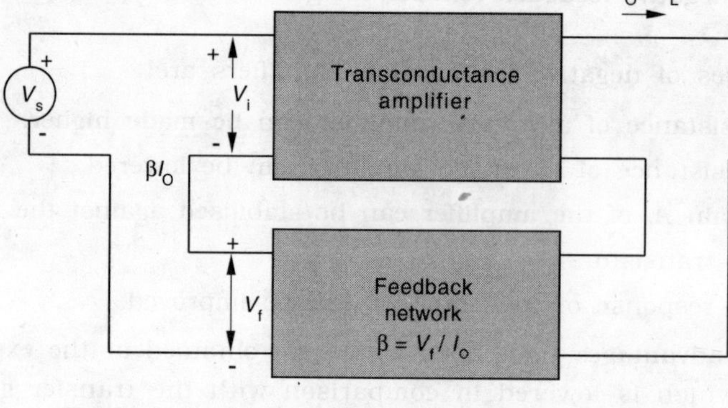


**(b) Transconductance amplifier with current-series feedback**

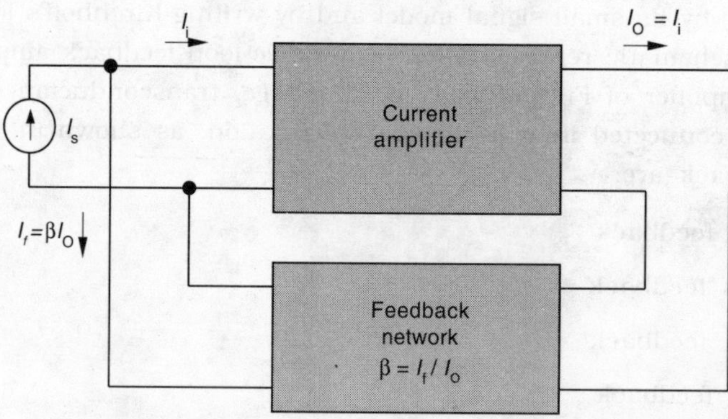


**(c) Current amplifier with current-shunt feedback**

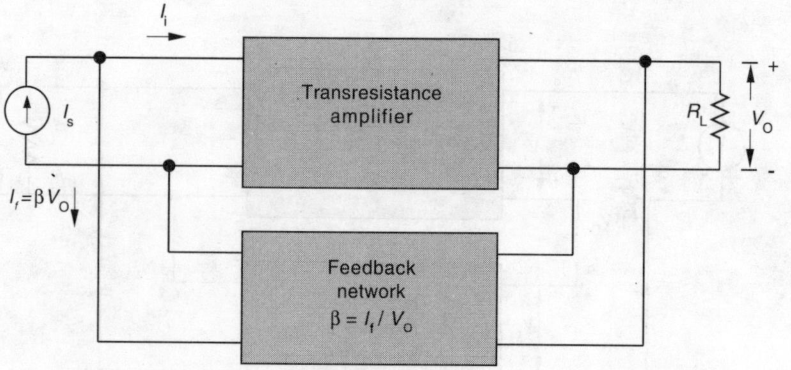


**(d) Transresistance amplifier with voltage-shunt feedback**

**Fig. 6.9 Types of amplifiers with negative feedback**

## TRANSFER GAIN WITH FEEDBACK

Consider the schematic representation of negative feedback amplifier as shown in Fig. 6.8. The source resistance $R_s$ is considered to be part of the amplifier and the transfer gain $A$ ($A_v$, $A_i$, $G_m$, $R_m$) includes the effect of the loading of the β network upon the amplifier. The input signal $X_s$, the output signal $X_o$, the feedback signal $X_f$, and the difference signal $X_d$, each represents either a voltage or a current. These signals and also the ratios $A$ and β are summarised in Table 6.2.

**Table 6.2 Voltage and Current signals in feedback amplifiers**

| Signal or ratio | Type of feedback | | | |
|---|---|---|---|---|
| | **Voltage-series** | **Current-series** | **Current-shunt** | **Voltage-shunt** |
| $X_o$ | Voltage | Current | Current | Voltage |
| $X_s$, $X_f$, $X_d$ | Voltage | Voltage | Current | Current |
| $A$ | $A_v$ | $G_m$ | $A_i$ | $R_m$ |
| β | $V_f/V_o$ | $V_f/I_o$ | $I_f/I_o$ | $I_f/V_o$ |

The transfer gain A is defined by

$$A = \frac{X_o}{X_s} \longrightarrow (1)$$

The output of the mixer is the sum of the inputs $X_s$ and $X_f$ and is given by

$$X_d = X_s + (-X_f) = X_i \longrightarrow (2)$$

Since $X_d$ represents the difference between the applied signal $X_s$ and that fed back ($X_f$) to the input, it is called **difference**, **error** or **comparison signal**.

If β be the feedback ratio and is defined by

$$\beta = \frac{X_f}{X_o} \longrightarrow (3)$$

The factor β is often a positive or a negative real number, but in general, β is a complex function of the signal frequency.

The transfer gain of the amplifier with feedback is given by

$$A_f = \frac{X_o}{X_s} \longrightarrow (4)$$

From Eq. (2), $X_s = X_i + X_f$, $A_f = \dfrac{X_o}{X_i + X_f}$

Dividing the numerator and denominator by $X_i$, we get

$$A_f = \frac{X_o/X_i}{(X_i + X_f)/X_i}$$

Since $A = \frac{X_o}{X_i}$, we have $$A_f = \frac{A}{1 + X_f / X_i}$$

or $$A_f = \frac{A}{1 + \left(\frac{X_f}{X_o}\right)\left(\frac{X_o}{X_i}\right)}$$

Since $\beta = \frac{X_f}{X_o}$, $$A_f = \frac{A}{1 + \beta A} \quad \text{(5)}$$

If $|A_f| < |A|$, the feedback is called **negative** or **degenerative**. If $|A_f| > |A|$, the feedback is termed **positive** or **regenerative**. From Eq. (5), we see that, in the case of negative feedback, the gain $A$ of the basic amplifier without feedback is always greater than gain with feedback $A_f$.

For voltage amplifier, gain with negative feedback is given by

$$A_{Vf} = \frac{A_V}{1 + A_V \beta} \quad \text{(6)}$$

### Loop Gain

In Fig. 6.8, the signal $X_d$ is multiplied by $A$ in passing through the amplifier, is multiplied by $\beta$ in transmission through the feedback network and is multiplied by -1 in the mixer network. The path taken from the input terminals around the loop consisting of the amplifier and feedback network back to the input forms a single loop and the gain around that loop is loop gain. The product $-A\beta$ is called the **loop gain** or return ratio. The difference between unity and the loop gain is called the **return difference**, $D = 1 + A\beta$.

The amount feedback introduced into an amplifier is also expressed in decibels. By definition,

$$N = dB \text{ of feedback}$$

$$N = 20 \log \left|\frac{A_f}{A}\right| = 20 \log \left|\frac{1}{1 + A\beta}\right| \quad \text{(7)}$$

If the feedback is negative, then $N$ is a negative number.

### Fundamental Assumptions

1. The input signal is transmitted to the output through the amplifier and not through the feedback network.
2. The feedback signal is transmitted from the output to the input through the feedback network and not through the amplifier.
3. The feedback ratio $\beta$ is independent of the load and the source resistances $R_L$ and $R_S$.

## CHARACTERISTICS OF NEGATIVE FEEDBACK AMPLIFIERS

### 1. Stability of Gain

Since the transfer gain of the amplifier depends on the factors such as temperature, operating point, aging, etc., it is not stable. The stability can be improved by introducing negative feedback.

The ratio of fractional change in amplification with feedback to the fractional change of gain without feedback is called the **sensitivity of the transfer gain**.

$\therefore$ Sensitivity of the transfer gain = $\left|\dfrac{dA_f}{dA}\right|$ ⟶(1)

We know that $$A_f = \frac{A}{1+A\beta}$$ ⟶(2)

Differentiating Eq. (2) with respect to $A$, we get

$$dA_f = \frac{(1+A\beta)1 - A\beta}{(1+A\beta)^2}.dA = \frac{1}{(1+A\beta)^2}.dA$$

Dividing both sides by $A_f$, we obtain

$$\frac{dA_f}{A_f} = \frac{1}{(1+A\beta)^2}.\frac{dA}{A_f}$$

$$= \frac{1}{(1+A\beta)^2}.\frac{dA}{\left(A/_{1+A\beta}\right)}$$

$$\frac{dA_f}{A_f} = \frac{dA}{A}\frac{1}{(1+A\beta)}$$ ⟶(3)

i.e. $$\left|\frac{dA_f}{A_f}\right| = \left|\frac{dA}{A}\right|.\frac{1}{|1+A\beta|}$$ ⟶(4)

where $\dfrac{dA_f}{A_f}$ = Fractional change in gain with feedback

$\dfrac{dA}{A}$ = Fractional change in gain without feedback

Here $1/_{1+A\beta}$ is **sensitivity**. The reciprocal of the sensitivity is called the **desensitivity *D***. The term desensitivity indicates the factor by which the voltage gain has been reduced due to feedback.

$\therefore$ Desensitivity, D = $1 + A\beta$ ⟶(5)

It may be proved that Desensitivity, $D = \dfrac{A}{A_f}$ ⟶(6)

For an amplifier with 20 $dB$ of negative feedback, $D$ = 10 and hence, for example, a 5 percent change in gain without feedback is reduced to a 0.5 percent variation after feedback is introduced.

$$A_f = \frac{A}{1+A\beta} = \frac{A}{D}$$ ⟶(7)

If $|A\beta| >> 1$, then $$A_f = \frac{A}{1+A\beta} = \frac{A}{A\beta}$$

$$\mathbf{A_f = \frac{1}{\beta}}$$ ⟶(8)

Hence the gain may be made to depend entirely on the feedback network. If the feedback network contains only stable passive elements, the improvement in stability may indeed be pronounced.

Since $A$ represents either $A_V$, $A_I$, $G_M$ or $R_M$, then $A_f$ represents the corresponding transfer gains with feedback: either $A_{Vf}$, $A_{If}$, $G_{Mf}$ or $R_{Mf}$. Equation (8) signifies that

For voltage-series feedback $A_{Vf} = \frac{1}{\beta}$ voltage gain is stabilised

For current-series feedback $G_{Mf} = \frac{1}{\beta}$ transconductance gain is stabilised

For voltage-shunt feedback $R_{Mf} = \frac{1}{\beta}$ transresistance gain is stabilised

For current-shunt feedback $A_{If} = \frac{1}{\beta}$ current gain is stabilised

## 2. Reduction in Frequency Distortion

If the feedback network is purely resistive, the overall gain is then not a function of frequency even though the basic amplifier gain is frequency dependent. Under these conditions a substantial reduction in frequency and phase distortion is obtained.

If a frequency-selective feedback network is used, the feedback factor $\beta$ then depends upon frequency. As a result, the amplification may also change with frequency. This fact is used in tuned amplifiers. In tuned amplifiers, the feedback network is so designed that at tuned frequency $\beta \to 0$ and at other frequencies $\beta \to \infty$. As a result, amplifier provides high gain for signal at tuned frequency and rejects all other frequencies.

## 3. Nonlinear Distortion

Negative feedback reduces the amount of noise signal (such as power supply hum) and nonlinear distortion. Nonlinear distortion results in the production of terms such as harmonics, which are present in the input.

Consider a non-feedback amplifier as shown in Fig. 6.10. Suppose that a large amplitude signal is applied to an amplifier so that the operation of the device extends slightly beyond its range of linear operation. As a result the output signal is slightly distorted.

The output of the amplifier is

$$V_O = A_V V_S + V_d \quad \text{(1)}$$

where $V_d$ represents the distortion signal such as harmonics.

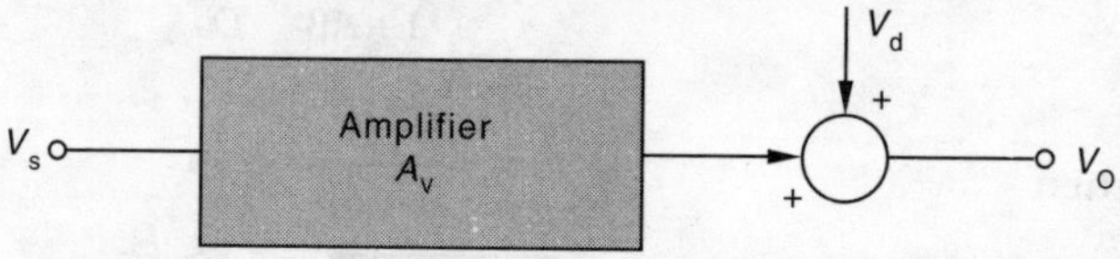


**Fig. 6.10 A model used to represent the effect of distortion in an amplifier**

The distortion is given by

$$D = \frac{V_d}{A_v V_s} \longrightarrow (2)$$

If a negative feedback is introduced, the input signal is increased by the same amount by which the gain is reduced, so that the amplitude of the output signal remains the same. Since the output signal operates at the same levels in both cases, the distortion signals are the same and $V_d' = V_d$. Analyzing this we obtain

$$V_o' = \frac{A_v}{1 + A_v\beta} V_s' + \frac{V_d'}{1 + A_v\beta} \longrightarrow (3)$$

Hence,

$$D' = \frac{V_d'}{A_v V_s'} \longrightarrow (4)$$

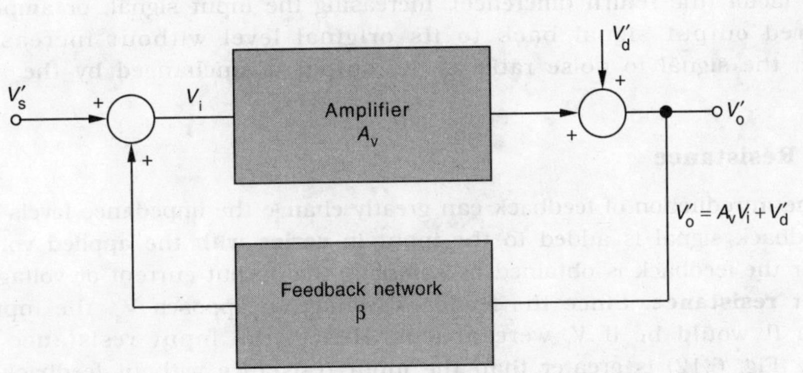


**Fig. 6.11 The model of Fig. 6.10 used in feedback configuration**

Taking the ratio of Eqs. (2) and (4), we get

$$\frac{D'}{D} = \frac{V_s}{V_s'} \quad \text{(because } V_d = V_d'\text{)} \longrightarrow (5)$$

The input signals $V_s$ and $V_s'$ (without and with feedback) must be adjusted so that the input signals are equal. The output signals are given in Eqs. (3) and (1) so that if the signal components are equal we must have

$$A_v V_s = \frac{A_v}{1 + A_v\beta} V_s' \longrightarrow (6)$$

Manipulating and substituting in Eq. (5), we get

$$\frac{D'}{D} = \frac{1}{1 + A_v\beta}$$

or

$$\boldsymbol{D' = \frac{D}{1 + A_v\beta}} \longrightarrow (7)$$

Thus, the distortion has been reduced by the amount $(1 + A_v\beta)$.

**4. Reduction of Noise**

There are other unwanted signals that occur in an amplifier's output which are not due to nonlinear distortion. These are called **noise**. Noise can manifest itself in many ways. Power supplies that operate from the 50 *Hz* power lines produce small sinusoidal voltages present at frequencies of 50 *Hz* and its harmonics. If these signals are significant, a noise called **hum** results.

The effect of feedback upon noise varies. Consider two cases here. Suppose that only significant amount of hum is produced by the power supply for the output stage. Let $V_d$ represent the hum voltage. Thus, provided that the input signal can be increased, the output signal to noise ratio will be reduced by the return difference and feedback can produce a significant advantage.

Note that the introduction of feedback reduces both the signal and the noise by the same factor (the return difference). Increasing the input signal, or amplifying it, brings the desired output signal back to its original level without increasing the noise. Therefore, the signal to noise ratio at the output is unchanged by the incorporation of feedback.

**5. Input Resistance**

The introduction of feedback can greatly change the impedance levels within a device. If the feedback signal is added to the input in **series** with the applied voltage (regardless of whether the feedback is obtained by sampling the output current or voltage), **it increases the input resistance.** Since the feedback voltage $V_f$ opposes $V_s$, the input current $I_i$ is less than it would be if $V_f$ were absent. Hence, the input resistance with feedback $R_{if} = V_s/I_i$ (Fig. 6.12) is greater than the input resistance without feedback $R_i$.

On the other hand, if the input signal is added to the input in **shunt** with the applied voltage (regardless of whether the feedback is obtained by sampling the output current or voltage), it **decreases the input resistance**. Since $I_s = I_i + I_f$, then the current $I_i$ is decreased from what it would be if there were no feedback current. Hence $R_{if} = V_i/I_s = I_i R_i/I_s$ (Fig. 6.13) is decreased because of feedback.

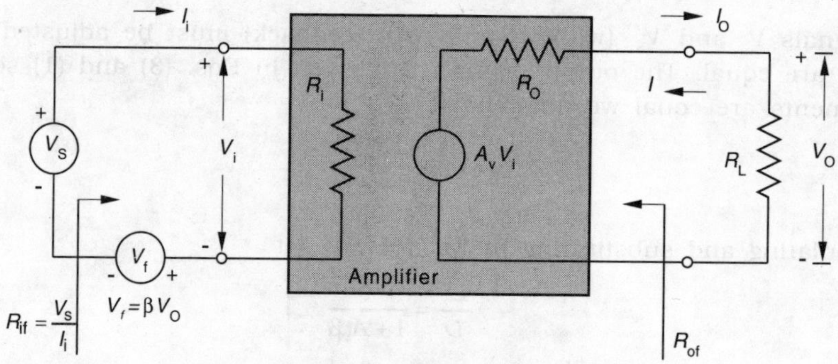


**Fig. 6.12 Voltage-series feedback circuit**

**(a) Voltage-series Feedback**

The topology of voltage-series feedback is shown in Fig. 6.12 with the amplifier replaced by Thevenin's model. Let $A_v$ represent the open-circuit voltage gain taking $R_s$ into account. Since throughout the discussion of feedback amplifiers we shall consider $R_s$ to be part of the amplifier, we shall drop the subscript $s$ on the transfer gain and input impedance ($A_v$ instead of $A_{vs}$, $R_i$ instead of $R_{is}$, $R_{if}$ instead of $R_{ifs}$, $G_m$ instead of $G_{ms}$, etc).

From Fig. 6.12 the input resistance with feedback is given as

$$R_{if} = \frac{V_s}{I_i} \longrightarrow (1)$$

Applying KVL to the input circuit, we get

$$V_s = I_i R_i + V_f$$

Since, $V_f = \beta V_o$,

$$V_s = I_i R_i + \beta V_o \longrightarrow (2)$$

By voltage divider rule, the output voltage $V_o$ is given by

$$V_o = \frac{A_V V_i R_L}{R_o + R_L} = A_V I_i R_i \longrightarrow (3)$$

$$= A_V V_i$$

where

$$A_V = \frac{V_o}{V_i} = \frac{A_V R_L}{R_o + R_L} \longrightarrow (4)$$

is the voltage gain without feedback taking the load $R_L$ into account.

Input resistance with feedback is given by

$$R_{if} = \frac{V_s}{I_i} \longrightarrow (5)$$

Substituting the value of $V_s$ from Eq. (2), we get

$$R_{if} = \frac{I_i R_i + \beta V_o}{I_i}$$

Since $V_o = A_V V_i$,

$$R_{if} = \frac{I_i R_i + \beta A_V V_i}{I_i} = R_i + \beta A_V R_i$$

$\therefore$

$$\mathbf{R_{if} = R_i\,(1 + \beta A_V)} \longrightarrow (6)$$

### (b) Current-series Feedback

The topology of current series feedback is shown in Fig. 6.13 with the amplifier input circuit replaced by Thevenin's model and output circuit by Norton's model.

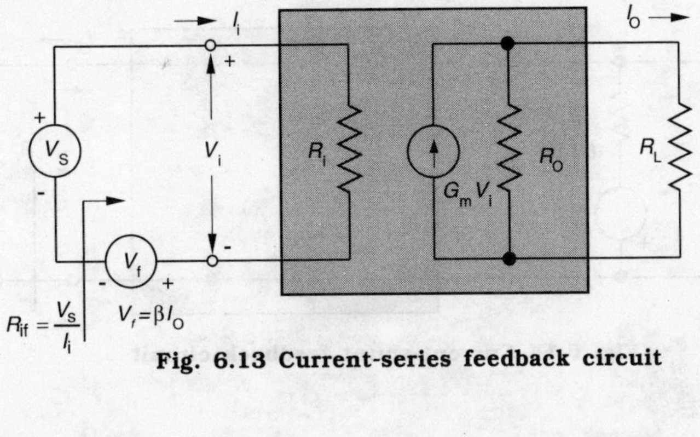


Fig. 6.13 Current-series feedback circuit

From Fig. 6.13, the input resistance with feedback is given by

$$R_{if} = \frac{V_s}{I_i} \longrightarrow (1)$$

Applying KVL to the input circuit, we get

$$V_s = I_i R_i + V_f$$

Since, $V_f = \beta I_o$,

$$V_s = I_i R_i + \beta I_o \longrightarrow (2)$$

By current divider rule, the output current $I_o$ is given by

$$I_o = \frac{G_m V_i R_o}{R_o + R_L} = G_M V_i \longrightarrow (3)$$

where

$$G_M = \frac{G_m R_o}{R_o + R_L} = \frac{I_o}{V_i} \longrightarrow (4)$$

Note that $G_m$ is the short-circuit transconductance, whereas $G_M$ is the transconductance without feedback taking the load $R_L$ into account.

Input resistance with feedback is given by

$$R_{if} = \frac{V_s}{I_i} \longrightarrow (5)$$

Substituting the value of $V_s$ from Eq. (2), we get

$$R_{if} = \frac{I_i R_i + \beta I_o}{I_i}$$

Since $I_o = G_M V_i$,

$$R_{if} = \frac{I_i R_i + \beta G_M V_i}{I_i} = R_i + \beta G_M R_i$$

$\therefore$

$$\boxed{R_{if} = R_i (1 + \beta G_M)} \longrightarrow (6)$$

Hence for series mixing $R_{if} > R_i$.

### (c) Current-shunt Feedback

The topology of current-shunt feedback is shown in Fig. 6.14, with the amplifier replaced by Norton's model. Let $A_i$ represent the short-circuit current gain taking $R_s$ into account.

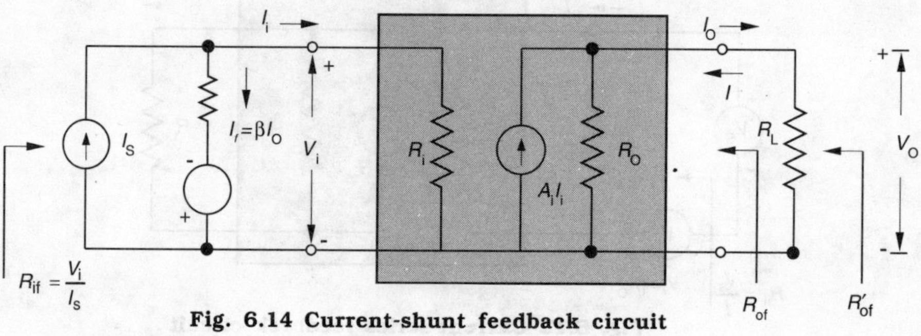


**Fig. 6.14 Current-shunt feedback circuit**

Applying KCL to the input node, we get

$$I_s = I_i + I_f$$

Since $I_f = \beta I_o$, $$I_s = I_i + \beta I_o \quad \longrightarrow (1)$$

The output voltage $V_o$ is given by

$$V_o = \frac{A_i I_i R_o}{R_o + R_L}$$

$$V_o = A_I I_i \quad \longrightarrow (2)$$

where $$A_I = \frac{A_i R_o}{R_o + R_L} = \frac{I_o}{I_i} \quad \longrightarrow (3)$$

Note that $A_I$ represents the current gain without feedback taking the load $R_L$ into account.

Input resistance with feedback is given by

$$R_{if} = \frac{V_i}{I_s} \quad \longrightarrow (4)$$

Substituting the value of $I_s$ from Eq. (1), we get

$$R_{if} = \frac{V_i}{I_i + \beta I_o}$$

Since $I_o = A_I I_i$, $$R_{if} = \frac{V_i}{I_i + \beta A_I I_i} = \frac{V_i}{I_i(1 + \beta A_I)}$$

$$\boldsymbol{R_{if} = \frac{R_i}{1 + \beta A_I}} \quad \longrightarrow (5)$$

**(d) Voltage-shunt Feedback**

The topology of voltage-shunt feedback is shown in Fig. 6.15 with amplifier input circuit replaced by Norton's model and output circuit replaced by Thevenin's model. Here $R_m$ is the open-circuit transresistance.

Applying KCL to the input node, we get

$$I_s = I_i + I_f$$

or $$I_s = I_i + \beta I_o \qquad \text{(because } I_f = \beta I_o\text{)} \quad \longrightarrow (1)$$

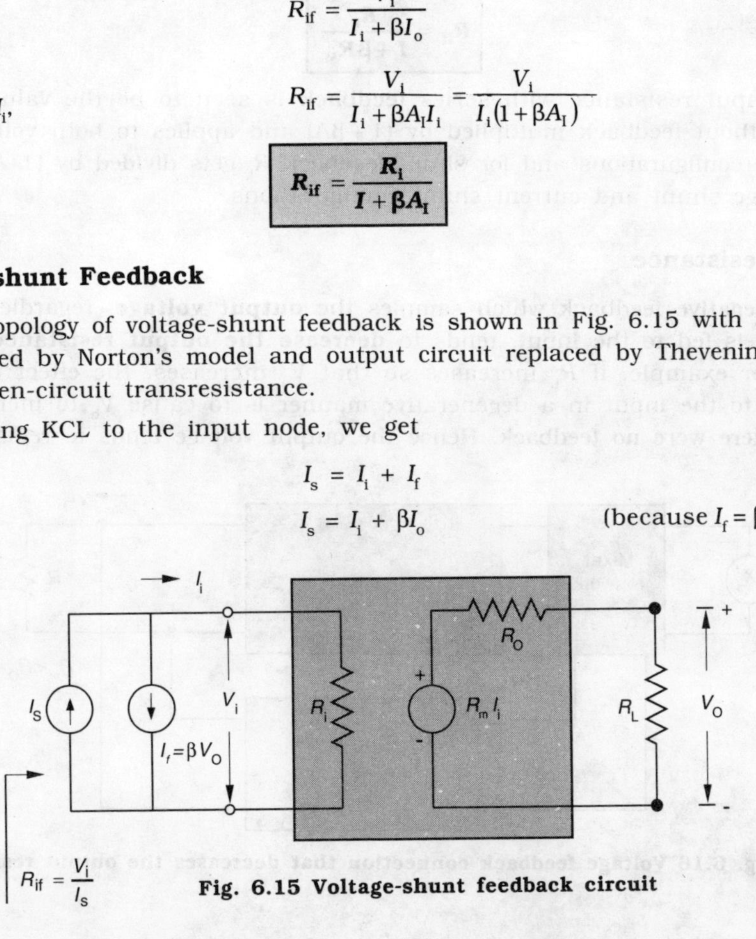


**Fig. 6.15 Voltage-shunt feedback circuit**

By voltage divider rule, the output voltage is given by

$$V_o = \frac{R_m I_i R_o}{R_o + R_L} \longrightarrow (2)$$

$$V_o = R_M I_i \longrightarrow (3)$$

where $$R_M = \frac{R_m R_o}{R_o + R_L} = \frac{I_o}{I_i} \longrightarrow (4)$$

is the transresistance without feedback taking the load $R_L$ into account.

Input resistance with feedback is given by

$$R_{if} = \frac{V_i}{I_s} \longrightarrow (5)$$

Substituting the value of $I_s$ from Eq. (1), we get

$$R_{if} = \frac{V_i}{I_i + \beta I_o} = \frac{V_i}{I_i + \beta R_M I_i} \qquad (\text{because } I_o = R_M I_i)$$

$$= \frac{V_i}{I_i(1 + \beta R_M)}$$

$$\boldsymbol{R_{if} = \frac{R_i}{I + \beta R_M}} \longrightarrow (6)$$

The input resistance with series feedback is seen to be the value of the input resistance without feedback multiplied by $(1 + \beta A)$ and applies to both voltage-series and current-series configurations and for shunt feedback $R_i$ gets divided by $(1+A\beta)$ and applies to both voltage shunt and current shunt configurations.

## 6. Output Resistance

The negative feedback which samples the **output voltage**, regardless of how this output signal is fed to the input, tends to **decrease the output resistance**, as shown in Fig. 6.16. For example, if $R_L$ increases so that $V_o$ increases, the effect of feeding this voltage back to the input in a degenerative manner is to cause $V_o$ to increase less than it would if there were no feedback. Hence the output voltage tends to remain constant as

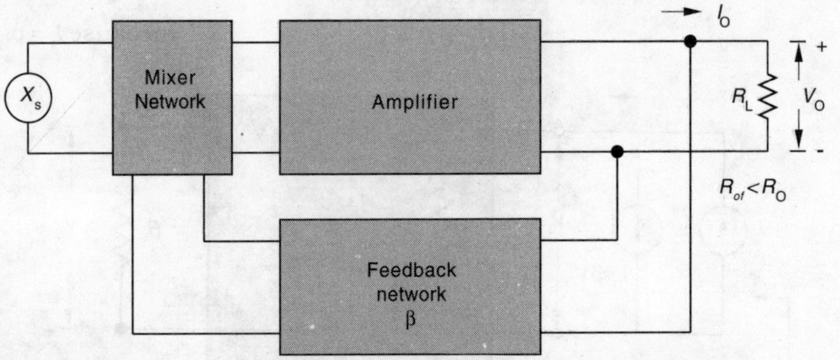


**Fig. 6.16 Voltage feedback connection that decreases the output resistance**

$R_L$ changes, which means that $R_{of} \ll R_L$.

On the other hand, the negative feedback which samples the **output current**, regardless of how this output signal is returned to the input, tends to **increase the output resistance**, as shown in Fig. 6.17.

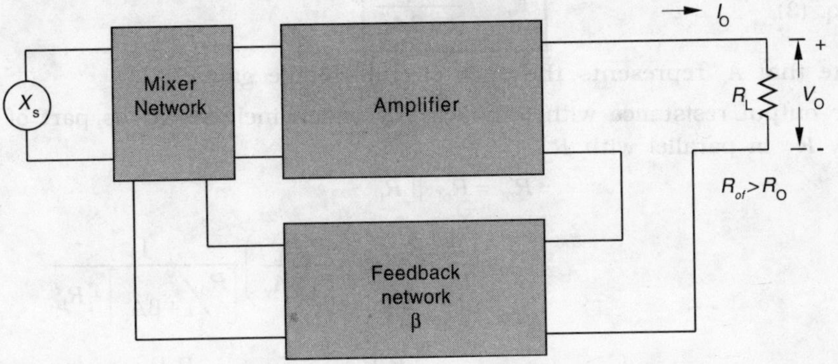


**Fig. 6.17 Current feedback connection that increases the output resistance**

## (a) Voltage-series Feedback

In this topology, the output resistance can be measured by shorting the input source (i.e., $V_s$ = 0 or $I_s$ = 0) and looking into the output terminals with $R_L$ disconnected, as shown in Fig. 6.18.

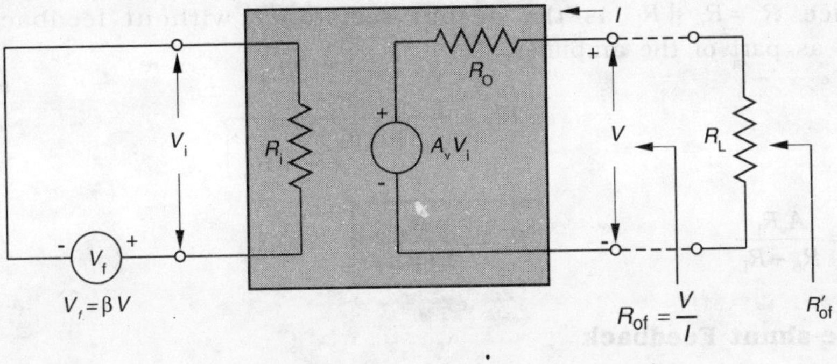


**Fig. 6.18 Voltage-series feedback connection**

Applying KVL to the output circuit, we have

$$A_v V_i + IR_o - V = 0$$

or
$$I = \frac{V - A_v V_i}{R_o} \qquad \text{(1)}$$

Since the input is shorted, $V_i = -V_f = -\beta V$ $\qquad$ (2)

Substituting the value of $V_i = \beta V$ in Eq. (1), we get

$$I = \frac{V - A_v(-\beta V)}{R_o} = \frac{V + \beta A_v V}{R_o}$$

or
$$I = \frac{V(1 + \beta A_v)}{R_o} \qquad \text{(3)}$$

The output resistance with feedback is given by

$$R_{of} = \frac{V}{I}$$

$\therefore$ From Eq. (3),

$$\boxed{R_{of} = \frac{R_o}{1+\beta A_v}} \longrightarrow (4)$$

Note that $A_v$ represents the open-circuit voltage gain.

The output resistance with feedback $R'_{of}$ which includes $R_L$ as part of the amplifier is given by $R_{of}$ in parallel with $R_L$.

$$R'_{of} = R_{of} \parallel R_L$$

$$= \frac{R_{of}R_L}{R_{of}+R_L} = \left(\frac{R_oR_L}{1+\beta A_v}\right) \cdot \frac{1}{\left[R_o \big/ 1+\beta A_v\right] + R_L}$$

$$R'_{of} = \frac{R_oR_L}{R_{of}+R_L(1+\beta A_v)} = \frac{R_oR_L}{R_o+R_L+\beta A_vR_L} \longrightarrow (5)$$

Dividing numerator and denominator by ($R_o$ + $R_L$), we get

$$R'_{of} = \frac{R_oR_L \big/ (R_o+R_L)}{1+\left(\beta A_vR_L \big/ R_o+R_L\right)} \longrightarrow (6)$$

Since $R'_o = R_o \parallel R_L$ is the output resistance without feedback but with $R_L$ considered as part of the amplifier.

$$R'_{of} = \frac{R'_o}{1+\left(\beta A_vR_L \big/ R_o+R_L\right)}$$

Since $A_V = \dfrac{A_vR_L}{R_o+R_L}$,

$$\boxed{R'_{of} = \frac{R'_o}{1+\beta A_V}} \longrightarrow (7)$$

**(b) Voltage-shunt Feedback**

In this topology, the output resistance can be measured by opening the input source (i.e., $I_s$ = 0) and looking into the output terminals with $R_L$ disconnected, as shown in Fig. 6.19.

Applying KVL to the output circuit, we have

$$R_m I_i + I R_o - V = 0$$

or

$$I = \frac{V - R_m I_i}{R_o} \longrightarrow (1)$$

Since the input is shorted, $I_i = -I_f = -\beta V$ $\longrightarrow$(2)

Substituting the value of $I_i$ in Eq. (1), we get

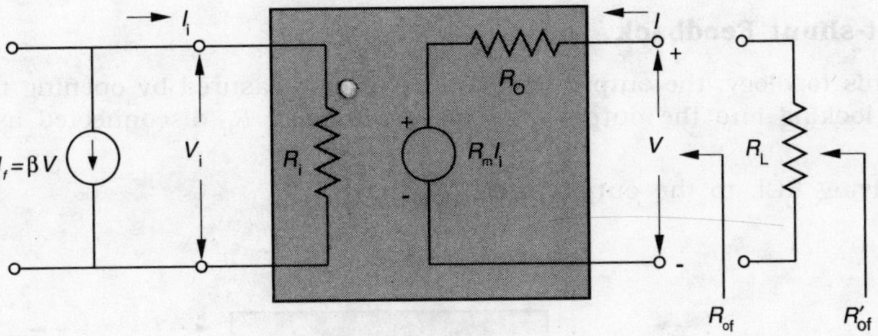


**Fig. 6.19 Voltage-shunt feedback connection**

$$I = \frac{V - R_m(-\beta V)}{R_o} = \frac{V + R_m \beta V}{R_o}$$

or
$$I = \frac{V(1+\beta R_m)}{R_o} \longrightarrow (3)$$

The output resistance with feedback is given by

$$R_{of} = \frac{V}{I}$$

From Eq. (3),
$$\boxed{R_{of} = \frac{R_o}{1+\beta R_m}} \longrightarrow (4)$$

$R_m$ is the open-loop transresistance without taking $R_L$ into account.

The output resistance with feedback which includes $R_L$ as part of the amplifier is given by

$$R'_{of} = R_{of} \,||\, R_L = \frac{R_{of} R_L}{R_{of} + R_L} \longrightarrow (5)$$

From Eq. (4),
$$R'_{of} = \frac{\left(R_o \big/ 1+\beta R_m\right) R_L}{\left(R_o \big/ 1+\beta R_m\right) + R_L} = \frac{R_o R_L}{R_o + R_L(1+\beta R_m)}$$

Dividing numerator and denominator by $(R_o + R_L)$, we get

$$= \frac{R_o R_L \big/ R_o + R_L}{1 + \left(\beta R_m R_L \big/ R_o + R_L\right)}$$

Since $R'_{of} = R_o \,||\, R_L$ is the output resistance without feedback,

$$R'_{of} = \frac{R'_o}{1 + \left(\beta R_m R_L \big/ R_o + R_L\right)}$$

or
$$\boxed{R'_{of} = \frac{R'_o}{1+\beta R_M}} \longrightarrow (6)$$

where $R_M = \dfrac{R_m R_L}{R_o + R_L}$ is the transresistance without feedback taking the load into account.

### (c) Current-shunt Feedback

In this topology, the output resistance can be measured by opening the input source $I_s = 0$ and looking into the output terminals, with load $R_L$ disconnected as shown in Fig. 6.20.

Applying KCL to the output node, we have

$$I = \frac{V}{R_o} - A_i I_i \qquad \longrightarrow (1)$$

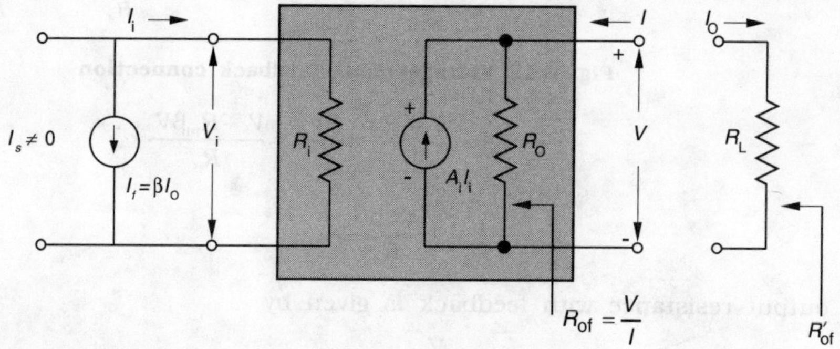


**Fig. 6.20 Current-shunt feedback connection**

The input current is given by

$$I_i = -I_f = -\beta I_o \qquad \text{(because } I_S = 0\text{)}$$

Since $I = -I_o$,

$$I_i = \beta I \qquad \longrightarrow (2)$$

Substituting the value of $I_i$ in Eq. (1), we get

$$I = \frac{V}{R_o} - A_i \beta I$$

or

$$I(1 + \beta A_i) = \frac{V}{R_o} \qquad \longrightarrow (3)$$

The output resistance with feedback is given by

$$R_{of} = \frac{V}{I}$$

From Eq. (3),

$$\mathbf{R_{of} = R_o(1 + \beta A_i)} \qquad \longrightarrow (4)$$

$A_i$ is the short-circuit gain without taking load $R_L$ into account.

The output resistance $R'_{of}$ which includes $R_L$ as part of the amplifier is given by

$$R'_{of} = R_{of} \,||\, R_L = \frac{R_{of} R_L}{R_{of} + R_L} \qquad \longrightarrow (5)$$

From Eq. (4),

$$R'_{of} = \frac{R_o(1 + \beta A_i) R_L}{R_o(1 + \beta A_i) + R_L} = \frac{R_o R_L (1 + \beta A_i)}{R_o + R_L + \beta A_i R_o}$$

Dividing numerator and denominator by $(R_o + R_L)$, we get

$$R'_{of} = \frac{(1+\beta A_i)R_o R_L / R_o + R_L}{1 + \left(\beta A_i R_o / R_o + R_L\right)}$$

Since $R'_o = R_o \,||\, R_L$ is the output resistance without feedback,

$$R'_{of} = \frac{R'_o(1+\beta A_i)}{1 + \left(\beta A_i R_o / R_o + R_L\right)}$$

$$\boldsymbol{R'_{of} = \frac{R'_o(1+\beta A_i)}{1+\beta A_I}} \longrightarrow (6)$$

where $A_I = \dfrac{A_i R_o}{R_o + R_L}$ is the current gain without feedback taking the load $R_L$ into account.

For $R_L = \infty$, $A_I = 0$ and $R'_o = R_o$ so that Eq. (6) reduces to

$$R'_{of} = R_o(1+\beta A_i) \longrightarrow (7)$$

**(d) Current-series Feedback**

In this topology, the output resistance can be measured by shorting the input source (i.e., $V_s = 0$) and looking into the output terminals with $R_L$ disconnected, as shown in Fig. 6.21.

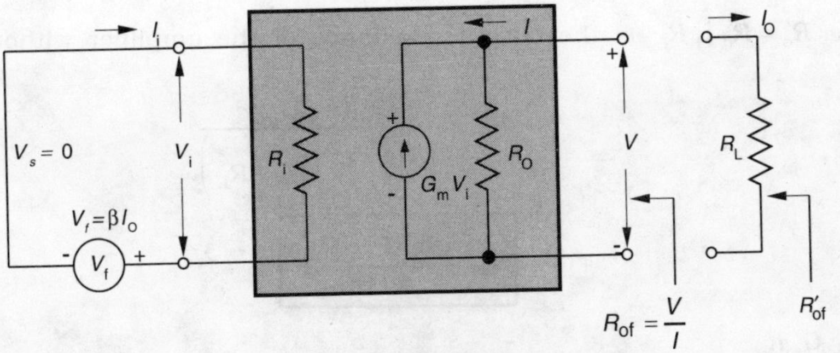


**Fig. 6.21 Current-series feedback connection**

Applying KCL to the output node, we get

$$I = \frac{V}{R_o} - G_m V_i \longrightarrow (1)$$

The input voltage is given by

$$V_i = -V_f = -\beta I_o$$

Since $I = -I_o$, $V_i = \beta I$ ⟶(2)

Substituting the value of $V_i$ from Eq. (2) in Eq. (1), we get

$$I = \frac{V}{R_o} - G_m \beta I$$

or $$I(1 + G_m \beta) = \frac{V}{R_o}$$ ⟶(3)

The output resistance with feedback is given by

$$R_{of} = \frac{V}{I}$$

From Eq. (3) $$\mathbf{R_{of} = R_o(1 + \beta G_m)}$$ ⟶(4)

where $G_m$ is the short-circuit transconductance without taking $R_L$ into account.

The output resistance $R'_{of}$ which includes $R_L$ as part of the amplifier is given by

$$R'_o = R_{of} \,||\, R_L = \frac{R_{of} R_L}{R_{of} + R_L}$$ ⟶(5)

From Eq. (4), $$R'_{of} = \frac{R_o(1 + \beta G_m) R_L}{R_o(1 + \beta G_m) + R_L} = \frac{(1 + \beta G_m) R_o R_L}{R_o + R_L + \beta G_m R_o}$$

Dividing numerator and denominator by $(R_o + R_L)$, we get

$$R'_{of} = \frac{(1 + \beta G_m) R_o R_L / (R_o + R_L)}{1 + \left( \beta G_m R_o / (R_o + R_L) \right)}$$

Since $R'_o = R_o \,||\, R_L$ is the output resistance of the amplifier without feedback,

$$R'_{of} = \frac{R'_o(1 + \beta G_m)}{1 + \left( \beta G_m R_o / (R_o + R_L) \right)}$$

$$\mathbf{R'_{of} = \frac{R'_o(1 + \beta G_m)}{1 + \beta G_M}}$$ ⟶(6)

where $G_M = \dfrac{G_m R_o}{R_o + R_L}$ is the transconductance without feedback taking the load $R_L$ into account.

From the above equation it is clear that $R_{of} > R_o$. For voltage feedback the output resistance is decreased, while current feedback increases the output resistance.

Table 6.3 summarises the characteristics of the four types of negative feedback configurations:

**Table 6.3 Effect of Negative Feedback on Amplifier Characteristics**

| Parameter | Type of feedback | | | |
|---|---|---|---|---|
| | **Voltage-series** | **Current-series** | **Current-shunt** | **Voltage-shunt** |
| 1. Stability | Improves | Improves | Improves | Improves |
| 2. Frequency distortion | Reduces | Reduces | Reduces | Reduces |
| 3. Gain with feedback | $A_{vf} = \dfrac{A_v}{1+\beta A_v}$ Decreases | $G_{mf} = \dfrac{G_m}{1+\beta G_m}$ Decreases | $A_{if} = \dfrac{A_i}{1+\beta A_i}$ Decreases | $R_{mf} = \dfrac{R_m}{1+\beta R_m}$ Decreases |
| 4. Bandwidth | Increases | Increases | Increases | Increases |
| 5. Nonlinear distortion and Noise | Decreases | Decreases | Decreases | Decreases |
| 6. Desensitizes | $A_{Vf}$ | $G_{Mf}$ | $A_{If}$ | $R_{Mf}$ |
| 7. Improves characteristics of | Voltage amplifier | Transconductance amplifier | Current amplifier | Transresistance amplifier |
| 8. Input resistance with feedback $R_{if}$ | Increases $R_{if} = R_i(1+\beta A_V)$ | Increases $R_{if} = R_i(1+\beta G_M)$ | Decreases $R_{if} = \dfrac{R_i}{1+\beta A_I}$ | Decreases $R_{if} = \dfrac{R_i}{1+\beta R_M}$ |
| 9. Output resistance with feedback ($R_{of}$) | Decreases $R_{of} = \dfrac{R_o}{1+\beta A_V}$ | Increases $R_{of} = R_o(1+\beta G_M)$ | Increases $R_{of} = R_o(1+\beta A_I)$ | Decreases $R_{of} = \dfrac{R_o}{1+\beta R_M}$ |

## IMPORTANT POINTS TO REMEMBER

1. Feedback is the process whereby a portion of the output signal of the amplifier is fed back to the input of the amplifier.
2. The two types of feedback used in electronic circuits are (i) positive feedback and (ii) negative feedback.
3. The positive feedback increases the gain of amplifier and produces excessive distortion.
4. The negative feedback reduces gain of the amplifier and improves the amplifier performance in many other respects.
5. Amplifiers are classified into (i) voltage (ii) current (iii) transconductance and (iv) transresistance amplifiers.
6. In voltage amplifier, the output voltage is proportional to the input voltage and the proportionality factor does not depend on the magnitudes of the source and load resistances.
7. In current amplifier, the output current is proportional to input current and the proportionality factor is independent of source and load resistances.
8. In transconductance amplifier, the output current is proportional to signal voltage and the proportionality factor is independent of the magnitudes of source and load resistances.
9. In transresistance amplifier, the output voltage is proportional to the signal current and the proportionality factor is independent of magnitudes of source and load resistances.

10. The two ways of mixing feedback signal with the input signal are (i) a fraction of output voltage is applied in series with the input voltage (ii) a fraction of output voltage is applied in parallel with the input through the feedback network.
11. The ratio of the output signal to the input signal of the amplifier is called the gain of the amplifier.
12. The four basic ways of connecting the feedback signal are: (i) voltage-series (ii) current-series (iii) current-shunt and (iv) voltage-shunt feedback.
13. Voltage-series/-shunt refers to connecting the output voltage as input to the feedback network.
14. Current-series/-shunt refers to tapping off some output current through the feedback network.
15. Series feedback connections tend to increase the input resistance.
16. Shunt feedback connections tend to decrease the input resistance.
17. Voltage feedback tends to decrease the output resistance.
18. Current feedback tends to increase the output resistance.
19. The feedback ratio is the ratio of feedback signal to the output signal.
20. The product -Aβ is called the loop gain or return ratio.
21. The difference between unity and the loop gain is called the return difference.
22. The term sensitivity (= 1 + Aβ) indicates the factor by which the voltage gain has been reduced due to feedback.
23. Negative feedback reduces the amount of noise signal and nonlinear distortion.
24. Negative feedback maintains the stability of the gain of an amplifier.
25. The introduction of feedback can greatly change the impedance levels within a device.

## KEY FORMULAE

1. Transfer characteristic i.e., voltage, $V_o = A_v V_s$; current, $I_L = A_i I_s$; transconductance, $I_L = G_m V_s$ and transresistance, $V_o = R_m I_s$
2. Gain with feedback, $A_f = \dfrac{A}{1 + \beta A}$
3. $dB$ of feedback, $N = 20 \log \left|\dfrac{A_f}{A}\right| = 20 \log \left|\dfrac{1}{1 + A\beta}\right|$
4. $\left|\dfrac{dA_f}{A_f}\right| = \left|\dfrac{dA}{A}\right| \dfrac{1}{|1 + A\beta|}$
5. Sensitivity $= \dfrac{1}{|1 + A\beta|}$ and desensitivity $D = 1 + A\beta$
6. Gain with feedback, $A_f = \dfrac{1}{\beta}$
7. Distortion with feedback, $D' = \dfrac{D}{1 + A_v \beta}$
8. In voltage-series feedback, $R_{if} = R_i (1 + \beta A_V)$
9. In current-series feedback $R_{if} = R_i (1 + \beta G_M)$

10. In current-shunt feedback $R_{if} = \frac{R_i}{1 + \beta A_I}$

11. In voltage-shunt feedback, $R_{if} = \frac{R_i}{1 + \beta R_M}$

12. In voltage-series feedback, $R_{of} = \frac{R_o}{1 + \beta A_V}$ and $R'_{of} = \frac{R'_o}{1 + \beta A_V}$

13. In voltage-shunt feedback, $R_{of} = \frac{R_o}{1 + \beta R_m}$ and $R'_{of} = \frac{R'_o}{1 + \beta R_M}$

14. In current-shunt feedback, $R_{of} = R_o\,(1 + \beta A_i)$ and $R'_{of} = \frac{R'_o(1 + \beta A_i)}{1 + \beta A_I}$

15. In current-series feedback, $R_{of} = R_o(1 + \beta G_m)$ and $R'_{of} = \frac{R'_o(1 + \beta G_m)}{1 + \beta G_M}$

## SOLVED PROBLEMS

**1. An amplifier has an open-loop gain of 500 and a feedback of 0.1. If open-loop gain changes by 20 % due to temperature find the percentage change in the closed-loop gain.**

**Data:** $A = 500$, $\beta = 0.1$ and $\frac{dA}{A} = 20$

**To find:** $\frac{dA_f}{A_f}$

**Solution:** Change in closed loop gain, $\frac{dA_f}{A_f} = \frac{dA}{A} \times \frac{1}{1 + \beta A}$

$$= 20 \times \frac{1}{1 + 500 \times 0.1} = \frac{20}{51}$$

Change in closed loop gain = **0.3921** or **39.21%**

**2. An amplifier has a voltage gain of 200. This gain is reduced to 50 when negative feedback is applied. Determine the reverse transmission factor and express the amount of feedback in *dB*.**

**Data:** A = 200, $A_f = 50$

**To find:** Reverse transmission factor β and feedback in *dB*

**Solution:** We know that $A_f = \frac{A}{1 + \beta A}$

$$50 = \frac{200}{1 + \beta \cdot 200}$$

Solving, $\beta = \mathbf{0.015}$

Feedback in *dB*, $N = 20 \log_{10}\left|\frac{A_f}{A}\right| = 20 \log_{10}\left(\frac{1}{1 + \beta A}\right)$

$$= 20 \log_{10}\left(\frac{1}{1 + 200 \times 0.015}\right) = 20 \log_{10}\left(\frac{1}{4}\right)$$

$$= 20 \log_{10} (0.25) = 20 [\bar{1}.3979]$$

$$= -20 + 7.958 = \mathbf{-12.042\ dB}$$

**3. For the circuit shown in Fig. 6.22, with $R_c$ = 4 kΩ, $R_L$ = 4 kΩ, $R_B$ = 20 kΩ, $R_s$ = 1 kΩ and the transistor parameters are:**
**$h_{ie}$ = 1.1 kΩ, $h_{fe}$ = 50, $h_{re}$ = 2.5 × $10^{-4}$ and $h_{oe}$ = 24 μS.**
**Find (a) the current gain (b) the voltage gain (c) the transconductance (d) the transresistance (e) the input resistance seen by the source and (f) the output resistance seen by the load. Neglect all capacitive effects.**

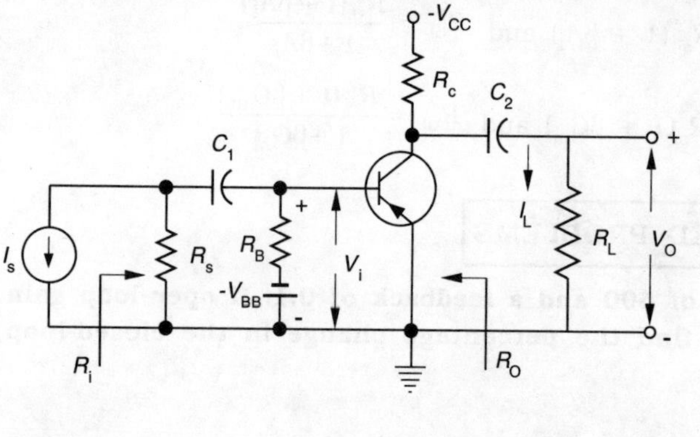


**Fig. 6.22**

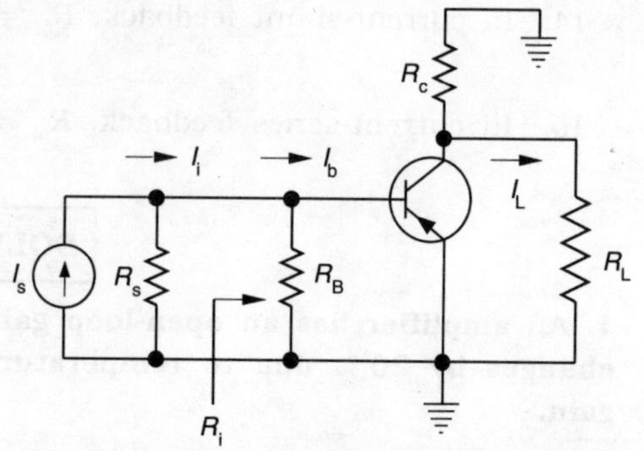


**Fig. 6.23**

**Data:** $R_c$ = 4 kΩ, $R_L$ = 4 kΩ, $R_B$ = 20 kΩ, $R_s$ = 1kΩ

**To find:** $A_i$, $A_v$, $G_m$, $R_m$, $R_i$ and $R_o$

**Solution:** The *ac* equivalent circuit of Fig. 6.22 is shown in Fig. 6.23.

From the circuit of Fig. 6.23, we have

(a) Current gain, $= \quad A_i = \dfrac{I_L}{I_s} = \dfrac{I_i}{I_s} \cdot \dfrac{I_b}{I_i} \cdot \dfrac{I_L}{I_b}$

where $\dfrac{I_i}{I_s} = \dfrac{R_s}{R_s + R_i}$

and Input resistance $R_i = R_b \parallel h_{ie} = 20\,k \parallel 1.1\,k$

$\therefore$ $R_i = \dfrac{20 \times 1.1}{20\ \ 1.1} = \dfrac{22}{21.1} = \mathbf{1.04\ k\Omega}$

Then $\dfrac{I_i}{I_s} = \dfrac{1k}{1k + 1.04k} = \dfrac{1}{2.04}$

and $\dfrac{I_b}{I_i} = \dfrac{R_B}{R_B + h_{ie}} = \dfrac{20}{20 + 1.1} = \dfrac{20}{21.1} = 0.95$

$\dfrac{I_L}{I_b} = -h_{fe} \dfrac{R_c}{R_c + R_L} = -50 \times \dfrac{4}{4+4} = \dfrac{-200}{8} = -25$

$$\therefore \quad A_i = \frac{I_L}{I_s} = \frac{1}{2.04} \times (0.95) \times (-25) = \mathbf{-11.65}$$

(b) Voltage gain, = $A_v = \frac{V_o}{V_s} = \frac{I_L R_L}{I_s R_s}$

$$= (-11.65)\frac{4k}{1k} = \mathbf{-46.6}$$

(c) Transconductance, $G_m = \frac{I_L}{V_s} = \frac{V_o}{R_L} \cdot \frac{1}{V_s} = \frac{V_o}{V_s} \times \frac{1}{R_L}$

$$= \frac{-46.6}{4k} = \mathbf{-11.65\ mA/V}$$

(d) Transresistance, $R_m = \frac{V_o}{I_s} = \frac{R_s}{V_s} \cdot V_o = R_s \cdot \frac{V_o}{V_s}$

$$= 1\ k \times -46.6 = -46.6\ k\Omega$$

(e) Input resistance, $R_i = R_B \parallel h_{ie} = 20\ k \parallel 1.1\ k = \mathbf{1.04\ k\Omega}$

(f) Output resistance, $R_o = R_C \parallel \frac{1}{h_{oe}} = 4k \parallel 40\ k$

$$= \mathbf{3.64\ k\Omega}$$

**4. Repeat Problem 3 for the circuit shown in Fig. 6.24, with $g_m$ = 5 *mA/V* and $r_d$ = 100 kΩ. Note that $V_s = I_s R_s$.**

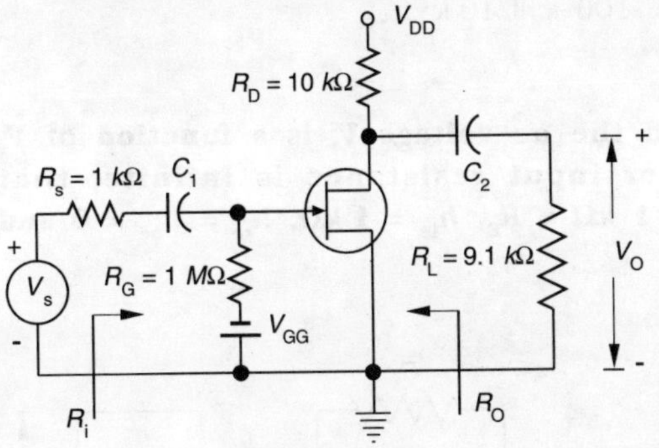


**Fig. 6.24**

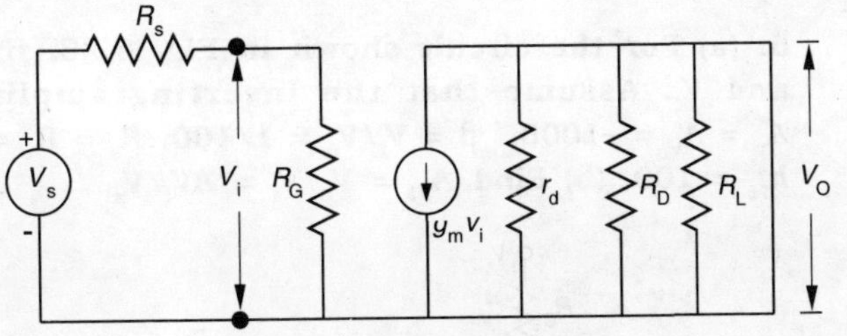


**Fig. 6.25**

**Data:** $R_D$ = 10 kΩ, $R_S$ = 1 kΩ, $R_G$ = 1 MΩ, $R_L$ = 9.1 kΩ, $g_m$ = 5 mS, $r_d$ = 100 kΩ

**To find:** $A_i$, $A_v$, $G_m$, $R_m$, $R_i$ and $R_o$

**Solution:** The *ac* equivalent circuit is shown in Fig. 6.25.

(a) Current gain, $A_i = \frac{I_L}{I_i} = \frac{R_g I_L}{V_i} = \frac{R_g}{V_i} \times \frac{V_o}{R_L} = \frac{R_g}{R_L} \cdot \frac{V_o}{V_i}$

From the *ac* equivalent circuit, we have

$$\frac{V_o}{V_i} = -g_m\,[r_d \parallel R_D \parallel R_L]$$

$$= -5 \times 10^{-3} \left[\left(\frac{100 \times 10}{110}\right) || \, 9.1\right] = -5 \times 10^{-3} \, [9.1||9.1]$$

$$= -5 \times 10^{-3} \times 4.55 = \mathbf{-22.75}$$

$$\therefore \qquad A_i = \frac{I_L}{I_i} = \frac{10^6}{9.1 \times 10^3} \times (-22.75) = \mathbf{-2500}$$

or

$$\frac{I_L}{I_s} = \frac{R_s}{R_s + R_G} \cdot \frac{I_L}{I_i} = \frac{1k}{1k + 1M}(-2500) \approx \mathbf{-2.5}$$

(b) Voltage gain,

$$A_v = \frac{V_o}{V_s} = \frac{I_L \; R_L}{I_i \; (R_s + R_G)}$$

$$A_v = -2500 \cdot \frac{9.1}{1001} \approx \mathbf{-22.7}$$

(c) Transconductance,

$$G_m = \frac{I_L}{V_s} = \frac{V_o}{R_L} \cdot \frac{1}{V_s} = \left(\frac{V_o}{V_s}\right) \cdot R_s = -22.75 \times \frac{1}{9 \times 10^3} = \mathbf{-2.5\; mS}$$

(d) Transresistance,

$$R_m = \frac{V_o}{I_s} = V_o \cdot \frac{R_s}{V_s} = \left(\frac{V_o}{V_s}\right) \cdot R_s$$

$$= -22.75 \times 10^3 = \mathbf{-22.75\; k\Omega}$$

(e) Input resistance,

$$R_i = \frac{V_i}{I_i} = R_s = \mathbf{1\, M\Omega}$$

(f) Output resistance,

$$R_o = r_d \; || \; R_D = 100\; k \; || \; 10\, k$$

$$= \mathbf{9.1\; k\Omega}$$

**5. (a) For the circuit shown in Fig. 6.26, find the *ac* voltage $V_i$ is a function of $V_s$ and $V_f$. Assume that the inverting amplifier input resistance is infinite, that $A_v = A_V = -1000$, $\beta = V_f/V_o = 1/100$, $R_s = R_c = 1\; k\Omega = R_E$, $h_{ie} = 1\; k\Omega$, $h_{re} = h_{oe} = 0$ and $h_{fe} = 100$. (b) Find $A_{Vf} = V_o/V_i = AV_i/V_s$.**

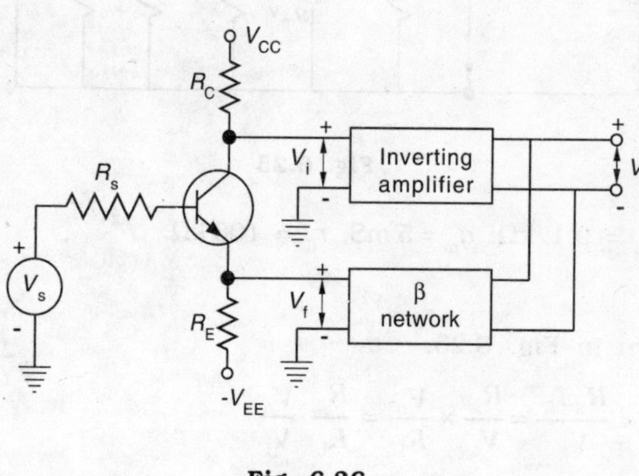


Fig. 6.26

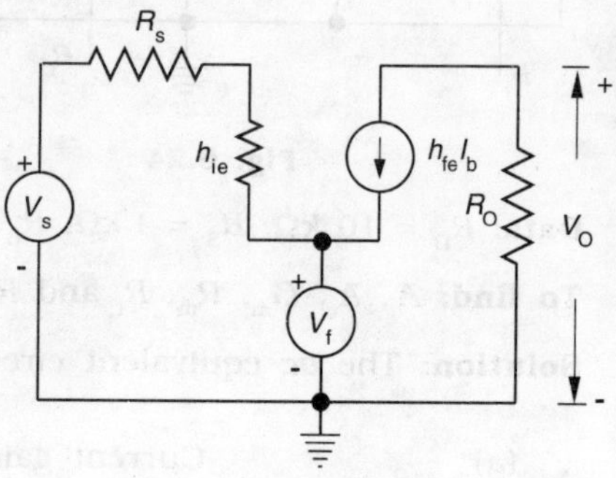


Fig. 6.27

**Data.:** $A_v = A_V = -1000$, $\beta = 0.01$, $R_s = R_C = R'_E = 1\; k\Omega$, $h_{ie} = 1\; k\Omega$, $h_{re} = h_{oe} = 0$ and $h_{fe} = 100$

**To find:** (a) AC voltage $V_i$ is a function of $V_s$ and $V_f$. (b) $A_{Vf}$

**Solution:** (a) The *ac* equivalent circuit is shown in Fig. 6.27. Assume that the β network can be represented by an ideal controlled voltage source with $V_f = \beta V_o$. From the equivalent circuit, we have

$$V_i = -h_{fe}\, I_b\, R_C \quad \text{where} \quad I_b = \frac{V_s - V_f}{R_s + h_{ie}}$$

$$\therefore \quad V_i = -h_{fe}\, R_C \frac{V_s - V_f}{R_s + h_{ie}}$$

$$V_i = -100.1k\ \frac{V_s - V_f}{(1+1)\,k} = \mathbf{-50\,(V_s - V_f)}$$

(b) With the output of the inverting amplifier connected to the input of the β network we have

$$V_f = \beta\, V_o \text{ and } V_o = A\, V_i = A_V\, V_i$$

$$\therefore \quad V_o = A_V \times -50\ (V_s - V_f) = -A_V \times 50\,(V_s - \beta\, V_o)$$

$$V_o = 5 \times 10^4\ (V_s - 0.01\, V_o)$$

or $$V_o\ (1 + 500) = 5 \times 10^4\ V_S$$

$$\therefore \quad A_{vf} = \frac{V_o}{V_s} = \frac{5 \times 10^4}{501} = \mathbf{100}$$

**6. An amplifier with open-loop voltage gain $A_V$ = 1000 ± 100 is available. It is necessary to have an amplifier whose voltage gain varies by no more than ± 0.1 percent. Find (a) the feedback ratio and (b) the gain with feedback.**

**Data:** $A_v = 1000 \pm 100$, $dA_f/A_f = 0.1\%$

**To find:** Feedback ratio β and $A_f$

**Solution:** (a) We know that $$\frac{d A_f}{A_f} = \frac{1}{1 + \beta A} \cdot \frac{dA}{A}$$

or $$\frac{0.1}{100} = \frac{1}{1 + \beta A} \times \frac{100}{1000} = \frac{1}{1 + \beta A} \times \frac{1}{10}$$

or $$1 + \beta A = 100$$

$$\beta = \frac{99}{1000} = \mathbf{0.099}$$

(b) Voltage gain with feedback is given by

$$A_f = \frac{A}{1 + \beta A} = \frac{1000}{1 + 0.099 \times 1000} = \frac{1000}{100} = \mathbf{10}$$

**7. An amplifier without feedback gives a fundamental output of 36 V with 7 percent second harmonic distortion when the input is 0.028 V.**

**(a) If 1.2 percent of the output is fed back into the input in a negative voltage-series feedback circuit, what is the output voltage?**

**(b) If the fundamental output is maintained at 36 V but the second harmonic distortion is reduced to 1 percent, what is the input voltage?**

**Data:** $V_o$ = 36 V, $V_i$ = 0.028 V, $D/D'$ = 7, $V_f$ = 1.2 % $V_o$ and $D/D'$ = 7

**To find:** Feedback output voltage, and input voltage $V_s$

**Solution:** (a) Voltage gain, $|A| = \frac{V_o}{V_i} = \frac{36}{0.028} = 1285$

Feedback ratio, $\beta = \frac{V_f}{V_o} = \frac{1.2}{100} = 0.012$

We know that $A_f = \frac{A}{1+\beta A} = \frac{1285}{1+0.012 \times 1285}$ **= 78.2**

Output voltage, $V'_o = A_f V_s = 78.2 \times 0.028 =$ **2.19 V**

(b) If the output is maintained constant at 36 *V* then the distortion generated by the device is unchanged. The reduction of the total distortion is caused by feedback.

We know that $D' = \frac{D}{1 + \beta A}$

or $1 + \beta A = \frac{D}{D'} = 7$

Hence $\beta A = 6$

And $A_f = \frac{A}{1 + \beta A} = \frac{1285}{7}$

and $V_s = \frac{V_o}{V_f} = \frac{36}{1285/7} = \frac{36 \times 7}{1285} =$ **0.196 V**

**8. The output resistance of voltage-series feedback amplifier is 10 Ω. If the gain of the basic amplifier is 100 and the feedback fraction is 0.01, what is the output resistance of the amplifier without feedback?**

**Data:** $R_{of} = 10\ \Omega$, $A = 100$, $\beta = 0.01$

**To find:** Output resistance $R_o$

**Solution:** We know that $R_{of} = \frac{R_o}{1 + \beta A}$

$R_o = R_{of}\ (1 + \beta A)$

$\therefore$ $R_o = 10\ (1 + 0.01 \times 100) =$ **20 Ω**

**9. The signal and output voltages of an amplifier are 1 *mV* and 1 *V* respectively. If the gain with negative feedback is 100 and input resistance without feedback (voltage-series) is 2 *k*Ω, find the feedback fraction and input resistance with feedback.**

**Data:** $V_s = 1\ mV$, $V_o = 1\ V$, $A_f = 100$, $R_i = 2\ k\Omega$

**To find:** Feedback fraction β and $R_{if}$

**Solution:** We know that $A = \frac{V_o}{V_s} = \frac{1V}{1mV} =$ **1000**

$A_f = \frac{A}{1 + \beta A}$

or $$1+\beta A = \frac{A}{A_f} = \frac{1000}{100} = 10$$

or $$\beta = \mathbf{0.009}$$

Input resistance with feedback, $$R_{if} = R_i\ (1 + \beta A)$$

$$= 2\ k\Omega \times 10 = \mathbf{20\ k\Omega}$$

**10. If an amplifier has a bandwidth of 200 *kHz* and voltage gain of 80, what will be the new bandwidth and gain if 5 percent of negative feedback is introduced?**

**Data:** $BW = 200\ kHz$, $A = 80$, $\beta = 5\% = 0.05$

**To find:** New bandwidth, $BW_f$

**Solution:** We know that $$A_f = \frac{A}{1+\beta A}$$

$$A_f = \frac{80}{1+0.05\times 80} = \frac{80}{5} = \mathbf{16}$$

With feedback, $$A\cdot BW = A_f\cdot BW_f$$

∴ Bandwidth with feedback, $$BW_f = \frac{A\cdot BW}{A_f} = \frac{80\times 200}{16} = \mathbf{1\ MHz}$$

## QUESTIONS

1. What is meant by feedback? Name the types of feedback.
2. Distinguish between voltage amplifier and current amplifier. Draw their equivalent circuits.
3. Distinguish between positive feedback and negative feedback.
4. Explain how feedback affects the overall gain of an amplifier.
5. (a) Draw the equivalent circuit for a voltage amplifier. (b) For an ideal amplifier, what are the values of $R_i$ and $R_o$? (c) What are the dimensions of the transfer gain?
6. Repeat Question no. 5 for a current amplifier.
7. Repeat Question no. 5 for a transconductance amplifier.
8. Repeat Question no. 5 for a transresistance amplifier.
9. Write topology for various types of feedback amplifiers.
10. Explain with block diagram the feedback amplifier. Identify each block and state its function.
11. (a) What are the four possible topologies of a feedback amplifier? (b) Identify the output signal $X_o$ and the feedback signal $X_f$ for each topology. (c) Identify the transfer gain $A$ for each topology.
12. Distinguish between open-loop gain and closed-loop gain.
13. Is distortion in an amplifier reduced by positive or by negative feedback?
14. What are the advantages of negative feedback?
15. Derive the relationship between the transfer gain with feedback $A_f$ and without feedback $A$.
16. What is meant by loop gain? Define the amount of feedback in decibels.
17. State the three conditions which are made in order that the expression $A_f = A/(1+\beta A)$ be satisfied exactly.
18. Define desensitivity. For large values of desensitivity what is $A_f$. What is the significance of this result?

19. Explain the effect of negative feedback on noise.
20. List five characteristics of an amplifier which are modified by negative feedback.
21. How does negative feedback affect the stability of an amplifier? Derive an expression $\left|\frac{dA_f}{A_f}\right| = \left|\frac{dA}{A}\right| \frac{1}{(1+A\beta))}$.
22. Explain the effect of negative feedback on frequency distortion.
23. Explain the effect of negative feedback on nonlinear distortion. Derive an expression $D' = D/(1+\beta A)$.
24. Explain how gain is stabilised with negative feedback.
25. State whether the input resistance $R_{if}$ is increased or decreased for each topology.
26. Explain the sampling and mixing networks.
27. Derive an expression for $R_{if}$ in the case of (i) voltage-series feedback (ii) current-series (iii) current-shunt and (iv) voltage-shunt feedback amplifiers.
28. State whether the output resistance $R_{of}$ is increased or decreased for each topology.
29. Derive an expression for $R_{of}$ in the case of (i) voltage-series (ii) current-series (iii) current-shunt (iv) voltage-shunt feedback amplifiers.

## EXERCISES

1. If an amplifier with gain 2000 has a 40% variation in its gain, calculate the change in gain when a 20 % feedback is employed. **[Ans: 0.1%]**
2. The voltage gain of an amplifier without feedback is 8000. Calculate the voltage gain if negative feedback is introduced. Given $\beta = 0.01$. **[Ans: 98.76]**
3. An amplifier has a gain of 600. When the negative feedback is applied, the gain is reduced to 200. Find the feedback ratio. **[Ans. 0.01]**
4. An amplifier has a gain of 100 and 5 % distortion with an input signal 1 *V*. When an input signal of 1 *V* is applied to the amplifier, calculate (a) output signal voltage (b) distortion (c) output voltage. **[Ans: (a) 100 *V*; (b) 5 *V* and (c) 105 *V*]**
5. An amplifier has the voltage gain of 1000. The gain is reduced to 200 with a negative feedback. Calculate the feedback fraction and express the amount of feedback in *dB*. **[Ans: 0.8 %; 13.98 *dB*]**
6. Identify topology, with justification for the circuit shown in Fig. 6.28. Transistors used are identical and have parameters $h_{ie} = 2\,k\Omega$, $h_{fe} = 50$ and $h_{oe} = h_{re} = 0$. Determine $A_{vf}$.
**[Ans: $h_{ie} = 2\,k\Omega$; $A_{v2} = -165$; $R_{i1} = 86.15\,k\Omega$; $A_{v1} = -1.114$; $A_v = 183.96$; $\beta = 0.5$; $D = 92.93$ and $A_{vf} = 1.978$]**
7. Identify the topology of feedback in the circuit of Fig. 6.29 giving justification. Two transistors are identical with $h_{ie} = 2\,k\Omega$, and $h_{fe} = 100$. Calculate (i) $R_{if}$ (ii) $A_{if}$ and (iii) $A_{vf}$. **[Ans: (i) 34.75 Ω; (ii) 18.92 and (iii) 75.68]**
8. Obtain *A* and β for a feedback amplifier that produces 100 *W* of output with 0.05% distortion. The load resistance is 8 Ω. When the output stage supplies 100 *W* to an 8 Ω load, in a nonfeedback case, 5 % harmonic distortion is produced. The input signal for 100 *W* of output is to be 1 *V*. **[Ans: 2830 and 0.035]**
9. An amplifier consists of 3 identical stages connected in cascade. The output voltage is sampled and returned to the input in series opposing. If it is specified that the relative change $dA_f/A_f$ in the closed-loop voltage gain $A_f$ must not exceed $\psi_f$, show

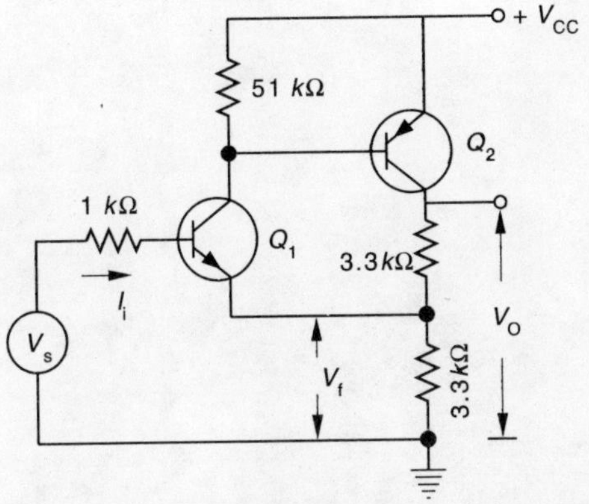


Fig. 6.28

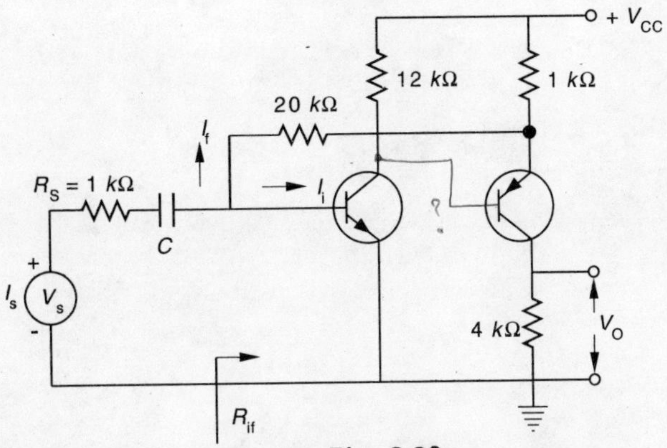


Fig. 6.29

that the minimum value of the open-loop gain $A$ of the amplifier is given by

$A = 3\ A_f \left|\frac{\psi_1}{\psi_f}\right|$ and $\psi_1 = dA_1 / A_1$ is the relative change in the voltage gain of each stage of the amplifier.

10. An amplifier with an open-loop voltage gain of 1000 delivers 10 $W$ of output power at 10 percent second-harmonic distortion when the input signal is 10 $mV$. If 40 $dB$ negative voltage series feedback is applied and the output power is to remain at 10 $W$, determine (a) the required input signal, (b) the percent harmonic distortion.

**[Ans: (a) 1V; (b) 0.1%]**

# 7 POWER CIRCUITS AND SYSTEMS

## Chapter Outline



## INTRODUCTION

An amplifying system usually has several cascaded stages. The input and intermediate stages are small-signal amplifiers operated in class-A. Their function is only to amplify the small input signal to a sufficiently large level to drive the final device. This output stage feeds a transducer such as a loudspeaker, cathode-ray tube, servometer, etc. Hence the output stage must be capable of handling large signals and deliver appreciable power to the load. Therefore, large-signal amplifiers are also called **power amplifiers.**

A power amplifier is more commonly known as audio amplifier. The audio amplifiers are used in public address systems, tape recorders, stereo systems, television receivers, radio receivers, broadcast transmitters, etc. In case of broadcast transmitters, power amplifiers capable of feeding large power to the antenna are required. It will be interesting to know that a power amplifier does not actually amplify the power. As a matter of fact, it takes power from the *dc* power supply connected to the output circuit and converts it into useful *ac* signal power. This power is fed to the load. The type of *ac* power developed, at the output of a power amplifier, is controlled by the input signal. Thus, we can say that actually a power amplifier is a *dc* to *ac* power converter, whose action is controlled by the input signal.

Large-signal amplifiers can be classified according to the class of operation, which is decided by the location of the quiescent point on the device characteristics. The different classes of operation are:

(i) Class A amplifier
(ii) Class B amplifier
(iii) Class AB amplifier and
(iv) Class C amplifier.

## CLASS A LARGE-SIGNAL AMPLIFIERS

The circuit diagram of a simple transistor amplifier that supplies power to a pure resistance load is shown in Fig. 7.1. Let $i_C$ represent the total instantaneous collector current, $i_c$ designate the instantaneous variation from the quiescent value $I_C$ of the collector current. Similarly, $i_B$, $i_b$ and $I_B$ represent corresponding base currents. $v_{CE}$ represents the total instantaneous collector-to-emitter voltage, $v_{ce}$ designates the instantaneous variation from the quiescent value $V_{CE}$ of the collector-to-emitter voltage.

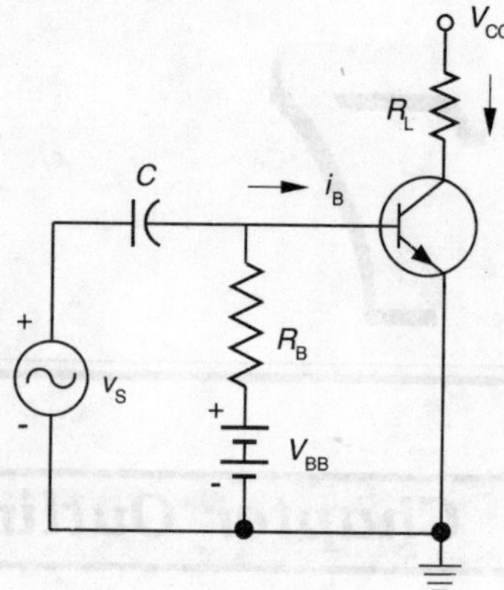

**Fig. 7.1 Series-fed transistor amplifier**

Consider the static output characteristics of a transistor, shown in Fig. 7.2. Assume that the characteristics are equidistant for equal increments of input base current $i_b$. Then, if the input signal $i_b$ is a sinusoid, the output current and voltage are also sinusoidal as shown in Fig. 7.2. The output power may be found graphically and is given by

$$P = V_{ce} I_C = I_C^2 R_L \longrightarrow (1)$$

where $V_{ce}$ and $I_C$ are the rms values of the output voltage $v_{ce}$ and current $i_c$ respectively. The numerical values of $V_{ce}$ and $I_C$ can be determined graphically in terms of the maximum and minimum voltage and currents, as shown in Fig. 7.2. If $V_m$ and $I_m$ represent the peak sinusoidal voltage and current swings, then

$$I_C = \frac{I_m}{\sqrt{2}} = \frac{I_{max} - I_{min}}{2\sqrt{2}} \longrightarrow (2)$$

and

$$V_{ce} = \frac{V_m}{\sqrt{2}} = \frac{V_{max} - V_{min}}{2\sqrt{2}} \longrightarrow (3)$$

$\therefore$

$$\textbf{Power, } P = \frac{V_m I_m}{2} = \frac{I_m^2 R_L}{2} = \frac{V_m^2}{2R_L} \longrightarrow (4)$$

or

$$P_{ac} = \frac{(V_{max} - V_{min})}{2\sqrt{2}} \cdot \frac{(I_{max} - I_{min})}{2\sqrt{2}}$$

$$P_{ac} = \frac{(V_{max} - V_{min})(I_{max} - I_{min})}{8} \longrightarrow (5)$$

and

$$P_{dc} = V_{CC} \cdot I_{CQ} \longrightarrow (6)$$

Efficiency is given by

$$\eta = \frac{(V_{max} - V_{min})(I_{max} - I_{min})}{8 V_{CC} \cdot I_{CQ}} \longrightarrow (7)$$

**Maximum Efficiency**

Referring to Fig. 7.2(b),

For a maximum swing, $V_{max} = V_{CC}$ and $V_{min} = 0$

and $I_{max} = 2I_{CQ}$ and $I_{min} = 0$

$$\therefore \quad \eta = \frac{V_{CC} \times 2I_{CQ}}{8V_{CC} \times I_{CQ}} = 0.25 = 25\ \%$$

Thus the maximum efficiency in class-A amplifier is 25%.

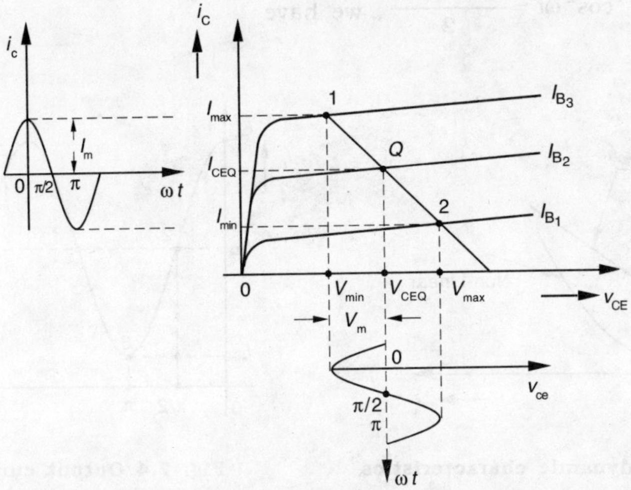

Fig. 7.2 (a) The output characteristics and the current and voltage waveforms

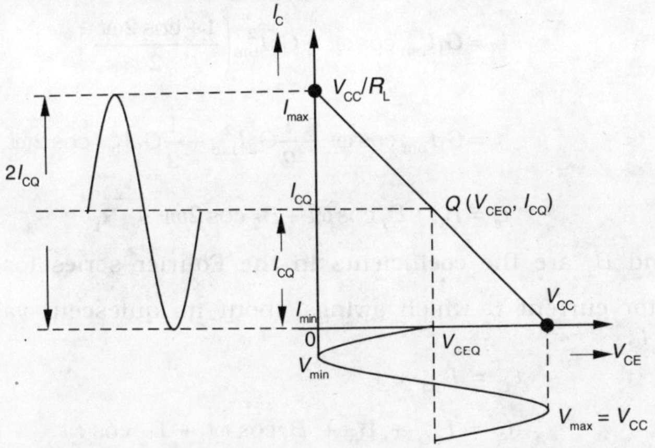

Fig. 7.2 (b) Maximum voltage and current swings

## SECOND HARMONIC DISTORTION

In practice, the static output characteristics are not equidistant straight lines for constant increments of input excitation resulting in nonlinearity and hence distortion of the output signal. This type of distortion is called **nonlinear** or **amplitude distortion**.

Under large-signal conditions the output current $i_c$ does not vary linearly with the input signal $i_b$. The transfer characteristic $i_c$ versus $i_b$ will be nonlinear. In terms of a Taylor's series expansion of $i_b$ we can write output current $i_c$ as

$$i_c = G_1 i_b + G_2 i_b^2 + G_3 i_b^3 + \ldots\ldots\ldots \longrightarrow (1)$$

where $G_1$, $G_2$, .... are constants.

If the input waveform is sinusoidal, then

$$i_b = I_{bm} \cos \omega t \longrightarrow (2)$$

Then $$i_c = G_1 I_{bm} \cos \omega t + G_2 I^2_{bm} \cos^2 \omega t + \ldots\ldots\ldots$$

Since $$\cos^2 \omega t = \frac{1 + \cos 2\omega t}{2}, \text{ we have}$$

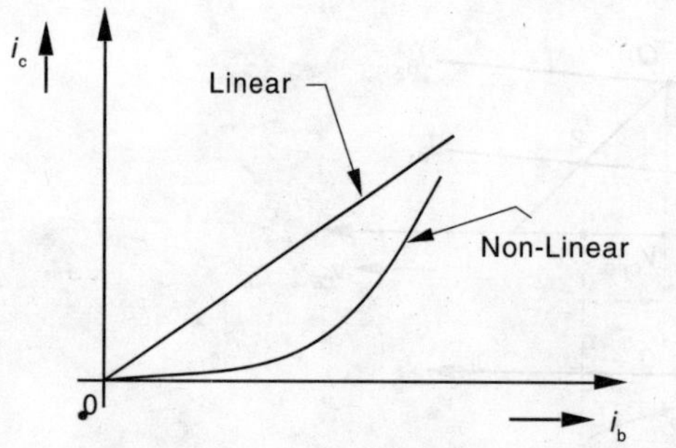


**Fig. 7.3 Nonlinear dynamic characteristics**

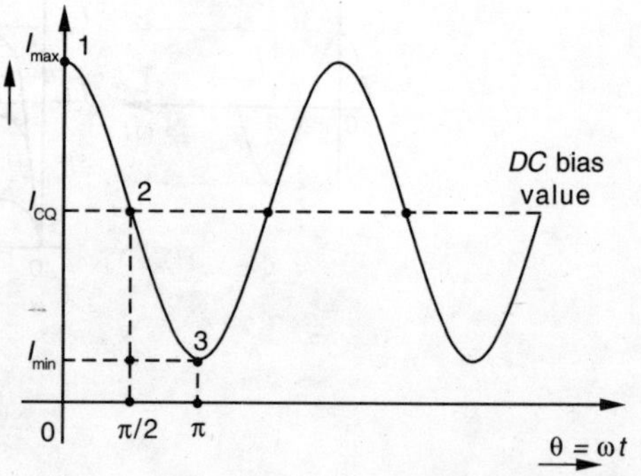


**Fig. 7.4 Output current waveform**

$$i_c = G_1 I_{bm} \cos \omega t + G_2 I_{bm}^2 \left( \frac{1 + \cos 2\omega t}{2} \right)$$

$$= G_1 I_{bm} \cos \omega t + \frac{1}{2} G_2 I_{bm}^2 + \frac{1}{2} G_2 I_{bm}^2 \cos 2\omega t$$

$$i_c = B_0 + B_1 \cos \omega t + B_2 \cos 2\omega t \longrightarrow (3)$$

where $B_0$, $B_1$ and $B_2$ are the coefficients in the Fourier series for the current $i_c$.

The total collector current $i_C$ which swings about its quiescent value $I_{CQ}$ [Fig. 7.4] is the sum of $I_{CQ}$ and $i_c$.

$\therefore$ $$i_C = I_{CQ} + i_c$$

$$i_C = I_{CQ} + B_0 + B_1 \cos \omega t + B_2 \cos \omega t \longrightarrow (4)$$

It shows that the application of a sinusoidal signal on a parabolic dynamic characteristic results in an output current which contains, in addition to the term of the same frequency ($B_1$ cos ωt) as the input, a second harmonic term ($B_2$ cos2ωt) and also a constant current ($I_{CQ}$). The constant term $B_0$ adds to the original *dc* value $I_{CQ}$ to yield a total *dc* component of current $I_{CQ} + B_0$. Parabolic nonlinear distortion introduces into the output a component whose frequency is **twice** that of the sinusoidal input excitation. Also, since a sinusoidal input signal changes the average value of the output current, rectification takes place.

The amplitudes $B_0$, $B_1$ and $B_2$ for a given load resistor can be determined from either the static or the dynamic characteristics. To evaluate these values we need output currents at three different values [Fig. 7.4].

At point 1, $\omega t = 0$; $i_C = I_{max}$

At point 2, $\omega t = \pi/2$; $i_C = I_{CQ}$ ⟶ (5)

At point 3, $\omega t = \pi$; $i_C = I_{min}$

Substituting these values in Eq. (4), we get

$$I_{max} = I_{CQ} + B_0 + B_1 + B_2 \quad \longrightarrow (6)$$

$$I_{CQ} = I_{CQ} + B_0 - B_2 \quad \longrightarrow (7)$$

$$I_{min} = I_{CQ} + B_0 - B_1 + B_2 \quad \longrightarrow (8)$$

From Eq. (7), we have $B_0 = B_2$ ⟶ (9)

Subtracting Eq. (8) from Eq. (6), we get

$$I_{max} - I_{min} = 2B_1$$

or $$B_1 = \frac{I_{max} - I_{min}}{2} \quad \longrightarrow (10)$$

Adding Eqs. (6) and (8), $I_{max} + I_{min} = 2I_{CQ} + 2B_0 + 2B_2$

Since $B_0 = B_2$, $= 2I_{CQ} + 4B_2$

or $$B_2 = B_0 = \frac{I_{max} + I_{min} - 2I_{CQ}}{4} \quad \longrightarrow (11)$$

As the amplitudes of the fundamental and second harmonic are known, the second harmonic distortion $D_2$ is defined as

$$\boxed{D_2 = \frac{|B_2|}{|B_1|}} \quad \longrightarrow (12)$$

or % second harmonic distortion, $D_2 = \frac{|B_2|}{|B_1|} \times 100\%$ ⟶ (13)

Since the method uses three points on the collector current waveform to obtain the amplitudes of the harmonics, the method is known as **"three-point method"** of determining second harmonic distortion.

## HIGHER-ORDER HARMONIC DISTORTION

For a power amplifier with a large input swing, the output current $i_c$ can be written in terms of a Taylor's series expansion as

$$i_c = G_1 i_b + G_2 i_b^2 + G_3 i_b^3 + G_4 i_b^4 + \ldots\ldots\ldots\ldots \quad \longrightarrow (1)$$

where $G_1$, $G_2$, ............. are constants.

If the input wave is a simple cosine function of time, then

$$i_b = I_{bm} \cos\omega t \longrightarrow (2)$$

Then $$i_c = G_1 I_{bm} \cos\omega t + G_2 I_{bm}^2 \cos^2\omega t + G_3 I_{bm}^3 \cos^3\omega t + \ldots$$

By using trigonometric identities for multiple angles and combining the resultant coefficients, $i_c$ can be rewritten as

$$i_c = B_0 + B_1\cos\omega t + B_2\cos 2\omega t + B_3\cos 3\omega t + B_4\cos 4\omega t + \ldots \longrightarrow (3)$$

where $B_0$, $B_1$, $B_2$, $B_3$, $B_4$, ......... are the coefficients in the Fourier series for the current $i_c$.

The total output current is given by

$$i_C = I_{CQ} + i_c$$

$$i_C = I_{CQ} + B_0 + B_1\cos\omega t + B_2\cos 2\omega t + B_3\cos 3\omega t + B_4\cos 4\omega t + \ldots \longrightarrow (4)$$

where $(I_{CQ} + B_0)$ is the *dc* component.

Since $i_C$ is an even function of time, the Fourier series in Eq. (4) representing a periodic function possessing the symmetry, contains only cosine terms.

Suppose we assume as an approximation that harmonics higher than the fourth are negligible in the above Fourier series, then we will have five unknown terms $B_0$, $B_1$, $B_2$, $B_3$ and $B_4$. To evaluate these we need output currents at five different values of $I_B$ [Fig. 7.5]. Let us assume that

$$i_c = 2\Delta i \cos\omega t \longrightarrow (5)$$

Hence $$I_B = I_{BQ} + 2\Delta i \cos\omega t \longrightarrow (6)$$

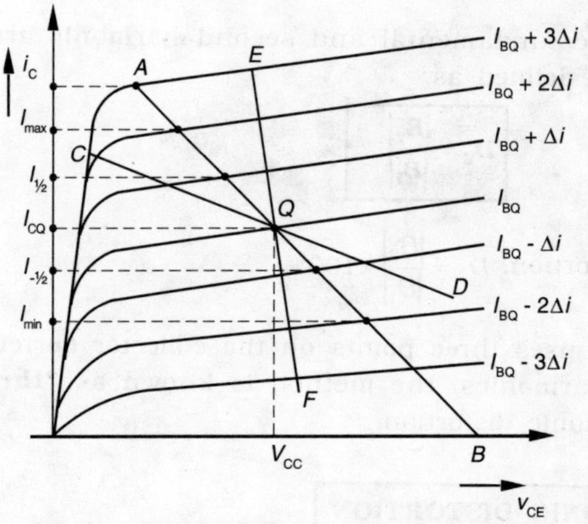


**Fig. 7.5 Graphical evaluation of harmonic distortion**

At $\omega t = 0$, $\quad I_B = I_{BQ} + 2\Delta i$, $\quad i_C = I_{max} \longrightarrow (7)$

At $\omega t = \pi/3$, $\quad I_B = I_{BQ} + \Delta i$, $\quad i_C = I_{1/2} \longrightarrow (8)$

At $\omega t = \pi/2$, $I_B = I_{BQ}$, $i_C = I_{CQ}$ $\longrightarrow$ (9)

At $\omega t = 2\pi/3$, $I_B = I_{BQ} - \Delta i$, $i_C = I_{-1/2}$ $\longrightarrow$ (10)

At $\omega t = \pi$, $I_B = I_{BQ} - 2\Delta i$, $i_C = I_{min}$ $\longrightarrow$ (11)

By combining Eqs. (4) and (7) to (11), we get five equations and solving them we obtain the following relations:

$$B_0 = \frac{1}{6}\left[I_{max} + 2I_{1/2} + 2I_{-1/2} + I_{min}\right] - I_{CQ} \longrightarrow (12)$$

$$B_1 = \frac{1}{3}\left[I_{max} + I_{1/2} - I_{-1/2} - I_{min}\right] \longrightarrow (13)$$

$$B_2 = \frac{1}{4}\left[I_{max} - 2I_{CQ} + I_{min}\right] \longrightarrow (14)$$

$$B_3 = \frac{1}{6}\left[I_{max} - 2I_{1/2} + 2I_{-1/2} - I_{min}\right] \longrightarrow (15)$$

$$B_4 = \frac{1}{12}\left[I_{max} - 4I_{1/2} + 6I_{CQ} - 4I_{-1/2} + I_{min}\right] \longrightarrow (16)$$

The harmonic distortion is defined as

$$D_2 = \frac{|B_2|}{|B_1|},\quad D_3 = \frac{|B_3|}{|B_1|},\quad D_4 = \frac{|B_4|}{|B_1|} \longrightarrow (17)$$

where $D_n$ represents the distortion of the $n^{th}$ harmonic. Since this method uses five points on the output waveform to obtain the amplitudes of harmonics, the method is known as **"five point method"** of determining the higher order harmonic distortion.

## Power Output Due to Distortion

If the distortion is not negligible, the power delivered to the load at the fundamental frequency is given by

$$P_1 = \frac{B_1^{\,2} R_L}{2} \longrightarrow (1)$$

Total *ac* power output is

$$P_{ac} = \frac{\left(B_1^2 + B_2^2 + B_3^2 + \ldots\ldots\right) R_L}{2} \longrightarrow (2)$$

$$= \left(1 + D_2^2 + D_3^2 + \ldots\ldots\right) P_1 \longrightarrow (3)$$

where $D_2$, $D_3$, etc., are the second, third, etc., harmonic distortions. Hence

$$\mathbf{P_{ac} = (1 + D^2)\, P_1} \longrightarrow (4)$$

where $D$ is the **total distortion factor** and is given by

$$D = \sqrt{D_2^2 + D_3^2 + D_4^2 + \ldots\ldots\ldots\ldots} \longrightarrow (5)$$

If $D$ = 10 % of the fundamental, then

$$P_{ac} = [1 + (0.1)^2]\,P_1$$

$$\boxed{P_{ac} = 1.01\,P_1} \qquad \longrightarrow (6)$$

When total distortion is 10 %, the power output is only 1% higher than the fundamental power. Thus, only a small error is made in using only the fundamental term $P_1$ for calculating the output power.

## THE TRANSFORMER-COUPLED AUDIO POWER AMPLIFIER

The main reason for the poor efficiency of a direct-coupled class A amplifier is the large amount of *dc* power that the resistive load in collector dissipates. This problem (distortion) can be solved by using a transformer for coupling the load (say, a speaker) to the amplifier stage as shown in Fig. 7.6. Transformer coupling permits impedance matching and also it keeps the *dc* power loss small because of the small resistance of the transformer primary winding.

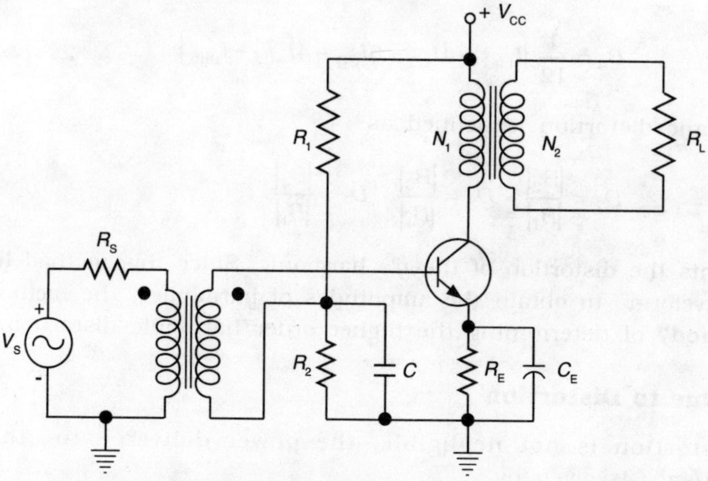


**Fig. 7.6 Transformer-coupled transistor amplifier**

### Transformer Impedance Matching

Assume that the transformer is ideal and there are no losses in the transformer. The resistance seen looking into the primary of the transformer is related to the resistance connected across the secondary. The impedance matching properties follow from the simple transformer relations:

$$V_1 = \frac{N_1}{N_2}V_2 \quad \text{and} \quad I_1 = \frac{N_2}{N_1}I_2 \qquad \longrightarrow (1)$$

where $V_1$ = primary voltage; $V_2$ = secondary voltage

$I_1$ = primary current; $I_2$ = secondary current

$N_1$ = number of primary turns and $N_2$ = number of secondary turns.

It may be noted that if $N_2 < N_1$, then the above equations show that the transformer reduces the voltage in proportion to the turns ratio $n = N_2/N_1$ and steps up the current in the same ratio. The ratio of these two equations gives

$$\frac{V_1}{I_1} = \frac{(N_1/N_2)V_2}{(N_2/N_1)I_2} = \left(\frac{N_1}{N_2}\right)^2 \frac{V_2}{I_2}$$

$$\frac{V_1}{I_1} = \left(\frac{1}{n^2}\right)\frac{V_2}{I_2} \longrightarrow (2)$$

Since $V_1/I_1$ represents the effective input resistance $R_L'$, whereas $V_2/I_2$ represents output resistance $R_L$, then

$$\boldsymbol{R_L' = \frac{1}{n^2} R_L = \left(\frac{N_1}{N_2}\right)^2 R_L} \longrightarrow (3)$$

Note that in an ideal transformer, there is no primary drop. As a result, all the supply voltage $V_{CC}$ appears as the collector-to-emitter voltage $V_{CE}$ of the transistor i.e., $V_{CC} = V_{CEQ}$. Now all the power supplied by the *dc* supply $V_{CC}$ is delivered to the transistor. Hence, the overall and collector efficiencies become equal i.e.,

$$\eta_{overall} = \frac{P_{out\,(ac)}}{V_{CC} I_{CQ}} = \frac{P_{out\,(ac)}}{V_{CEQ} I_{CQ}} = \eta_{collector} \longrightarrow (4)$$

The maximum value of overall or collector efficiency of a transformer-coupled class-A transformer is 50 % (proved in the next section).

**Maximum Power Output**

The *ac*-power output and efficiency of transformer-coupled amplifier may be calculated by plotting the *dc* and *ac* load lines on the family of output (collector) characteristics of the transistor. Since the *ac* and the *dc* load resistances of a transformer-coupled amplifier are different, it is essential to construct two separate load lines.

**DC Load Line:** The transformer *dc* (winding) resistance determines the *dc* load line for the circuit. Typically, the *dc* resistance is very small. Thus, the *dc* load line passes through the point ($V_{CEQ} = V_{CC}$, 0) on the collector voltage axis and is almost a straight (vertical) load line. This is the ideal load line for the transformer. Practical transformer windings would provide a slight slope for the load line. Since there is no voltage drop across the *dc* load resistance,

$$V_{CC} = V_{CEQ} \longrightarrow (5)$$

When the values of the resistance $R_B$ $(= R_1 \| R_2)$ and the collector supply voltage $V_{CC}$ are known, the base current at the operating point may be calculated by the equation,

$$I_B = \frac{V_{CC} - V_{BE}}{R_B} \approx \frac{V_{CC}}{R_B} \longrightarrow (6)$$

**Quiescent Operating Point:** The operating point is obtained graphically at the point of intersection of the *dc* load line and the transistor base current curve.

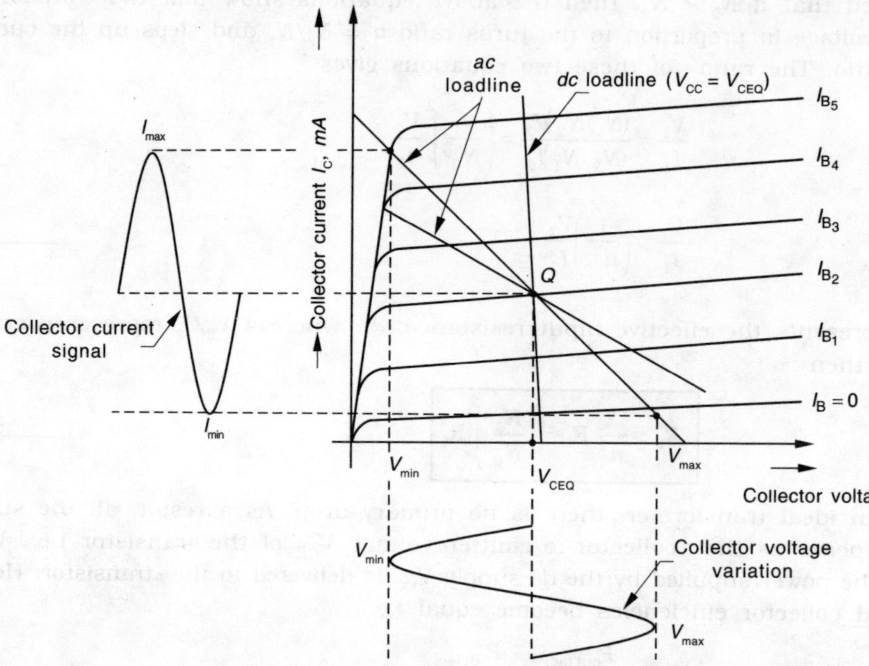


**Fig. 7.7 Collector characteristics of a power transistor showing the *dc* and the *ac* load lines**

After the operating point is determined the next step is to construct the *ac* load line passing through this point.

## AC Load Line

In order to draw the *ac* load line first calculate the *ac* load resistance seen looking into the primary side of the transformer. The effective load resistance is calculated using Eq. (3) from the values of the secondary load resistance and transformer turns ratio. Having obtained the value of $R_L'$ the *ac* load line must be drawn so that it passes through the operating point $Q$ and has a slope equal to $-1/R_L'$, the load-line slope being the negative reciprocal of the *ac* load resistance. The *dc* and the *ac* load lines along the operating point $Q$ are shown in Fig. 7.7. Two *ac* load lines are drawn through $Q$ for different values of $R_L'$.

For $R_L'$ very small, the voltage swing, and hence the output power *P*, approaches zero. For $R_L'$ very large, the current swing is small , and again *P* approaches zero. Therefore, in Fig. 7.8 the plot of *P* versus $R_L'$ has a maximum. Note also that this maximum is quite broad.

## EFFICIENCY

Consider that the stage is supplying power to a pure resistance load. Then the average power input from the *dc* supply is $V_{CC}I_C$. The power absorbed by the output circuit is $I_C^2R_1 + I_cV_{ce}$, where $I_c$ and $V_{ce}$ are the rms output current and voltage respectively and $R_1$ is the static load resistance. If $P_D$ is the average power dissipated by the active device

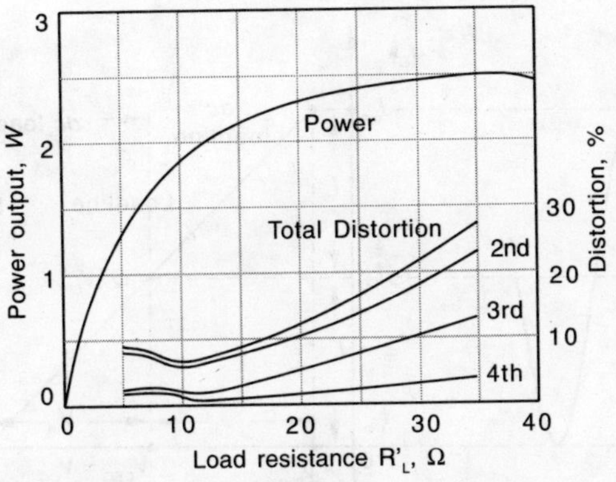


**Fig. 7.8 Output power and distortion as a function of load resistance**

then, in accordance with the principle of the conservation of energy,

$$V_{CC}I_C = I_C^2R_1 + I_cV_{ce} + P_D \longrightarrow (1)$$

Since $V_{CC} = V_{CEQ} + I_CR_1$, $P_D$ may be written in the form

$$P_D = V_{CEQ}\,I_C - V_{ce}I_c \longrightarrow (2)$$

If the load is not pure resistance, then $V_{ce}I_c$ must be replaced by $V_{ce}I_c$ cosϕ, where cosϕ is the power factor of the load.

The above equation expresses the amount of power that must be dissipated by the active device. It represents the kinetic energy of the electrons which is converted into heat upon the bombardment of the collector by these electrons. If the *ac* output power is zero, i.e., if no applied signal exists, then

$$P_D = V_{CEQ}\,I_C \longrightarrow (3)$$

Hence a device is cooler when delivering power to a load than when there is no such *ac* power transfer.

**Conversion Efficiency: Conversion efficiency is the measure of ability of an active device to convert the *dc* power of the supply into the *ac* (signal) power delivered to the load.**

This figure of merit is also called the **collector-circuit efficiency** for a transistor and is denoted by η.

Percentage efficiency, $$\eta = \frac{ac \text{ output power}}{dc \text{ power input}} \times 100\,\% \longrightarrow (4)$$

In general, $$\eta = \frac{(1/2)B_1^2R'_L}{V_{CC}(I_C + B_o)} \times 100\,\% \longrightarrow (5)$$

If the distortion components are neglected, then

$$\eta = \frac{(1/2)V_mI_m}{V_{CC}I_C} \times 100\,\%$$

$$= 50\frac{V_mI_m}{V_{CC}I_C} \longrightarrow (6)$$

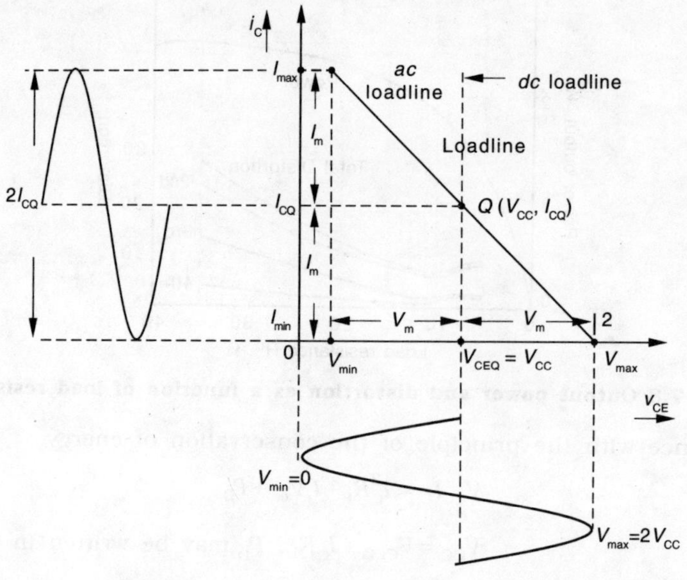


**Fig. 7.9 Maximum voltage and current**

## Maximum Value of Efficiency

An approximate expression for efficiency $\eta$ can be obtained by assuming ideal characteristic curves. Referring to Fig. 7.9, the maximum value of the sine wave output voltage is

$$V_m = \frac{V_{max} - V_{min}}{2} \qquad \longrightarrow (7)$$

and

$$I_m = I_{CQ} \qquad \longrightarrow (8)$$

The rms value of *ac* collector voltage is given by

$$V_{rms} = \frac{V_m}{\sqrt{2}} = \frac{V_{max} - V_{min}}{2\sqrt{2}} \qquad \longrightarrow (9)$$

Similarly,

$$I_{rms} = \frac{I_{max} - I_{min}}{2\sqrt{2}} \qquad \longrightarrow (10)$$

The *ac* output power is given by

$$P_{ac} = V_{rms} \cdot I_{rms} = \frac{(V_{max} - V_{min})}{2\sqrt{2}} \cdot \frac{(I_{max} - I_{min})}{2\sqrt{2}}$$

$$P_{ac} = \frac{(V_{max} - V_{min}) \cdot (I_{max} - I_{min})}{8} \qquad \longrightarrow (11)$$

The power supplied to the circuit is given by

$$P_{dc} = V_{CC} \cdot I_{CQ}$$

∴ Collector efficiency, $\eta = \frac{P_{ac}}{P_{dc}} = \frac{(V_{max} - V_{min}) \cdot (I_{max} - I_{min})}{8V_{CC}I_{CQ}} \times 100\,\%$ $\longrightarrow$ (12)

The efficiency of a transformer-coupled class A amplifier can also be expressed by

$$\eta = 50\left(\frac{V_{max} - V_{min}}{V_{max} + V_{min}}\right)\% \longrightarrow (13)$$

The maximum efficiency will be obtained when $V_{min} = 0$, $I_{min} = 0$, $V_{max} = 2\,V_{CC}$ and $I_{max} = 2\,I_{CQ}$. Substituting these values in Eq. (12), we get

$$\eta_{max} = \frac{2V_{CC} \cdot 2I_{CQ}}{8V_{CC} \cdot I_{CQ}} \times 100\,\% = 50\,\% \longrightarrow (14)$$

Hence, class-A power has a maximum efficiency of 50 % i.e., maximum of 50 % of *dc* supply power is converted into *ac* power output. In practice, the efficiency of class-A power amplifier is less than 50 % (about 35%) due to power loss in the primary of transformer. Note that this is double that of case with resistive load.

**Disadvantages**

1. The total harmonic distortion is very high.
2. Maximum output power is developed at a load resistance that is different from the value for maximum harmonic distortion.
3. The output transformer is subject to saturation problems due to the *dc* current in the primary.

## PUSH-PULL AMPLIFIERS

The distortion introduced by the nonlinearity of the dynamic transfer characteristic may be eliminated by the circuit known as a **push-pull configuration**. It employs two active devices and requires two input signals 180º out of phase with each other.

Figure 7.10 shows a transformer coupled class-B push-pull amplifier. The circuit consists of two centre-tapped transformers $T_1$ and $T_2$ and two identical transistors $Q_1$ and $Q_2$. The input transformer $T_1$ is called phase splitter. It provides opposite polarity signals to the transistor inputs. These two signal voltages, with opposite polarity drive the input of transistors $Q_1$ and $Q_2$. The output transformer $T_2$ is required to couple the *ac* output signal from the collector to the load.

On application of a sinusoidal signal one transistor amplifies the positive half-cycle of the sinusoidal signal and the other amplifies the negative half-cycle of the same signal. When a transistor is operated as class-B amplifier, the bias point should be fixed at cut-off so that practically no base current flows without an applied signal.

Consider an input signal (base current) of the form $i_{b_1} = I_{bm} \cos \omega t$ applied to transistor $Q_1$. The output current of transistor $Q_1$ is given by

$$i_1 = I_C + B_0 + B_1 \cos \omega t + B_2 \cos \omega t + B_3 \cos \omega t + \ldots\ldots \longrightarrow (1)$$

The input signal to transistor $Q_2$ is

$$i_{b_2} = -i_{b_1} = I_{bm} \cos(\omega t + \pi)$$

The output current of transistor $Q_2$ is then given by

$$i_2 = I_C + B_0 + B_1 \cos(\omega t + \pi) + B_2 \cos 2(\omega t + \pi) + \ldots\ldots \longrightarrow (2)$$

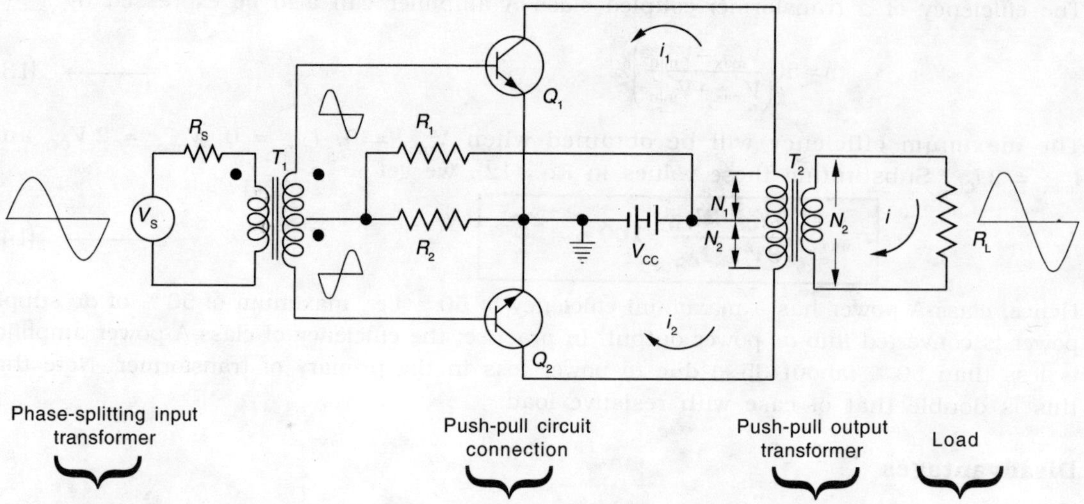


**Fig. 7.10 Transformer-coupled class-B push-pull amplifier**

or $$i_2 = I_C + B_0 - B_1 \cos \omega t + B_2 \cos 2\omega t - B_3 \cos 3\omega t + \text{- - - - - - - - -} \longrightarrow (3)$$

These two output currents $i_1$ and $i_2$ flow in opposite directions through the output-transformer $T_2$ primary windings, shown in the figure. The total output current is then proportional to the difference between the collector currents in the two transistors.

Total output current, $i = k(i_1 - i_2)$

where $k$ is some constant of proportionality.

$$i = 2k\ (B_1 \cos \omega t + B_3 \cos 3\omega t + \text{- - - - - - - - - -}) \longrightarrow (4)$$

The above equation shows that the output of a push-pull circuit will balance out even harmonics such as $B_0$, $B_2$, etc., and contains only the third harmonic term $B_3$ as the principal source of distortion.

Since the output current contains no even-harmonic terms the push-pull system possesses "half-wave" or "mirror" symmetry, in addition to the zero-axis symmetry. Half-wave symmetry requires that the bottom loop of the wave, when shifted 180º along the axis, will be a mirror image of the top loop. The condition of mirror symmetry is represented mathematically by the relation,

$$i(\omega t) = -i(\omega t + \pi) \longrightarrow (5)$$

Since the percent third harmonic distortion is given by

$$D_3 = \frac{2B_3}{2B_1} \times 100\ \% = \frac{B_3}{B_1} \times 100\ \% \longrightarrow (6)$$

Total harmonic distortion, $D = \sqrt{D_3^2 + D_5^2} \text{- - - - - -} \longrightarrow (7)$

### Advantages

A push-pull amplifier possesses many advantages over a single-ended power amplifier. The following are the advantages of the push-pull amplifier:

1. The *dc* components of the collector currents oppose each other in the transformer. As a result, there is no possibility of core saturation of the transformer, even at the peak value of the signals. Thus we can use smaller sized cores in the transformers, without affecting the circuit performance.
2. The use of push-pull system in the class-B amplifier eliminates even order harmonics in the output of a push-pull amplifier. This reduces the harmonic distortion.
3. Because of the absence of even harmonics, the circuit gives more output per active device under given distortion.
4. The effect of ripple voltages that may be contained in the power supply because of inadequate filtering will be balanced out. This cancellation results because the currents produced by the ripple voltages are in opposite directions in the transformer winding, and so will not appear in the load.
5. Due to the transformer, impedance matching is possible.
6. The efficiency is much higher than the class-A amplifier.

## CLASS-B AMPLIFIERS

When an amplifier is biased at cutoff so that it operates in the linear region for 180º of the input cycle and is in cutoff for 180º, it is a class-B amplifier. The transistor circuit of Fig. 7.10 operates as class-B if $R_2 = 0$ because a silicon transistor is essentially at cutoff if the base is shorted to the emitter.

### Advantages of Class-B over the Class-A Amplifier

1. In class-B operation, the transistor dissipation is zero under no-signal condition, and it increases with increased input signals. It is possible to obtain greater output power.
2. Since the average current in class-B operation is less than that in class-A, the amount of power dissipated by the transistor is also less. Maximum theoretical efficiency is 78.5% to class-B operation.

### Disadvantages

1. Harmonic distortion is higher.
2. Self-bias cannot be used.
3. Supply voltages must have good regulation.

### Power Considerations

The graphical construction for determining the output waveforms of a single class-B transistor stage is shown in Fig. 7.11. We know that the *ac* output power in a class-B amplifier is developed only during one-half cycle of the input signal. The effective load resistance is given by

$$R'_L = \left(\frac{N_1}{N_2}\right)^2 R_L \qquad \longrightarrow (1)$$

The waveforms illustrated in Fig. 7.11 represent one transistor $Q_1$ only. The output of transistor $Q_2$ is a series of sine loop pulses that are 180° out of phase with those of transistor $Q_1$. The load current, which is proportional to the difference between the two collector currents, is therefore a perfect sine wave. The output power is given by

$$P_o = \frac{V_m I_m}{2} = \frac{I_m}{2}(V_{cc} - V_{min}) \qquad \longrightarrow (2)$$

From the figure $\quad V_m = V_{CC} - V_{min}$

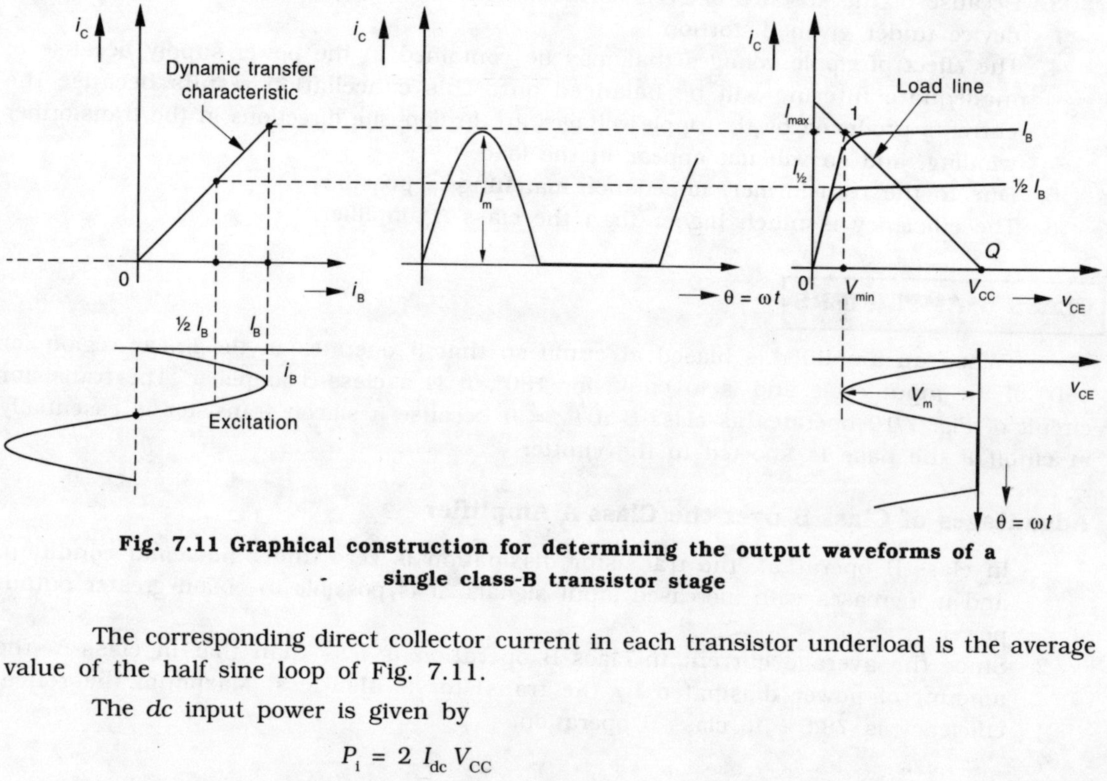


**Fig. 7.11 Graphical construction for determining the output waveforms of a single class-B transistor stage**

The corresponding direct collector current in each transistor underload is the average value of the half sine loop of Fig. 7.11.

The *dc* input power is given by

$$P_i = 2\, I_{dc}\, V_{CC}$$

Since $\quad I_{dc} = \dfrac{I_m}{\pi},\ P_i = \dfrac{2\, I_m\, V_{cc}}{\pi} \qquad \longrightarrow (3)$

The factor 2 in the above expression appears because two transistors are used in the push-pull system.

Collector-circuit efficiency, $\eta = \dfrac{P_o}{P_i} \times 100 = \dfrac{I_m (V_{cc} - V_{min})/2}{2 I_m V_{cc}/\pi} \times 100$

$$= \frac{\pi}{4}\left(\frac{V_{cc} - V_{min}}{V_{cc}}\right) \times 100 = \frac{\pi}{4}\frac{V_m}{V_{cc}} \times 100$$

or
$$\eta = \frac{\pi}{4}\left(1 - \frac{V_{min}}{V_{cc}}\right) \times 100\,\%$$

Since $V_{min} << V_{cc}$,
$$\eta = \frac{\pi}{4} \times 100\% = \mathbf{78.5\%} \qquad \longrightarrow (4)$$

This large value of $\eta$ results from the fact that there is no current in a class-B system if there is no excitation, whereas there is a drain from the power supply in a class-A system even at zero signal.

## Power Dissipation

The power dissipation $P_C$ in both transistors is the difference between the *ac* power output and *dc* power input.

$$P_C = P_{dc} - P_{ac} = P_i - P_o$$

$$= \frac{2}{\pi} V_{cc}\, I_m - \frac{V_m I_m}{2}$$

$$P_C = \frac{2}{\pi} V_{cc} \frac{V_m}{R'_L} - \frac{V_m^2}{2R'_L} \qquad \longrightarrow (5)$$

This equation shows that the collector dissipation is zero at no signal ($V_m = 0$), rises as $V_m$ increases and passes through a maximum at $V_m = 2\,V_{cc}/\pi$.

**Maximum Power Dissipation:** The condition for maximum power dissipation can be found by differentiating Eq. (5) with respect to $V_m$ and equating it to zero.

$$\frac{d\,P_C}{dV_m} = \frac{2}{\pi}\frac{V_{cc}}{R'_L} - \frac{2V_m}{2R'_L} = 0$$

or
$$\frac{V_m}{R'_L} = \frac{2}{\pi}\frac{V_{cc}}{R'_L}$$

or
$$V_m = \frac{2}{\pi} V_{cc} \qquad \longrightarrow (6)$$

Substituting the value of $V_m$ in Eq. (5), we get

$$P_{C,\,max} = \frac{2}{\pi}\frac{V_{cc}}{R'_L} \cdot \left(\frac{2}{\pi} V_{cc}\right) - \left(\frac{2}{\pi} V_{cc}\right)^2 \times \frac{1}{2R'_L}$$

$$= \frac{4\,V_{cc}^2}{\pi^2 R'_L} - \frac{2\,V_{cc}^2}{\pi^2 R'_L}$$

$$P_{C,\,max} = \frac{2}{\pi^2}\frac{V_{cc}^2}{R'_L} \qquad \longrightarrow (7)$$

For maximum efficiency, $V_m = V_{CC}$, hence the power dissipation is not maximum when the efficiency is maximum.

Now $$P_o = \frac{V_m^2}{2R'_L}$$

When $V_m = V_{CC}$, $$P_{o,\,max} = \frac{V_{cc}^2}{2R'_L} \longrightarrow (8)$$

Equation (7) can be written as

$$P_{C,\,max} = \frac{4}{\pi^2}\left(\frac{V_{CC}^2}{2R'_L}\right)$$

$$= \frac{4}{\pi^2} \cdot P_{o,\,max}$$

∴ $$\boldsymbol{P_{C,\,max} = \frac{4}{\pi^2} \cdot P_{o,\,max} \approx 0.4\ P_{o,\,max}} \longrightarrow (9)$$

The above equation shows that the maximum power is dissipated by both the transistors and therefore the maximum power dissipation per transistor is $(P_{C,\,max})/2$.

∴ $$P_{C,\,max} \text{ per transistor} = \frac{4}{\pi^2} P_{o,\,max} \Big/ 2$$

$$= \frac{2}{\pi^2} P_{o,\,max} = 0.2\ P_{o,\,max} \longrightarrow (10)$$

If, for example, 10 *W* maximum power is to be delivered from a class-B push-pull amplifier to the load, then power dissipation rating of each transistor should be 0.2 × 10 *W* = 2 *W*.

### Harmonic Distortion

The output of a push-pull system always possesses mirror symmetry, so that $I_C = 0$, $I_{max} = -I_{min}$ and $I_{½} = -I_{-½}$.

We know that $$B_0 = \frac{1}{6}\left[I_{max} + 2\,I_{½} + 2I_{-½} + I_{min}\right] - I_C$$

$$B_1 = \frac{1}{3}\left[I_{max} + I_{½} - I_{-½} - I_{min}\right]$$

$$B_2 = \frac{1}{4}\left[I_{max} - 2I_C + I_{min}\right]$$

$$B_3 = \frac{1}{6}\left[I_{max} - 2I_{½} + 2I_{-½} - I_{min}\right]$$

$$B_4 = \frac{1}{12}\left[I_{max} - 4I_{½} + 6I_C - 4I_{-½} + I_{min}\right]$$

When $I_C = 0$, $I_{max} = -I_{min}$ and $I_{½} = -I_{-½}$, the above equations reduce to

$$B_o = B_2 = B_4 = 0 \longrightarrow (11)$$

$$B_1 = \frac{2}{3}\left(I_{max} + I_{1/2}\right) \quad \longrightarrow (12)$$

$$B_3 = \frac{1}{3}\left(I_{max} - 2I_{1/2}\right) \quad \longrightarrow (13)$$

Note that there is no even-harmonic distortion. The principal contribution to distortion is the third harmonic and is given by

$$D_3 = \frac{|B_3|}{|B_1|} \times 100\,\% \quad \longrightarrow (14)$$

The values of $I_{max}$ and $I_{1/2}$ are found as follows: A load line corresponding to $R'_L = (N_1/N_2)^2 R_L$ is drawn on the collector characteristics through the point $I_C = 0$ and $V_{CE} = V_{CC}$. If the peak base current is $I_B$, then the intersection of the load line with the $I_B$ curve is $I_{max}$ and with the $I_B/2$ characteristics is $I_{1/2}$ as shown in Fig. 7.11. The output power taking distortion into account is given by

$$P_0 = (1 + D_3)^2 \; \frac{B_1^2 R'_L}{2} \quad \longrightarrow (15)$$

## Other Class-B Push-pull Amplifiers

A circuit that dispenses with the output transformer is shown in Fig. 7.12. This arrangement requires a power supply whose centre tap is grounded. In this arrangement high powered transistors are used. They have a collector-to-emitter output impedance in the order of 4 Ω to 8 Ω. This allows the "single-ended push-pull" operation, in which the collector current of each transistor flows through the load directly. The voltage developed across the load is again due to the difference in collector currents, $i_1 - i_2$, so this is a true push-pull application.

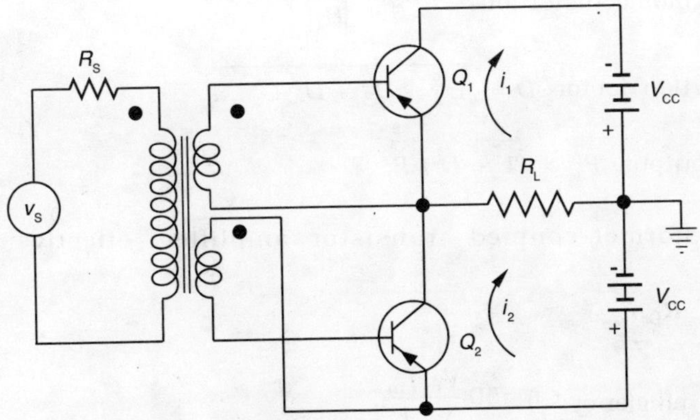


**Fig. 7.12 Circuit of a class-B push-pull amplifier that does not require an output transformer**

## IMPORTANT POINTS TO REMEMBER

1. A large-signal amplifier must operate efficiently and be capable of handling large amounts of power.
2. Large-signal amplifiers are called power amplifiers.

3. Large-signal amplifiers are classified according to the class of operation, which is decided by the location of the quiescent point on the device characteristics.
4. The $Q$ point must be centered on the load line for maximum class-A output signal swing.
5. In a class-A amplifier utilizing a direct coupled resistive load, the maximum conversion efficiency is 25 percent.
6. When a signal is applied to a class-A amplifier, the power dissipated in the transistor decreases by an amount equal to the *ac* power delivered to the load.
7. A transformer-coupled class A amplifier has a maximum theoretical conversion efficiency of 50 %, but its maximum *ac* power delivering capability is still one half of its maximum dissipation capability.
8. Conversion efficiency is defined as the measure of ability of an active device to convert the *dc* power of the supply into the *ac* power delivered to the load.
9. The operation of a push-pull amplifier is such as to cause a cancelling of all the even harmonics without a change in the fundamental and odd harmonics, so distortion is greatly reduced.
10. A class-B amplifier operates in the linear region for half of the input cycle (180º); and it is in cutoff for the other half.
11. Class-B amplifiers are normally operated in a push-pull configuration in order to produce an output that is a replica of the input.
12. The maximum efficiency of a class-B amplifier is 78.5 percent.

## KEY FORMULAE

1. In class A amplifier, power, $P = \dfrac{(V_{max} - V_{min})(I_{max} - I_{min})}{8}$

2. Second harmonic distortion, $D_2 = \dfrac{|B_2|}{|B_1|} \times 100\,\%$

3. Total distortion factor $D = \sqrt{D_2^2 + D_3^2 + D_4^2 + \ldots\ldots}$

4. AC power output, $P_{ac} = (1 + D^2)\, P_1$

5. In a transformer-coupled transistor amplifier, effective input resistance,

$$R'_L = \left(\frac{N_1}{N_2}\right)^2 R_L$$

6. Conversion efficiency, $\eta = 50 \dfrac{V_m I_m}{V_{cc} I_c}$

7. Maximum value of efficiency, $\eta_{max} = \dfrac{(V_{max} - V_{min})(I_{max} - I_{min})}{8\, V_{CC}\, I_{CQ}} \times 100\,\%$

8. Efficiency of a transformer-coupled class-A amplifier is $\eta = 50 \left(\dfrac{V_{max} - V_{min}}{V_{max} + V_{min}}\right) \%$

9. Collector efficiency of class-B amplifier is $\eta = \frac{\pi}{4}\frac{V_m}{V_{cc}} \times 100\,\%$

10. Maximum power dissipation, $P_{c\,max} = \frac{4}{\pi^2} P_{o,\,max}$

## SOLVED PROBLEMS

**1. Calculate the input power, output power, and efficiency of the amplifier in Fig. 7.13 for an input voltage resulting in a base current of 10 *mA* peak. Also calculate the power dissipated by the transistor.**

**Data:** $V_{cc} = 20\ V,\ R_B = 1\ k\Omega,\ \beta = 25,\ I_{B\,(max)} = 10\ mA$

**To find:** Input power $P_i$, output-power $P_o$, efficiency $\eta$ and $P_C$.

**Solution:** The *Q*-point for the circuit of Fig. 7.13 is determined as follows:

$$I_B = \frac{V_{cc} - V_{BE}}{R_B} = \frac{20 - 0.7}{1 \times 10^3} = 19.3\ mA$$

$$I_C = \beta I_B = 25 \times 19.3\ mA = 482.5\ mA$$

$$V_{CE} = V_{CC} - I_C R_C = 20 - 482.5 \times 10^{-3} \times 20 = 10.35\ V$$

$$I_{C\,(peak)} = \beta I_{B\,(peak)} = 25 \times (10\ mA\ \text{peak}) = 250\ mA\ \text{peak}$$

$$P_{ac} = \left(\frac{I_{C\,peak}}{\sqrt{2}}\right)^2 R_c = \left(\frac{250 \times 10^{-3}}{\sqrt{2}}\right)^2 \times 20 = \mathbf{0.625\ W}$$

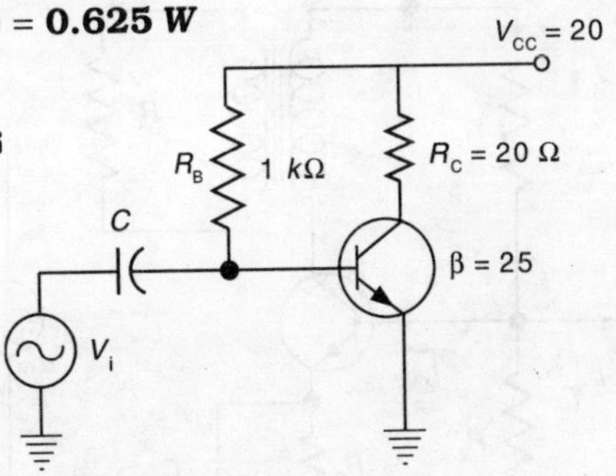


Fig. 7.13

Input power, $P_{ac} = V_{CC} I_C = 20 \times 482.5 \times 10^{-3} = \mathbf{9.6}$

Efficiency, $\eta = \frac{P_{ac}}{P_{dc}} = \frac{0.625}{9.65} \times 100\ \% = \mathbf{6.48\ \%}$

Power dissipated by the transistor is then

$$P_C = P_{dc} - P_{ac}$$

$$= 9.65\ W - 0.625\ W = \mathbf{9.025\ W}$$

**2. Calculate the effective resistance $R'_L$ seen looking into the primary of a 15:1 transformer connected to an output load of 8 Ω.**

**Data:** $N_1 : N_2 = 15 : 1$ and $R_L = 8\ \Omega$

**To find:** Effective resistance, $R'_L$

**Solution:** Effective resistance is given by

$$R'_L = \left(\frac{N_1}{N_2}\right)^2 R_L = (15)^2 \times 8 = \mathbf{1.8\ k\Omega}$$

**3. A class-A power amplifier with a direct coupled load has a collector efficiency of 15 % and delivers a power input of 5 *W*. Find (a) the *dc* power input (b) the power dissipation of full output and (c) the desirable power dissipation rating for the BJT.**

**Data:** $P_{ac} = 5\ W$, $\eta = 15\ \% = 0.15$

**To find:** The *dc* power input $P_{dc}$, $P_C$ at full output, and $P_C$ at no input

**Solution:** (a) We know that $$\eta = \frac{P_{ac}}{P_{dc}}$$

$$\therefore \quad P_{dc} = \frac{P_{ac}}{\eta} = \frac{5}{0.15} = \mathbf{33.33\ W}$$

(b) Dissipation at full output $\quad P_C = 33.33 - 5 = \mathbf{28.33\ W}$

(c) Dissipation at no output, $\quad P'_c = P_{dc} = \mathbf{33.33\ W}$

Hence, $\quad$ BJT rating $= \mathbf{33.33\ W}$

**4. For the class-A transformer-coupled power amplifier circuit given in Fig. 7.14, assume $R_1 = 2\ k\Omega$, $V_{cc} = 20\ V$, $R_2 = 200\ \Omega$, $R_L = 5\ \Omega$, $R_E = 10\ \Omega$ and $C_E = 100\ \mu F$. The turns ratio is 14:1. Transformer efficiency = 0.9. The *dc* primary resistance of the transformer is 10 Ω. The BJT has β = 20 and $V_{BE} = 0.5\ V$. Determine (a) the maximum power output to the load (b) the circuit efficiency (c) the dissipation of BJT under no signal conditions and (d) the operating point that yields maximum circuit efficiency.**

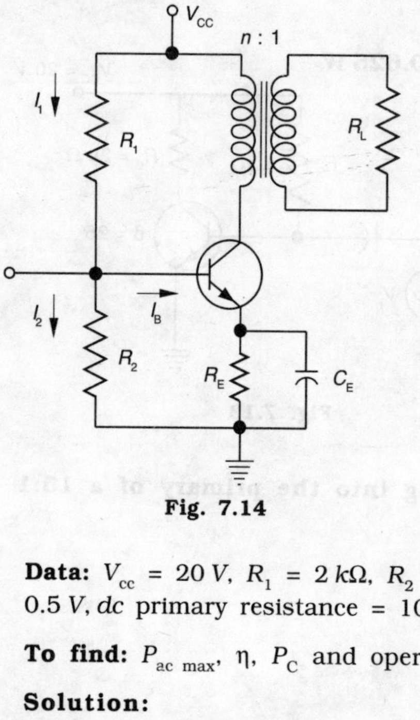


**Fig. 7.14**

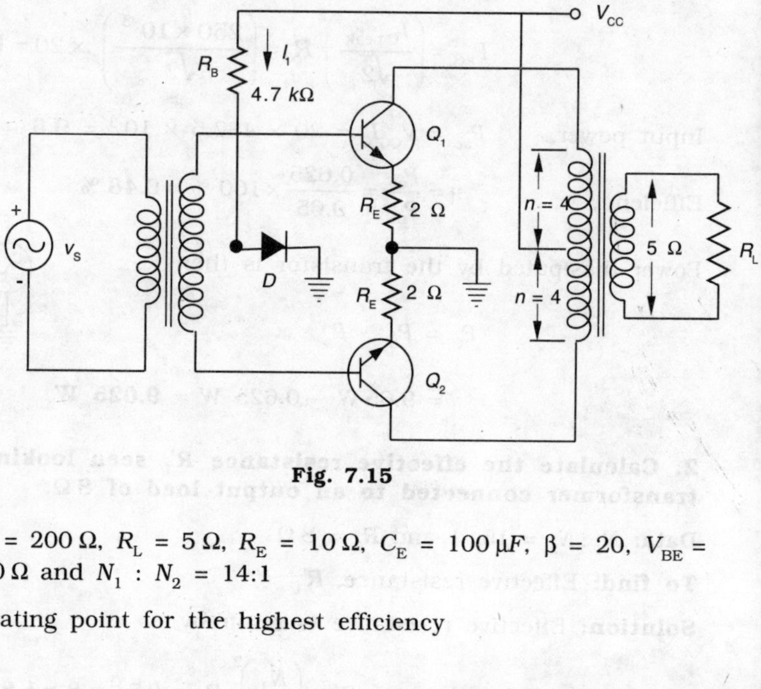


**Fig. 7.15**

**Data:** $V_{CC} = 20\ V$, $R_1 = 2\ k\Omega$, $R_2 = 200\ \Omega$, $R_L = 5\ \Omega$, $R_E = 10\ \Omega$, $C_E = 100\ \mu F$, $\beta = 20$, $V_{BE} = 0.5\ V$, *dc* primary resistance = 10 Ω and $N_1 : N_2 = 14:1$

**To find:** $P_{ac\ max}$, η, $P_C$ and operating point for the highest efficiency

**Solution:**

(a) For achieving the maximum power output, the *Q*-point should be so set as to allow an equal voltage swing on each side of this point.

$$I_{CQ} = \frac{V_{CQ}}{R_{ac}} \text{ (ac condition)}$$

$$V_{CQ} = V_{CC} - I_{CQ} \cdot R_{dc} \text{ (dc condition)}$$

Solving, we get $$I_{CQ} = \frac{V_{CC}}{R_{dc} + R_{ac}}$$

Maximum power that can be delivered to the load is

$$P_{ac\ max} = \eta_t \frac{I_{CQ}^2}{2} n^2 R_L$$

where $\eta_t$ = transformer efficiency.

In the circuit of Fig. 7.14,

$$I_{BQ} = I_1 - I_2 = \frac{I_{CQ}}{\beta}$$

$$I_1 = \frac{V_{CC} - V_B}{R_1}$$

$$I_2 = \frac{V_B}{R_2}$$

Hence $$\frac{I_{CQ}}{20} = \frac{20 - V_B}{2k\Omega} - \frac{V_B}{0.2k\Omega}$$

Also $$V_B = I_{EQ} R_E + V_{BE}$$
$$= 1.05\ I_{CQ}\ (0.01\ k\Omega) + 0.5$$

Solving the two relations simultaneously, we get

$$I_{CQ} = 67.3\ mA$$

$$P_{ac\ max} = \frac{0.9 \times (0.0673)^2 \times 80}{2} = \mathbf{0.163\ W}$$

(b) The power drawn from the source including the bias circuit power is

$$P_{dc} = V_{CC}\ (I_{CQ} + I_1)$$

and $$I_1 = \frac{V_{CC} - V_B}{R_1}$$

$$V_B = 1.05 \times 67.3 \times 10^{-3} \times 0.01 \times 10^3 + 0.5$$
$$= 1.2\ V$$

$\therefore$ $$I_1 = \frac{20 - 1.2}{2 \times 10^3} = \frac{18.8}{2k} = 9.4\ mA$$

Hence $$I_{CQ} + I_1 = 67.3\ mA + 9.4\ mA = 76.7\ mA$$

$$P_{dc} = 20\ (76.7\ mA) = \mathbf{1.534\ W}$$

Circuit efficiency, $$\eta = \frac{P_{ac}}{P_{dc}} \times 100 = \frac{0.163}{1.534} \times 100 = \mathbf{10.6\ \%}$$

(c) No signal dissipation is given by

$$P_D = V_{CQ} \cdot I_{CQ}$$

$$V_{CQ} = V_{CC} - I_{CQ} \cdot R_{primary} - I_{EQ} \cdot R_E$$
$$= 20 - 67.3 \times 0.01 - 70.7 \times 0.01$$
$$= 20 - 1.38 = 18.62\ V$$

∴ $$P_D = 18.62 \times 67.3 \times 10^{-3} = \mathbf{1.253\ W}$$

(d) The operating point for the highest efficiency is given by

$$I_{CQ} = \frac{V_{CC}}{R_{ac} + R_{dc}}$$

$$R_{ac} = n^2 R_L + R_{primary} = (4)^2 \times 5 + 10 = 90\ \Omega$$

$$R_{dc} = R_E + R_{primary} = 10 + 10 = 20\ \Omega$$

Hence $$I_{CQ} = \frac{20}{90 + 20} = \frac{20}{110} = \mathbf{0.182\ A}$$

**5. For the class-B power amplifier given in Fig. 7.15, assume the total *dc* primary resistance to be 20 Ω. Assume a transformer efficiency of 90 %. Estimate (a) the maximum output power, (b) the circuit efficiency under the condition of maximum power output, and (c) the maximum transistor dissipation that can occur in this circuit.**

**Data:** $V_{CC} = 30\ V$, $R_B = 4.7\ k\Omega$, $R_E = 2\ \Omega$, $R_L = 5\ \Omega$, $n = 4$, $R_{primary} = \frac{20\ \Omega}{2}$ and $\eta_t = 90\ \%$

**To find:** Maximum output power $P_{ac,\ max}$, η under $P_{ac,\ max}$ and $P_{D\ max}$

**Solution:** (a) When a transistor is conducting, the equivalent circuit of the stage will be as shown in Fig. 7.16.

The peak value of current is given by

$$I_m = \frac{V_{cc}}{R_{ac}} = \frac{V_{cc}}{n^2 R_L + R_{primary} + R_E}$$

$$= \frac{30}{16 \times 5 + 10 + 2} = \frac{30}{92} = 0.326\ A$$

The maximum power delivered to the load is

$$P_{ac,\ max} = \eta_t \frac{I_m^2}{2} n^2 R_L$$

$$= 0.9 \times \frac{(0.326)^2}{2} \times 4^2 \times 5 = \mathbf{3.826\ W}$$

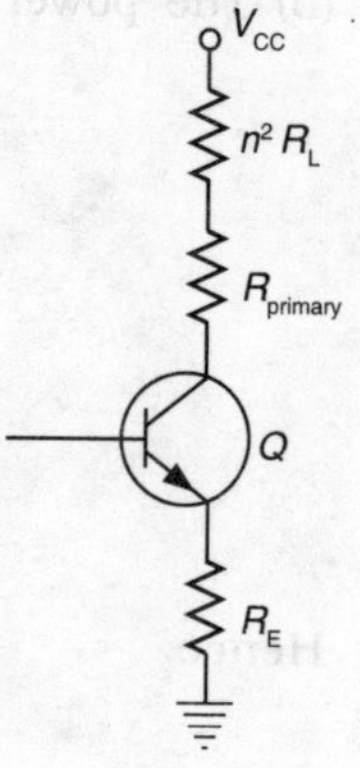

**Fig. 7.16**

(b) Under this condition, $$P_{dc} = V_{cc} \cdot \frac{2 I_m}{\pi}$$

$$= \frac{30 \times 2 \times 0.326}{3.142} = 6.23\ W$$

Bias circuit power $$= I_1 \times V_{cc} = V_{cc} \left( \frac{V_{cc} - V_D}{R_B} \right)$$

Let $V_D = 0.5\,V$, $I_1 V_{cc} = 30\left(\dfrac{30 - 0.5}{4.7k}\right) = 0.188W$

Total source power, $P_{source} = 6.23 + 0.188 = 6.42\ W$

Circuit efficiency, $\eta = \dfrac{P_{ac\,max}}{P_{source}} \times 100 = \dfrac{3.826}{6.42} \times 100$

$= \mathbf{59.6\ \%}$

(c) $P_{D\,max}$ per BJT $= \dfrac{0.1\,V_{cc}^2}{R_{ac}}$

$= \dfrac{0.1 \times (30)^2}{92} = \mathbf{0.978\ W}$

**6. A transistor supplies 0.85 *W* to a 4 *k*Ω load. The zero-signal *dc* collector current is 31 *mA*, and the *dc* collector current with signal is 34 *mA*. Determine the percent second harmonic distortion.**

**Data:** $P = 0.85\ W$, $R_L = 4\ k\Omega$, $I_{C,\ zero\text{-}signal} = 31\ mA$ and $I_{C,\ signal} = 34\ mA$

**To find:** Percent second-harmonic distortion, $D_2$

**Solution:** Using the dynamic characteristics of $i_c = G_1 i_b + G_2 i_b^2$, we have

$$B_o = I_{C\ signal} - I_{C\ no\ signal} = 34 - 31 = 3\ mA$$

We know that $B_o = B_2$

$$\text{Power, } P = \frac{B_1^2 R_L}{2} \quad \text{or} \quad B_1^2 = \frac{2P}{R_L}$$

$$\therefore \quad B_1^2 = \frac{2 \times 0.85}{4k} = 0.43 \times 10^{-3} = 430 \times 10^{-6}$$

or $B_1 = 20.6\ mA$

The second-harmonic distortion $D_2$ is given by

$$D_2 = \frac{|B_2|}{|B_1|} \times 100\ \% = \frac{3\ mA}{20.6\ mA} \times 100\ \%$$

$= \mathbf{14.6\ \%}$

**7. The *p-n-p* transistor whose input and output characteristics in Fig. 7.17 is used in the circuit of Fig. 7.6, with $R_s$ = 0 and $R_L' = (N_1/N_2)^2\ R_L$ = 10 Ω. The quiescent point is $I_C$ = –1.1 *A* and $V_{CE}$ = – 7.5 *V*. The peak-to-peak 2000 *Hz* sinusoidal base-to-emitter voltage is 140 *mV*.**

**(a) What is the fundamental current output? (b) What is the percent second, third- and fourth-harmonic distortion? (c) What is the output power? (d) What is the rectification component $B_o$ of the collector current? Neglect any changes in the operating point.**

**Data:** $R_S = 0$, $R_L' = 10\ \Omega$, $I_C = -1.1\ A$, $V_{CE} = -7.5\ V$, $V_{be} = 140\ mV$

**To find:** Fundamental current output $B_1$, $D_2$, $D_3$, $D_4$, $P_{ac}$ and $B_o$.

**Solution:** From Fig. 7.17, we have $I_{B\,max} = -30\,mA$, $I_{BQ} = -15\,mA$, $I_{B\,min} = -6.5\,mA$, $I_{B½} = -20.5\,mA$ $I_{B-½} = -10.5\,mA$, $I_{c\,max} = -1.77\,A$, $I_{C½} = -1.4\,A$, $I_{CQ} = -1.1\,A$, $I_{C-½} = -0.8\,A$, $I_{C\,min} = -0.53\,A$

(a) We know that $B_1 = \frac{1}{3}\left(I_{max} + I_{½} - I_{½} - I_{min}\right)$

$$= \frac{1}{3}\left[-1.77 - 1.4 + 0.8 + 0.53\right] = -\mathbf{0.613\,A}$$

(b) $B_2 = \frac{1}{4}\left[I_{max} - 2I_{CQ} + I_{min}\right]$

$$= \frac{1}{4}\left[-1.77 + 2\times 1.1 - 0.53\right] = -0.025\,A$$

$$B_3 = \frac{1}{6}\left[I_{max} - 2I_{½} + 2I_{-½} - I_{min}\right]$$

$$= \frac{1}{6}\left[-1.77 + 2\times 1.4 - 2\times 0.8 + 0.53\right] = -6.67\,mA$$

$$B_4 = \frac{1}{12}\left[I_{max} - 4I_{½} + 6I_{CQ} - 4I_{-½} + I_{min}\right]$$

$$= \frac{1}{12}\left[-1.77 + 4\times 1.4 - 6\times 1.1 + 4\times 0.8 - 0.53\right] = -8.32\,mA$$

Harmonic distortion, $D_2 = \frac{B_2}{B_1} = \frac{0.025}{0.613}\times 100\,\% = \mathbf{4.07\,\%}$

$$D_3 = \frac{B_3}{B_1} = \frac{6.67}{613}\times 100\,\% = \mathbf{1.09\,\%}$$

and $D_4 = \frac{B_4}{B_1} = \frac{8.32}{613}\times 100\,\% = \mathbf{1.36\,\%}$

(c) Output power, $P_o = (1 + D^2)\,P_1$ where $P_1 = \frac{B_1^2 R_L}{2}$

$\therefore$ $P_1 = \frac{(0.613)^2\times 10}{2} = 1.88W$

$$D = \sqrt{D_2^2 + D_3^2 + D_4^2}$$

$$D^2 = (4.07)^2 + (1.09)^2 + (1.36)^2 = 19.54\times 10^{-4} \approx 0.002$$

$\therefore$ $P_o = (1 + 0.002)\,1.88 = 1.89\,W$

(d) $B_o = \frac{1}{6}\left[I_{max} + 2I_{½} + 2I_{-½} + I_{min}\right] - I_{CQ}$

$$= \frac{1}{6}\left[-1.77 - 2\times 1.4 + 2\times 0.8 - 0.53\right] + 1.1 = \mathbf{-15\,mA}$$

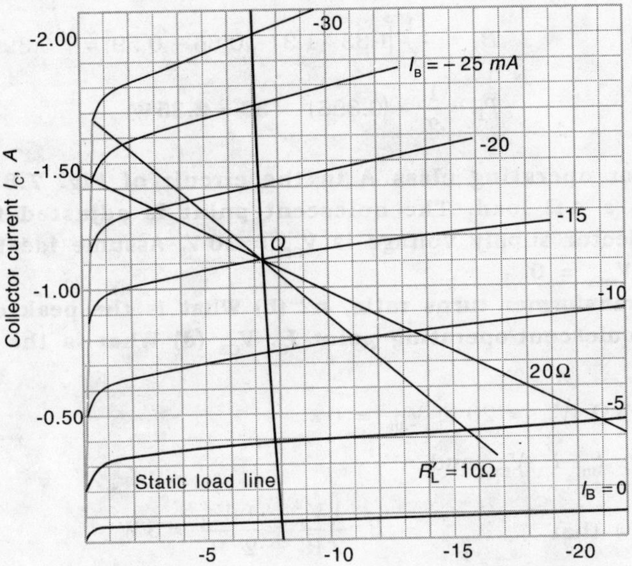


**Fig. 7.17 (a) The collector characteristics for a power transistor. A static load line for a transformer-load is indicated. Also shown are load lines for dynamic resistances of 10 and 20Ω.**

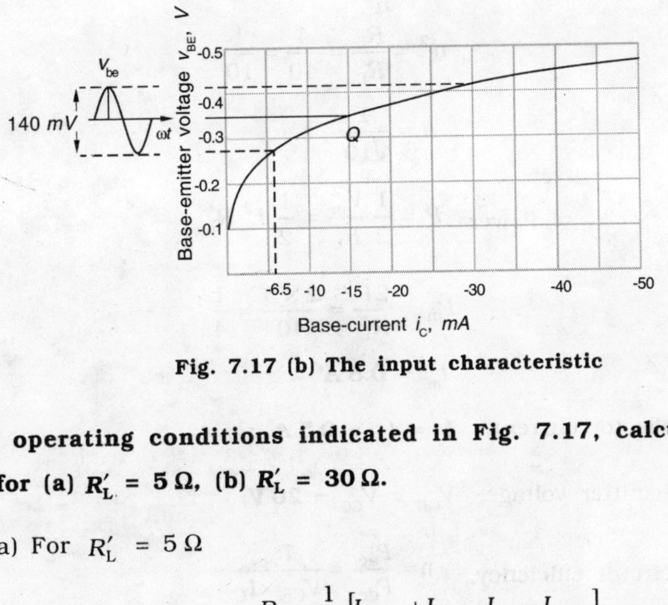


**Fig. 7.17 (b) The input characteristic**

**7. For the operating conditions indicated in Fig. 7.17, calculate the fundamental power, $P_1$ for (a) $R'_L$ = 5 Ω, (b) $R'_L$ = 30 Ω.**

**Solution:** (a) For $R'_L$ = 5 Ω

We know that
$$B_1 = \frac{1}{3}\left[I_{max} + I_{1/2} - I_{-1/2} - I_{min}\right]$$

$$B_1 = -\frac{1}{3}\left[1.92 + 1.43 - 0.78 - 0.51\right] = -\mathbf{0.686\,A}$$

Thus
$$P_1 = \frac{1}{2} B_1^2 R'_L = \frac{1}{2} \times (0.686)^2 \times 5 = \mathbf{1.18W}$$

(b) For $R'_L = 30\,\Omega$ $\quad B_1 = \frac{-1}{3}[1.33 + 1.31 - 0.86 - 0.59] = -0.396\,A$

$$P_1 = \frac{1}{2} \times (0.396)^2 \times 30 = \mathbf{2.35\,W}$$

**8. A power transistor operating class A in the circuit of Fig. 7.6 is to deliver a maximum of 5 *W* to a 4 Ω load. The quiescent point is adjusted for symmetrical clipping and the collector supply voltage is $V_{CC}$ = 20 V. Assume ideal characteristics, as in Fig. 7.9, with $V_{min}$ = 0.**

**(a) What is the transformer turns ratio, *n*? (b) What is the peak collector current $I_m$? (c) What is the quiescent-operating point $I_C$, $V_{CE}$ (d) What is the collector-circuit efficiency?**

**Data:** $P_C = 5\,W,\ R'_L = 4\,\Omega,\ V_{CC} = 20\,V,\ V_{min} = 0$

**To find:** Turns ratio $n$, $I_m$, $I_c$, $V_{CE}$ and $n$

**Solution:** (a) We know that
$$P = \frac{1}{2}\frac{V_m^2}{R'_L} = \frac{1}{2}\frac{V_{CC}^2}{R'_L} = 5\,W$$

∴
$$R'_L = \frac{V_{CC}^2}{2 \times 5} = \frac{(20)^2}{10} = 40\,\Omega$$

We also know that
$$R'_L = \frac{R_L}{n^2}$$

or
$$n^2 = \frac{R_L}{R'_L} = \frac{4}{40} = \frac{1}{10}$$

∴
$$n = \frac{\mathbf{1}}{\sqrt{\mathbf{10}}}$$

(b) Power
$$P = \frac{1}{2}\frac{V_m^2}{R'_L} = \frac{1}{2} I_m^2 R'_L$$

or
$$I_m^2 = \frac{2P}{R'_L} = \frac{2 \times 5}{40} = \frac{1}{4}$$

Thus
$$I_m = \mathbf{0.5\,A}$$

(c) Collector current, $I_C = I_m = \mathbf{0.5\,A}$

and Collector-emitter voltage, $V_{CE} = V_{CC} = \mathbf{20\,V}$

(d) Collector-circuit efficiency,
$$\eta = \frac{P_{ac}}{P_{dc}} = \frac{P_{ac}}{V_{CE} \times I_C}$$

$$\eta = \frac{5}{20 \times 0.5} = 0.5 = \mathbf{50\,\%}$$

**9. Design a class-B push-pull circuit to deliver 200 *mW* to a 4 Ω load. Output transformer efficiency is 70%, $V_{CE}$ = 25 V. Average rating of the transistor to be used is 165 *mW* at 25° C. Determine $V_{CC}$ collector to collector resistance $R_{C\text{-}C}$. Assuming $R_E$ = 10 Ω, find the voltage divider resistances $R_1$ and $R_2$. [Aug. 2000, VTU]**

**Data:** $P_{ac} = 200\,mW$, $R_L = 4\,\Omega$, $\eta = 0.7$, $V_{CE\,(max)} = 25\,V$, $P_{transistor} = 165\,mW$ at $25°\,C$, $R_E = 10\,\Omega$

**To find:** $V_{CC}$, $R_{C\text{-}C}$, and voltage divider resistances $R_1$ and $R_2$

**Solution:** Assume that the given power delivered to the load is maximum.

Then
$$P_{ac} = \frac{P_{ac\,max}}{\eta} = \frac{200}{0.7}$$
$$P_{ac} = \mathbf{285.714\ mW} \text{ on primary of transformer}$$

Maximum voltage rating per transistor is $2\,V_{CC}$.
$$V_{CE\,max} = 2\,V_{CC} = 25\,V$$
∴
$$V_{CC} = 12.5\,V$$
Let us choose
$$V_{CC} = 12\,V$$
Now
$$P_{ac\,primary} = \frac{1}{2}\frac{V_{CC}^2}{R_L'}$$
$$R_L' = \frac{1}{2}\frac{V_{CC}^2}{P_{ac\,primary}} = \frac{1}{2} \times \frac{(12)^2}{285.714 \times 10^{-3}}$$
$$R'_L = \mathbf{252\ \Omega}$$
We know that
$$R_L' = \frac{R_L}{n^2} = \left(\frac{N_1}{N_2}\right)^2 R_L$$
$$n^2 = \frac{R_L}{R_L'} = \frac{4}{252} = \mathbf{0.125}$$
∴
$$N_1/N_2 = 8$$
The turns ratio of output transformer is $2\,N_1 : N_2 = 16{:}1$

Collector-to-collector resistance, $R_{C\text{-}C} = \left(\frac{2\,N_1}{N_2}\right)^2 R_L$
$$= (16)^2 \times 4 = \mathbf{1024\ \Omega}$$

To find voltage divider resistances $R_1$ and $R_2$, consider the circuit as shown in Fig. 7.18.

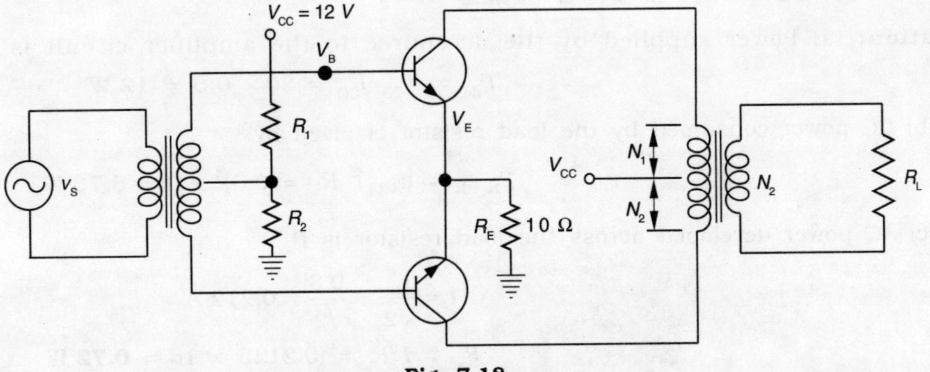


**Fig. 7.18**

$$I_m = \frac{V_m}{R'_L} = \frac{V_{CC}}{R'_L} = \frac{12}{252} = \mathbf{47.61\ mA}$$

From Fig. 7.18, $V_E = I_{dc} R_E = \frac{2I_m}{\pi} \times R_E$

$$= \frac{2 \times 47.61 \times 10^{-3}}{3.142} \times 10 = \mathbf{303.15\ mV}$$

Let $V_{BE} = 0.6\ V$, $V_{BE} = V_B - V_E$

$$0.6 = V_B - 303.15\ mV$$

$\therefore$ $V_B = \mathbf{903.75\ mV}$

Neglecting base current, $V_B = \frac{V_{CC} R_2}{R_1 + R_2}$

$$\frac{R_1 + R_2}{R_2} = \frac{V_{CC}}{V_B} = \frac{12}{903.75} \times 10^3 = 13.28\ \Omega$$

$$\frac{R_1}{R_2} + 1 = 13.28\ \Omega \text{ or } \frac{R_1}{R_2} = 12.28\ \Omega$$

$\therefore$ $R_1 = 12.28 R_2$

Choose $R_2 = 100\ \Omega$, $R_1 = 1.228\ k\Omega$

$$P_{dc} = I_{dc} \times V_m = 303.15 \times 10^{-3} \times 12$$

$$= \mathbf{363.78\ mW}$$

Power dissipation per transistor, $P = \frac{P_{dc} - P_{ac}}{2} = \frac{363.78 - 200}{2} = \mathbf{81.89\ mW}$

**10. A single-stage class-A amplifier has $V_{CC}$ = 20 V, $V_{CEQ}$ = 10 V, $I_{CQ}$ = 600 *mA* and $R_L$ = 16 Ω. The *ac* output current varies by ± 300 *mA* with the *ac* input signal. Determine (a) the power supplied by the *dc* source to the amplifier circuit, (b) *dc* power consumed by the load resistor (c) *ac* power developed across the load resistor, (d) the *dc* power delivered to the transistor, (e) *dc* power wasted in transistor collector (f) overall efficiency and (g) collector efficiency.**

**Data:** $V_{CC} = 20\ V$, $V_{CEQ} = 10\ V$, $I_{CQ} = 600\ mA$, $R_L = 16\ \Omega$, $I_{max} = 300\ mA$

**To find:** $P_{dc}$, $P_{R_L(dc)}$, $P_{ac}$, $P_{tra\ (dc)}$, $P_{dc}$, $\eta_{overall}$ and $\eta$

**Solution:** (a) Power supplied by the *dc* source to the amplifier circuit is given by

$$P_{dc} = V_{CC} \cdot I_{CQ} = 20 \times 0.6 = \mathbf{12\ W}$$

(b) *DC* power consumed by the load resistor is given by

$$P_{R_L(dc)} = (I_{CQ})^2 R_L = (0.6)^2 \times 16 = \mathbf{5.76\ W}$$

(c) *AC* power developed across the load resistor is $P_{ac}$.

$$I = \frac{I_m}{\sqrt{2}} = \frac{0.3}{\sqrt{2}} = 0.212\ A$$

$$P_{ac} = I^2 R_L = (0.212)^2 \times 16 = \mathbf{0.72\ W}$$

(d) DC power delivered to the transistor is

$$P_{tr(dc)} = P_{dc} - P_{dc(R_L)}$$

$$= 12 - 5.76 = \mathbf{6.24\ W}$$

(e) DC power wasted in the transistor collector is

$$P_{dc} = P_{tr(dc)} - P_{ac}$$

$$= 6.24 - 0.72 = \mathbf{5.52\ W}$$

(f) Overall efficiency, $\eta = \dfrac{P_{ac}}{P_{dc}} = \dfrac{0.72}{12} = 0.06 = \mathbf{6\ \%}$

(g) Collector efficiency, $\eta_c = \dfrac{P_{ac}}{P_{tr(dc)}} = \dfrac{0.72}{6.24} = \mathbf{11.5\ \%}$

## QUESTIONS

1. Why is a power amplifier called a large-signal amplifier? Explain it briefly.
2. Explain why a centered $Q$-point is important for a class A amplifier.
3. Explain the meaning of class-A, class-B and class-C as applied to power amplifiers.
4. Explain how the active device in a class-A direct-coupled resistive load amplifier dissipates less power, when a signal is applied than with no signal. Is this also true in a transformer-coupled-load power amplifier?
5. Derive an expression for the output power of a class-A large-signal amplifier in terms of $V_{max}$, $V_{min}$, $I_{max}$ and $I_{min}$.
6. Show that the maximum collector efficiency of class-A amplifier is 25 percent.
7. Explain how rectification takes place in a power amplifier.
8. When is the power dissipation maximum in class-A amplifiers? What is the power dissipation rating of a transistor?
9. Define intermodulation distortion.
10. What is harmonic distortion? Explain how the output signal gets distorted due to the harmonic distortion.
11. Describe the three-point method of computing harmonic distortion.
12. Explain the five-point method of computing harmonic distortion.
13. What is impedance transformation in case of a transformer? Explain the procedure of calculating a reflected load on primary for a step down transformer.
14. Explain with a diagram the operation of a transformer-coupled transistor amplifier. Also explain the need for impedance matching.
15. Prove that the maximum efficiency of a transformer coupled class-A amplifier is 50%.
16. What are the different possible distortions in an AF power amplifier? Which is the most significant? Why?
17. Explain why the circuit of transformer-coupled class-A amplifier exhibits a maximum in the power-output versus load resistance curve.
18. (a) Define the conversion efficiency $\eta$ of a power stage. Derive a simple expression for $\eta$.
    (b) Compare the maximum efficiency of a series-fed and transformer coupled class-A single transistor power stage.
19. Derive an expression for efficiency of a class-B power amplifier in terms of $V_{max}$ and $V_{min}$.
20. Explain why even harmonics are not present in a push-pull amplifier.
21. Write two additional advantages of a push-pull amplifier over that of a single-stage transistor amplifier.

22. Derive an expression for the output power of an idealized class-B push-pull circuit.
23. Prove that the maximum conversion efficiency of the idealized class-B push-pull amplifier is 78.5 percent.
24. Explain with a circuit diagram the working of a class-B push-pull amplifier.
25. What is required to provide push-pull operation? How is transistor saturation avoided in this type of operation?
26. Explain briefly why an output transformer is used in a power amplifier.
27. What are the design criteria for a power amplifier?
28. Write the advantages and disadvantages of push-pull configuration in power amplifiers.

## EXERCISES

1. The load of 4 Ω is connected to the secondary of a transformer having primary turns of 200 and secondary turns of 20. Calculate the reflected load impedance on primary.
**[Ans: 400 Ω]**

2. For a transformer, the load connected to the secondary has an impedance of 8 Ω. Its reflected impedance on primary is observed to be 648 Ω. Calculate the turns ratio.
**[Ans: 0.1]**

3. What turns ratio is required to match a 16 Ω speaker load to an amplifier, so that the effective load resistance is 10 *k*Ω?
**[Ans: 25:1]**

4. A single-ended class-A amplifier has a transformer coupled load of 8 Ω. If the transformer turns ratio is 10, find the maximum power output delivered to the load. Take the zero signal collector current of 500 *mA*.
**[Ans: 100 *W*]**

5. A transistor rated for a maximum collector dissipation of 100 *mW* operates a single-ended class-A output stage from a 10 *V* supply. Calculate the approximate values of (a) maximum undistorted *ac* output power, (b) the quiescent current and (c) turns-ratio of the output transformer, if the load resistance is 16 Ω. Assume overall and collector efficiency = 0.5.
**[Ans: (a) 50 *mW*; (b) 0.01 *A* and (c) 8]**

6. The loudspeaker of 8 Ω is connected to the secondary of the output transformer of a class-A amplifier circuit. The quiescent collector current is 140 *mA*. The turns ratio of the transformer is 3:1. The collector supply voltage is 10 *V*. If *ac* power delivered to the loudspeaker is 0.48 *W*, assuming ideal transformer, calculate (i) *ac* power developed across primary; (ii) rms value of load voltage; (iii) rms value of primary voltage; (iv) rms value of load current; (v) rms value of primary current; (vi) the *dc* power input; (vii) the efficiency (viii) the power dissipation.
**[Ans: (i) 0.48 *W*; (ii) 1.96 *V*; (iii) 5.88 *V*; (iv) 0.25 *A*; (v) 81.64 *mA*; (vi) 1.4 *W*; (vii) 34.28 %; (viii) 0.92 *W*]**

7. Use a five-point method to calculate harmonic components which gives the following results: $D_2$ = 0.1, $D_3$ = 0.02, $D_0$ = 0.01, with $I_1$ = 4 *A* and $R_C$ = 8 Ω. Calculate the total distortion, fundamental power component and total power. **[Ans: 0.1; 64 *W* and 64.64 *W*]**

8. Calculate the efficiency of the following amplifier classes and voltages:
(a) Class-A operation with $V_{CE\ max}$ = 24 *V* and $V_{CE\ min}$ =2 *V*.
(b) Class-B transformer operation with $V_{CE\ max}$ = 4 *V* and $V_{CC}$ = 22 *V*.
**[Ans: (a) 35.8%; (b) 64.2%]**

9. A power amplifier supplies 3 *W* to a load of 6 *k*Ω. The zero signal *dc* collector current is 55 *mA* and the collector current with signal is 60 *mA*. How much is the percentage of second harmonic distortion?
**[Ans: 16.01%]**

10. A single transistor is operating as an ideal class-B amplifier with a 1 $k\Omega$ load. A *dc* meter in the collector circuit reads 10 $mA$. How much signal power is delivered to the load?

**[Ans: 0.246 *W*]**

11. Given an ideal class-B transistor amplifier whose characteristics are as in Fig. 7.11. The collector supply voltage $V_{CC}$ and the effective load resistance $R'_L = n^2 R_L$ are fixed as the base current excitation is varied. Show that the collector dissipation $P_C$ is zero at no signal ($V_m = 0$), rises as $V_m$ increases and passes through a maximum given by

$$P_C = \frac{2V_{CC}^2}{\pi R'_L} \text{ at } V_m = \frac{2V_{CC}}{\pi}.$$

12. The power transistor whose characteristics are shown in Fig. 7.7 is used in the class-B push-pull circuit of Fig. 7.10 with $R_2 = 0$ and $-V_{CC} = -20\,V$. If the base current is sinusoidal, with a peak value of 20 $mA$ and $R'_L = 15\,\Omega$, calculate (a) the third harmonic distortion, (b) the power output, (c) the collector-circuit efficiency.

**[Ans: (a) 6%; (b) 13.5 *W* and (c) 74.5%]**

13. Repeat Prob. 12, using $-V_{CC} = -15\,V$, $R'_L = 7.5\,\Omega$ and a peak base current of 30 $mA$.

**[Ans: (a) 6.1%; (b) 14.4 *W* and (c) 71.2%]**

# 8 OPERATIONAL AMPLIFIER

## Chapter Outline



## INTRODUCTION

An operational amplifier is basically a direct coupled high gain differential amplifier with high input impedance and low output impedance. It consists of one or more differential amplifiers and followed by a level translator and an output stage. The output stage is either a push-pull or a push-pull complementary-symmetry pair. An operational amplifier amplifies the difference between two input signals, and is also known as **difference amplifier.**

The name **operational amplifier** is derived from the fact that the amplifier was originally used to perform electronically the mathematical operations such as addition, subtraction, integration, multiplication, differentiation, etc. However, operational amplifiers are now used in several applications such as wave shaping, impedance transformation, analog-to-digital conversion, generation of waveforms, logarithmic amplifiers, active filters, oscillators, comparators, regulators, and others.

The integrated operational amplifier now available as a monolithic IC, has gained wide acceptance as a versatile, predictable and economic system building block. It offers all the advantages of monolithic integrated circuits: small size, high reliability, reduced cost, temperature tracking and low offset voltage and current.

## THE BASIC OPERATIONAL AMPLIFIER

The schematic symbol of an op-amp is an arrow head and shown in Fig 8.1(a). The arrow head (triangle) signifies amplification and points from input to output, in the direction of signal flow. All operational amplifiers have differential inputs. The differential inputs are marked by (+) and (–) notations to indicate **non-inverting input,** ($V_1$) and **inverting input,** ($V_2$) respectively. $V_o$ represents the output of op-amp.

The differential input applied to op-amp is

$$V_{in} = V_1 - V_2 \quad \longrightarrow (1)$$

Note that voltages $V_1$, $V_2$ and $V_o$ are node voltages, measured with respect to ground. The differential input $V_{in}$ is the difference of two node voltages, $V_1$ and $V_2$ and $A_V$ is the voltage gain. The equivalent circuit of operational amplifier is shown in Fig. 8.1(b).

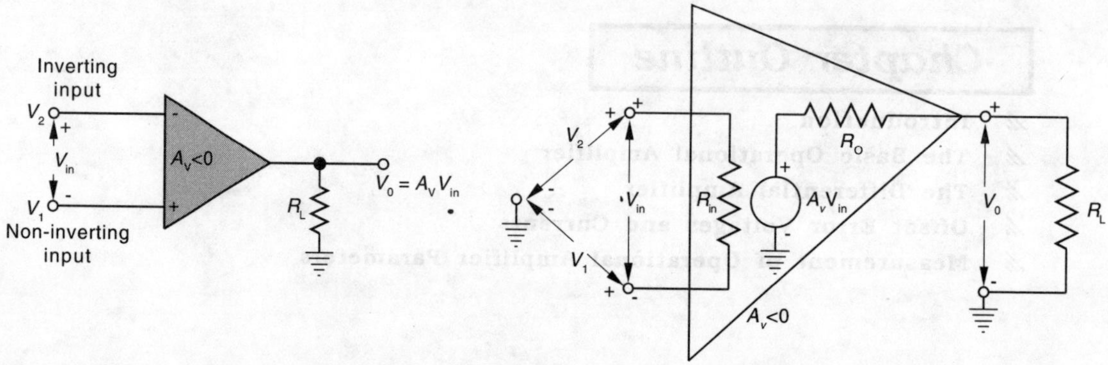


**(a) Basic operational amplifier** **(b) Low frequency equivalent circuit**

**Fig. 8.1**

**Ideal Operational Amplifier:** Ideal op-amp should provide the following characteristics/parameters:

1. Voltage gain, $A_v = \infty$
2. Input resistance, $R_{in} = \infty$
3. Output resistance, $R_o = 0$
4. Bandwidth = $\infty$
5. Perfect balance i.e., $V_o = 0$ when $V_1 = V_2$
6. Characteristics do not drift with temperature
7. Slew rate = $\infty$
8. PSRR = 0
9. CMRR = $\infty$

## Ideal Inverting Operational Amplifier

Figure 8.2 shows an ideal inverting operational amplifier, $Z$ is the input impedance and $Z_f$ feedback impedance. The non-inverting terminal is grounded.

$$V_o = (V_2 - V_1)\, A_v$$

For ideal value of $A_v$, $V_2 - V_1 = 0$

When $V_2$ is grounded $V_1 = 0$ i.e., $V_1$ will be at ground potential, without actually it being grounded. Therefore $G$ is called **virtual ground**.

**Virtual ground is defined as a node that has zero voltage but is not physically grounded.**

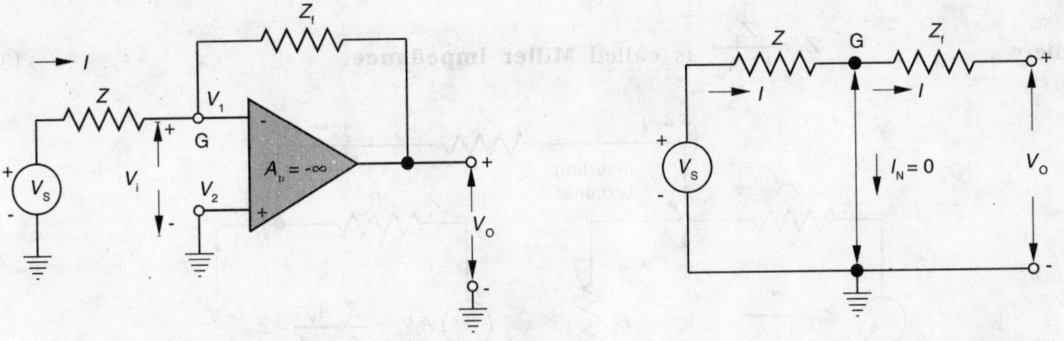


**(a) Inverting operational amplifier with added voltage-shunt feedback**

**(b) Virtual ground of operational amplifier**

**Fig. 8.2**

Ideally, since the input resistance of the op-amp is infinite no current enters the op-amp. Hence the current $I$ through $Z$ also flows through $Z_f$, as shown in Fig. 8.2(b).

From the figure, $$V_i = \frac{V_o}{A_\upsilon}$$

Since $|A_\upsilon| = \infty$ $\quad V_i = \frac{V_o}{\infty} = 0,$ so that the input is effectively shorted.

Voltage gain with feedback is given by

$$A_{Vf} = \frac{V_o}{V_s} = \frac{-IZ_f}{IZ}$$

$$\boxed{A_{Vf} = \frac{-Z_f}{Z}} \longrightarrow (2)$$

The overall voltage gain $A_{vf}$ is independent of the open loop gain $A_v$. The negative sign shows that output voltage is out of phase by 180º with respect to input voltage.

## Practical Inverting Operational Amplifier

Figure 8.3 shows the equivalent circuit of practical inverting operational amplifier. In a practical operational amplifier $|A_\upsilon| \neq \infty$, $R_{in} \neq \infty$ and $R_o \neq 0$. The impedances $Z_1$ and $Z_2$ indicate the effect of feedback impedance $Z_f$ on the input and output of the amplifier. The current $I$ drawn from inverting terminal through $Z_f$ can be obtained by disconnecting inverting terminal from $Z_f$ and bridging an impedance $Z_f/(1 - A_v)$ from inverting terminal to ground.

The current $I$ is given by

$$I = \frac{V_{in} - V_o}{Z_f} = \frac{V_{in}\left(1 - \frac{V_o}{V_{in}}\right)}{Z_f} = \frac{V_{in}(1 - A_\upsilon)}{Z_f} = \frac{V_{in}}{Z_f/(1 - A_\upsilon)}$$

$$I = \frac{V_{in}}{Z_1}$$

where $Z_1 = \frac{Z_f}{1 - A_v}$ is called **Miller impedance.** $\longrightarrow$ (3)

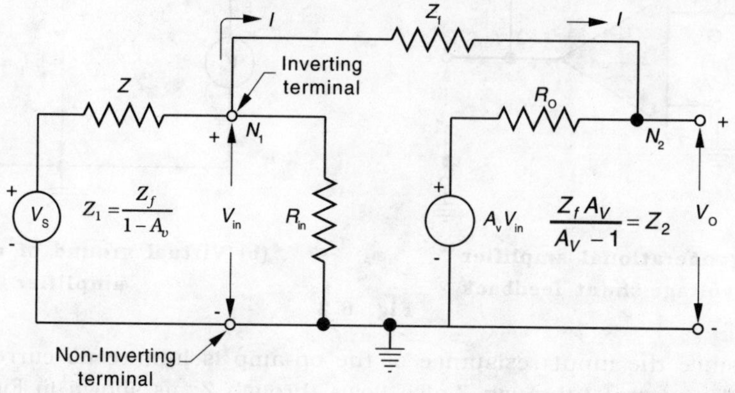


**Fig. 8.3 (a) Equivalent circuit of practical inverting operational amplifier**

Therefore, if $Z_1 = Z_f/(1-A_v)$ were shunted across inverting and ground terminals, the current $I$ drawn from inverting terminal would be the same as that from the original circuit.

The current $I$ drawn from terminal $N_2$ may be calculated by removing $Z_f$ and connecting an impedance $Z_2 = \frac{Z_f A_v}{A_v - 1}$ between $N_2$ and ground. It is given by

$$I = \frac{V_o - V_{in}}{Z_f} = \frac{V_o\left(1 - \frac{V_{in}}{V_o}\right)}{Z_f} = \frac{V_o\left(1 - \frac{1}{A_v}\right)}{Z_f} = \frac{V_o}{Z_f \Big/ \left(1 - \frac{1}{A_v}\right)}$$

$$\therefore \quad I = \frac{V_o}{Z_2}$$

where $Z_2 = \frac{Z_f}{1 - \frac{1}{A_v}} = \frac{Z_f A_v}{A_v - 1}$ is called **Miller impedance.** $\longrightarrow$ (4)

The equivalent circuit shown in Fig. 8.3(b) uses Miller's theorem with $A_v = V_o / V_i$ taking the loading of Z` into account. From the output circuit, we have

$$V_o = A_v V_i \frac{\left(Z'A_v \Big/ A_{v-1}\right)}{R_o + \left(Z'A_v \Big/ A_{v-1}\right)} = \frac{A_v V_i}{1 + R_o Y'\left(\frac{A_v - 1}{A_v}\right)}$$

$$A_V = \frac{V_o}{V_i} = \frac{A_v}{1 + R_o Y'\left(1 - \frac{1}{A_v}\right)} = \frac{A_v}{1 + R_o Y' - \left(R_o Y' \Big/ A_v\right)}$$

or $A_V(1+R_oY') = R_oY' + A_v$

$$\therefore \quad A_V = \frac{A_v + R_oY'}{1+R_oY'}$$

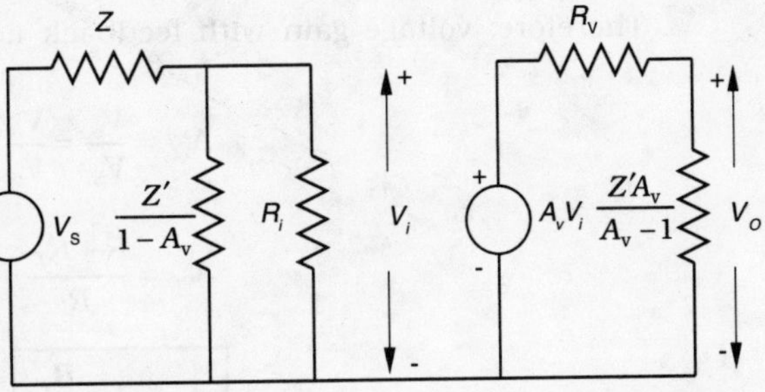


**Fig. 8.3 (b) Practical inverting operational amplifier**

From the input circuit, we can write

$V_i = I_{(\text{short-circuited})} \times \text{impedance}$

$$V_i = \frac{V_sY}{Y+(1-A_V)Y'+Y_i} = \frac{-V_sY}{-(Y+Y'+Y_i)+A_VY}$$

and
$$A_{Vf} = \frac{V_o}{V_s} = \frac{A_VV_i}{V_s} = \frac{-A_VY}{A_VY-(Y+Y'+Y_i)}$$

$$\boldsymbol{A_{Vf} = \frac{-Y}{Y - \frac{1}{A_v}(Y+Y'+Y_i)}}$$

## Ideal Non-inverting Operational Amplifier

Figure 8.4 shows the circuit of an ideal non-inverting operational amplifier. In an ideal op-amp, input resistance $R_{in} = \infty$ and output resistance $R_0 = 0$.

Since $R_{in} = \infty$, we have

$$V_2 = \frac{R}{R+R_f}V_0$$

Output voltage, $V_0 = A_V(V_2 - V_S)$

For a finite output voltage $V_0$ and voltage gain $A_v = \infty$, we have

$$\frac{V_0}{\infty} = V_2 - V_S = 0$$

or $V_2 = V_S$

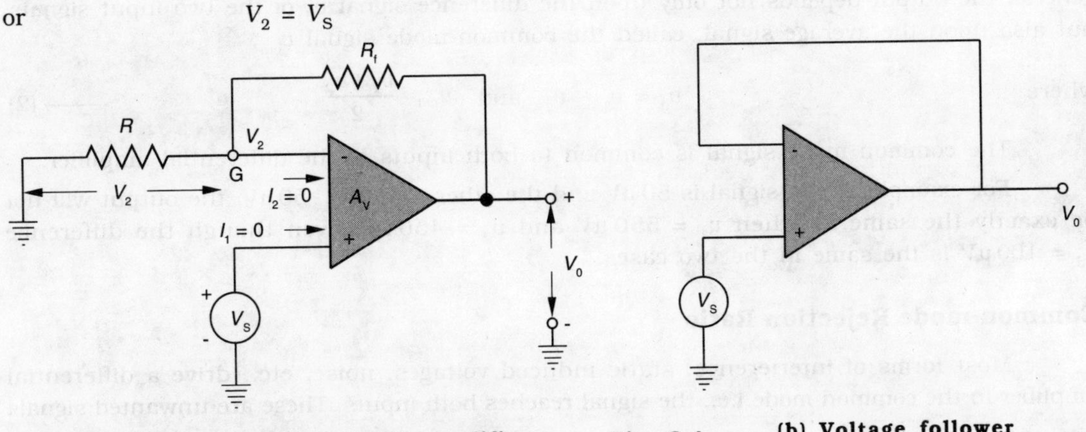


**(a) Non-inverting operational amplifier** **Fig. 8.4** **(b) Voltage follower**

Therefore, voltage gain with feedback i.e., overall voltage gain is given by

$$A_{Vf} = \frac{V_o}{V_S} = \frac{V_o}{V_2} = \frac{V_o}{V_o R/(R+R_f)}$$

$$A_{Vf} = \frac{R+R_f}{R}$$

$$\boxed{A_{Vf} = 1 + \frac{R_f}{R}} \longrightarrow (5)$$

Hence the closed-loop gain is greater than unity.

If $R = \infty$ and $R_f = 0$, then $A_{Vf} = +1$ and the amplifier acts as a **voltage follower** [Fig. 8.4(b)].

## THE DIFFERENTIAL AMPLIFIER

**An amplifier which amplifies the difference between two input signals is known as differential amplifier.**

Differential amplifier is the basic stage of an integrated operational amplifier and hence its study sets the groundwork for analysis and design procedures for operational amplifers.

The block diagram of differential amplifier is shown in Fig. 8.5. It has two input signals $v_1$ and $v_2$ and one output signal $v_0$, each measured with respect to ground. In an ideal differential amplifier, the output signal $v_0$ is given by

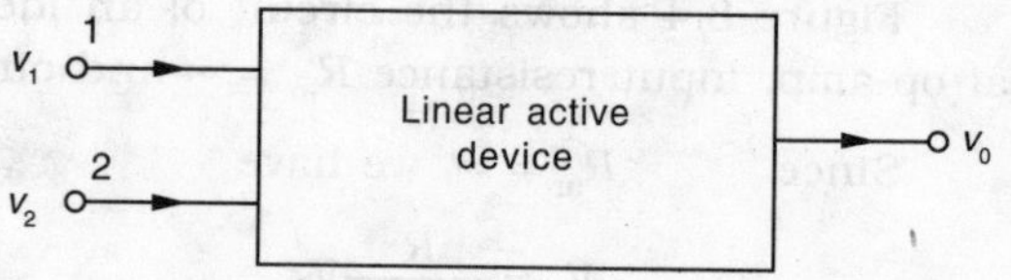


**Fig. 8.5 Block diagram of differential amplifier**

$$v_0 = A_d (v_1 - v_2) \longrightarrow (1)$$

where $A_d$ is the gain of the differential amplifier. It is thus referred to as the **differential mode gain**. But a practical differential amplifier cannot be described by Eq. (1), because, in general, the output depends not only upon the difference signal $v_d$ of the two input signals, but also upon the average signal, called the common-mode signal $v_{cm}$,

where $$v_d = v_1 - v_2 \text{ and } v_{cm} = \frac{v_1 + v_2}{2} \longrightarrow (2)$$

The common mode signal is common to both inputs of the differential amplifier.

For example, if one signal is 50 $\mu V$ and the other signal is –50 $\mu V$, the output will not be exactly the same as when $v_1$ = 550 $\mu V$ and $v_2$ = 450 $\mu V$, even though the difference $v_d$ = 100 $\mu V$ is the same in the two cases.

### Common-mode Rejection Ratio

Most forms of interference, static induced voltages, noise, etc., drive a differential amplifier in the common mode i.e., the signal reaches both inputs. These are unwanted signals which distort the output. The differential amplifier has the property of rejecting these common mode signals. Therefore common mode rejection ratio is the ability of a differential amplifier to reject a common-mode signal such as noise.

**Common-mode rejection ratio is defined as the ratio of differential voltage gain to common-mode voltage gain.** It is abbreviated as CMRR.

$$\text{CMRR} = \frac{\text{Differential voltage gain}}{\text{Common-mode voltage gain}} = \frac{A_d}{A_{cm}} \longrightarrow (3)$$

Ideally, $A_{cm} = 0$ and $A_d = A_v$ is very large therefore, the CMRR is very large. CMRR is also expressed in decibels ($dB$) and is denoted by $\rho$. For the 741 $C$, CMRR is around 90 $dB$.

$$\rho = 20 \log_{10} \text{CMRR} = 20 \log_{10} \left| \frac{A_d}{A_{cm}} \right| \longrightarrow (4)$$

The higher the CMRR, the better the ability of the differential amplifier to eliminate common mode signals. If a differential amplifier were ideal, CMRR would be infinite because $A_{cm}$ would be zero.

Because of the mismatch between the two transistors constituting the basic differential amplifier circuit, the gain encountered by the two input signals $V_1$ and $V_2$ are different. Therefore, the output voltage can be expressed as a linear combination of two input voltages.

$$v_0 = A_1 v_1 + A_2 v_2 \longrightarrow (5)$$

where $A_1$ ($A_2$) is the voltage amplification from input 1 (2) to the output under the condition that input 2 (1) is grounded.

From Eqs. (2), $$v_1 = v_{cm} + \frac{1}{2} v_d \quad \text{and} \quad v_2 = v_{cm} - \frac{1}{2} v_d \longrightarrow (6)$$

Substituting the above equations in Eq. (5), we get

$$v_0 = A_1 (v_{cm} + \frac{1}{2} v_d) + A_2 (v_{cm} - \frac{1}{2} v_d)$$

$$= A_1 v_{cm} + \frac{A_1 v_d}{2} + A_2 v_{cm} - \frac{A_2 v_d}{2}$$

$$= \frac{v_d}{2}(A_1 - A_2) + v_{cm}(A_1 + A_2)$$

$$v_0 = A_d v_d + A_{cm} v_{cm} \longrightarrow (7)$$

$$\text{where} \quad A_d = \frac{1}{2}(A_1 - A_2) \quad \text{and} \quad A_{cm} = A_1 + A_2 \longrightarrow (8)$$

$A_d$ can be measured directly by setting $v_1 = -v_2 = 0.5\,V$, then $v_d = 1$, $v_{cm} = 0$, and $v_0 = A_d$. This means the output voltage is a direct measurement of the differential voltage gain, $A_d$. Similarly, we can measure $A_{cm}$ directly by making $v_1 = v_2 = 1\,V$, then $v_d = 0$, $v_{cm} = 1\,V$ and $v_0 = A_{cm}$. Then the output voltage is a direct measurement of common-mode gain, $A_{cm}$.

Dividing Eq. (7) by $A_d$, we get

$$\frac{v_0}{A_d} = v_d + \frac{A_{cm}}{A_d} v_{cm}$$

$$= v_d + \frac{1}{\rho} v_{cm}$$

$$v_0 = A_d v_d + \frac{A_d}{\rho} v_{cm}$$

$$\boxed{v_0 = A_d v_d \left[1 + \frac{1}{\rho}\frac{v_{cm}}{v_d}\right]} \longrightarrow (9)$$

From the above equation it is clear that ρ should be large compared to the ratio of the common-mode signal to the difference mode signal.

## OFFSET ERROR VOLTAGES AND CURRENTS

In an ideal operational amplifier, for input to be zero, output voltage $v_0$ should be zero. In practice, an operational amplifier exhibits an unbalance caused by a mismatch of the input transistors, that is, the output may not be zero when the inputs are connected to ground. The mismatch results in unequal bias currents flowing through the input terminals. Hence it is necessary to apply a small *dc* bias voltage called input offset voltage between the input terminals to make the output voltage zero.

**Input Bias Current ($I_B$)** Input bias current is defined as the average of the currents that flow into the inverting and non-inverting input terminals of the op-amp [Fig. 8.6]. It is given by

$$I_B = \frac{I_{B1} + I_{B2}}{2} \quad \text{when} \quad V_0 = 0$$

For typical operational amplifiers, each input current is in the range of 10 to 100 *nA*.

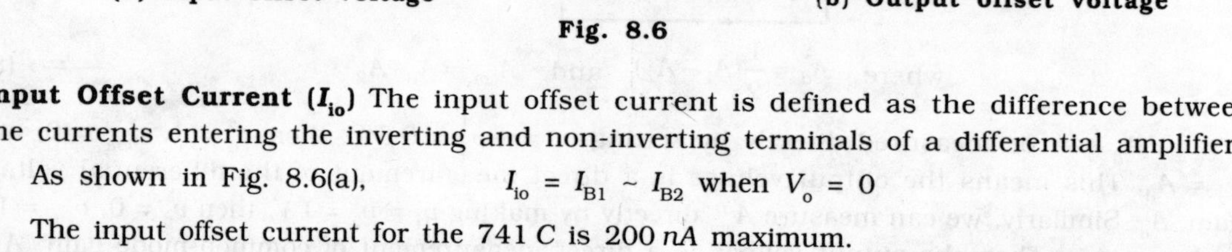

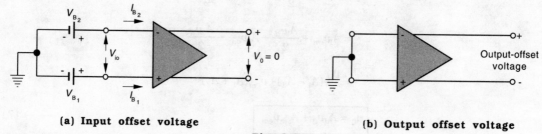

**(a) Input offset voltage** **(b) Output offset voltage**

**Fig. 8.6**

**Input Offset Current ($I_{io}$)** The input offset current is defined as the difference between the currents entering the inverting and non-inverting terminals of a differential amplifier.

As shown in Fig. 8.6(a), $I_{io} = I_{B1} \sim I_{B2}$ when $V_o = 0$

The input offset current for the 741 *C* is 200 *nA* maximum.

**Input Offset Current Drift** The input offset current drift is defined as the ratio of the change in input offset current to the change in temperature.

$$\text{Input offset current drift} = \frac{\text{Change in input offset current } (\Delta I_{io})}{\text{Change of temperature } (\Delta T)}$$

**Input Offset Voltage ($V_{io}$)** Input offset voltage is the voltage which must be applied between the two input terminals of an op-amp to null the output [Fig. 8.6(a)]. Typical value is around 5–6 *mV*. It is temperature sensitive.

**Input Offset Voltage Drift** The input offset voltage drift is defined as the ratio of the change in input offset voltage to the change in temperature.

$$\text{Input offset voltage drift} = \frac{\text{Change in input offset voltage } (\Delta V_{io})}{\text{Change in temperature } (\Delta T)}$$

**Output Offset Voltage** The output offset voltage is defined as the difference between the *dc* voltages present at the two output terminals (or at the output terminal and ground for an amplifier with one output) when the two input terminals are grounded [Fig. 8.6(b)].

**Power Supply Rejection Ratio (PSRR)** The power supply rejection ratio is defined as the ratio of the change in input offset voltage to the corresponding change in one power supply voltage, when all other power supply voltages remain constant.

For 741 IC, PSRR = 150 $\mu V/V$.

**Slew Rate** Slew rate is defined as the time rate of change of closed loop amplifier output voltage under large-signal conditions.

$$\text{Slew rate} = \left(\frac{dV_o}{dt}\right)_{max} V/\mu s$$

For an ideal op-amp it is infinite. Slew rate also indicates how rapidly the output of an op-amp can change in response to high frequency. It changes with change in voltage gain and is normally specified at unity (+1) gain. For 741 IC, slew rate = 0.5 $V/\mu s$.

**Table 8.1 Typical parameters of 741**

| Sl. No. | Parameters | Value |
|---|---|---|
| 1. | Open loop gain, $A_v$ | $10^3$ to $2 \times 10^5$ |
| 2. | Input offset voltage, $V_{io}$ | 6 *mV* |
| 3. | Input offset current, $I_{io}$ | 20 *nA* (200 *nA*) |
| 4. | Input bias current $I_B$ | 500 *nA* |
| 5. | Common-mode rejection ratio, CMRR | 30,000 (90 *dB*) |
| 6. | Power supply rejection ratio, PSRR | 150 $\mu V/V$ |
| 7. | Input offset current drift | 0.1 $nA/°C$ |
| 8. | Input offset voltage drift | 1.0 $\mu V/°C$ |
| 9. | Slew rate | 0.5 $V/\mu s$ |

### Universal Balancing Techniques

When an operational amplifier is used, it is often necesary to balance the offset voltage. Hence a small *dc* voltage must be applied in the input of an op-amp so as to cause the *dc* output voltage to become zero. The universal offset-voltage balancing circuits for inverting and non-inverting operational amplifiers are shown in Fig. 8.7. The circuit shown in Fig. 8.7(a) supplies a small voltage effectively in series with the non-inverting input terminal in the range $\pm V(R_2/(R_1+R_2)) = \pm 7.5\ mV$ if $\pm 15\ V$ power supplies are used and $R_1 = 200\ k\Omega$, $R_2 = 100\ \Omega$. Thus the circuit shown in Fig. 8.7(a) is useful for balancing inverting amplifiers. The feedback element can be either a resistor (capacitor) or a nonlinear element. The circuit shown in Fig. 8.7(b) is used for balancing the offset voltage when operational amplifier is used as a non-inverting amplifier.

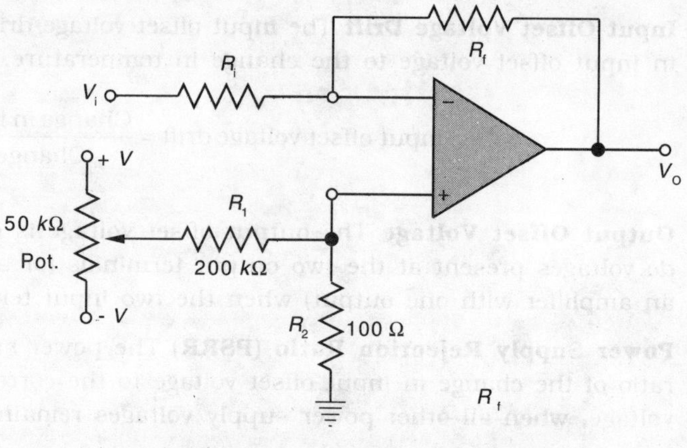


**(a) Inverting operational amplifier**

**Fig. 8.7 Universal offset-voltage balancing circuits**

Some integrated op-amps are provided with in-built offset null adjustment. Ex: μA741C has pins 1 and 5 to nullify the offset voltage as shown in Fig. 8.7(c).

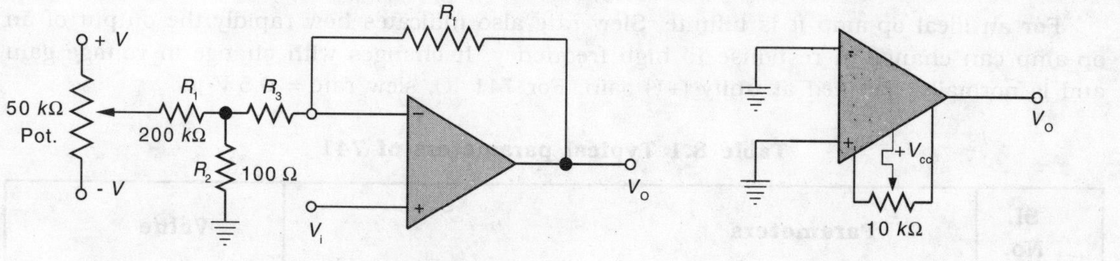


**(b) Non-inverting operational amplifier** **(c) Balancing in 741C**

**Fig. 8.7 Universal offset-voltage balancing circuits**

## MEASUREMENT OF OPERATIONAL AMPLIFIER PARAMETERS

Let us understand the practical methods of measuring some of the important parameters of operational amplifier such as (1) open-loop voltage gain ($A_v$), (2) output resistance $R_o$ without feedback, (3) differential input resistance $R_i$ (4) input offset voltage $V_{io}$ (5) input bias current $I_B$ and input offset current $I_{io}$, (6) common-mode rejection ratio CMRR, and (7) slew rate.

## 1. Open-loop Differential Voltage Gain ($A_v = A_d$)

**The open-loop voltage gain is defined as the ratio of the output signal voltage $V_o$ to the input differential signal voltage** $V_i$. That is, it is ratio $V_o/V_{in}$ with the feedback path opened.

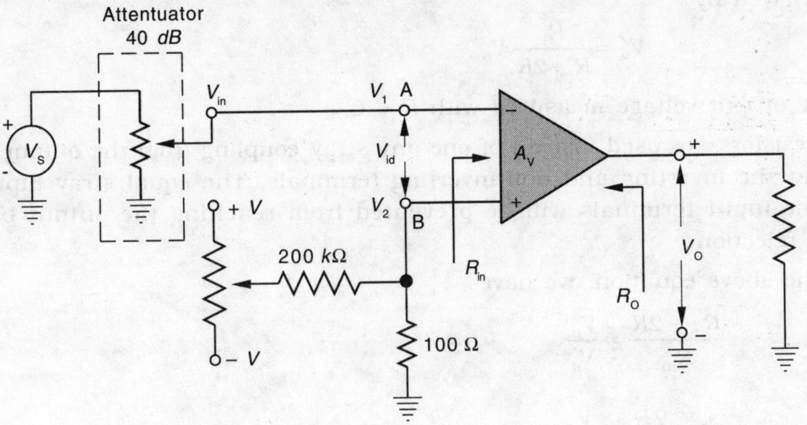


**Fig. 8.8 Circuit diagram for measuring $A_V$, $R_o$ and $R_{in}$**

Figure 8.8 shows the circuit diagram of measuring the open-loop voltage gain. It is essential to apply a small *dc* voltage at the input so as to cause the *dc* output voltage $V_o$ to become zero. Otherwise, the output will go to saturation [Fig. 8.7].

Let $V_s$ be the input *ac* signal. The input attenuator is essential so that $V_{in}$ can be at a sufficiently low level for the output swings to be not greater than about 30 per cent of the output voltage rating.

Open-loop voltage gain (or gain without feedback), $A_V = \dfrac{V_o}{V_{id}}$ where $V_{id} = V_1 - V_2$.

## 2. Output Resistance ($R_o$)

The output resistance $R_o$ of the operational amplifier can be obtained using the circuit of Fig. 8.8. Measure the decrease in the low-frequency gain $A_V$ caused by a load resistance $R_L$. From Fig. 8.1 (b), we can write

$$A_V = \frac{R_L}{R_L + R_o} \cdot A_\upsilon$$

where $A_v$ and $A_V$ represent the open-loop voltage gains with $R_L = \infty$ and $R_L \neq \infty$, respectively. From the above equation, we have

$$R_L + R_o = \frac{A_\upsilon}{A_V} \cdot R_L$$

or

$$\boxed{R_o = \left(\frac{A_\upsilon}{A_V} - 1\right) \cdot R_L}$$

**3. Differential Input Resistance ($R_{in}$)**

The differential input resistance can be measured by forming a voltage divider at the input of the amplifier in Fig. 8.8. The voltage divider is formed by inserting two equal resistors $R$ at points A and B in series with the inverting and non-inverting terminals. The output voltage is then given by

$$V_o' = \frac{R_i}{R_i + 2R} \cdot V_o$$

where $V_o$ is the output voltage measured with $R = 0$.

If two resistors are used instead of one any stray coupling from the output generates equal signals at the inverting and non-inverting terminals. The equal stray input signals produced at the input terminals will be prevented from reaching the output because of common-mode rejection.

From the above equation, we have

$$\frac{R_{in} + 2R}{R_{in}} = \frac{V_o}{V_o'}$$

or

$$1 + \frac{2R}{R_{in}} = \frac{V_o}{V_o'}$$

or

$$\frac{2R}{R_{in}} = \frac{V_o - V_o'}{V_o'}$$

or

$$\boldsymbol{R_{in} = 2R\left(\frac{V_o'}{V_o - V_o'}\right)}$$

A capacitor $C$ can be placed across each of the resistors $R$ to reduce high frequency noise. Notice that the input signal frequency must be much less than $1/2\pi RC$.

If $R_{in}$ is very high, then $V_o' \simeq V_o$ and the measurement is not practical.

**4. Input Offset Voltage ($V_{io}$)**

Figure 8.9 shows the closed-loop circuit used for the measurement of input offset voltage. $R'$ is the feedback resistor connected between the inverting and output terminals of op-amp. The output voltage is given by

$$V_o = \left(\frac{R + R'}{R}\right) V_{io} \qquad \longrightarrow (1)$$

or

$$V_{io} = \left(\frac{R}{R + R'}\right) V_o \qquad \longrightarrow (2)$$

If $R = 100\,\Omega$ and $R' = 100\,k\Omega$, then

$$V_{io} = \left(\frac{100}{100 + 10000}\right) V_o = \left(\frac{100}{100100}\right) V_o = \frac{V_o}{1001}$$

or

$$V_o = 1001\, V_{io}$$

Thus the small input offset voltge is multiplied at the output by a factor 1001, which can easily be measured.

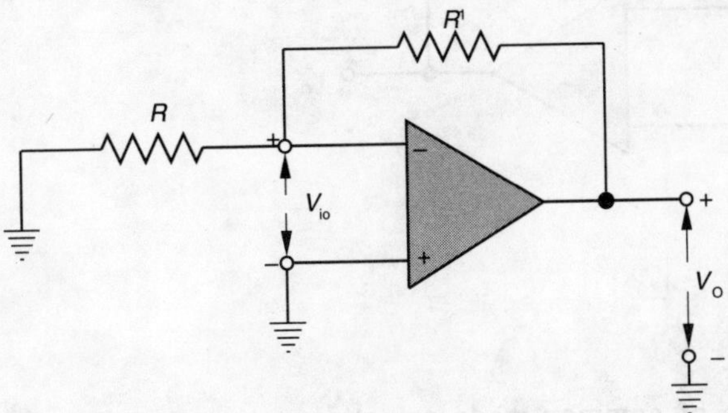


**Fig. 8.9 Measurement of input offset voltage**

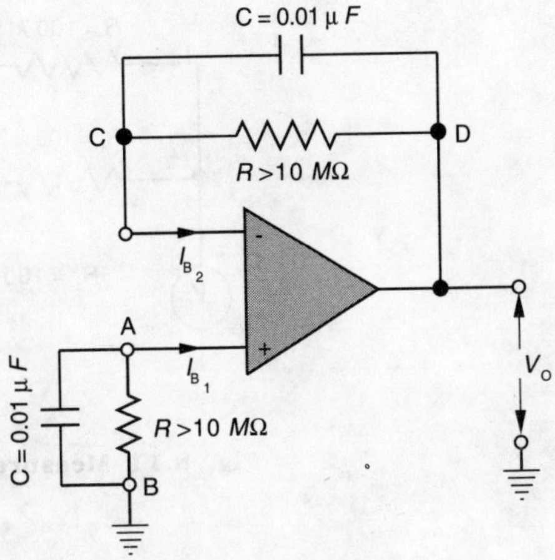


**Fig. 8.10 Measurement of input bias current**

## 5. Input Bias Current ($I_B$)

Figure 8.10 shows the circuit diagram for measuring input bias current. Two equal large resistors $R > 10\ M\Omega$ are connected at the input terminals as shown in the figure. The amplifier is connected as a unity gain non-inverting amplifier. Let $I_{B_1}$ and $I_{B_2}$ be the input bias currents flowing through the resistors. The large resistors $R$ are bypassed with 0.01 μF capacitor to reduce high-frequency noise.

If the terminals A and B are shorted, the output voltage is given by $V_o = I_{B_2} R$

Similarly, if terminals C and D are shorted, the output voltage is $V_o = -I_{B_1} R$

From the above equations, $I_{B_1} = -\dfrac{V_o}{R}$ and $I_{B_2} = \dfrac{V_o}{R}$

Since the input bias current is the average of $I_{B_1}$ and $I_{B_2}$ the input offset current $I_{io}$ is the difference of $I_{B_1}$ and $I_{B_2}$.

Input bias current, $$I_B = \frac{I_{B_1} + I_{B_2}}{2}$$

and Input offset current, $$I_{io} = I_{B_1} - I_{B_2}$$

## 6. Common-mode Rejection Ratio (CMRR)

Figure 8.11 shows the circuit diagram for the measurement of common-mode rejection ratio. The signal at A or B is common-mode signal and is given by

$$V_{cm} = \frac{R_1'}{R_1 + R_1'} V_s \longrightarrow (1)$$

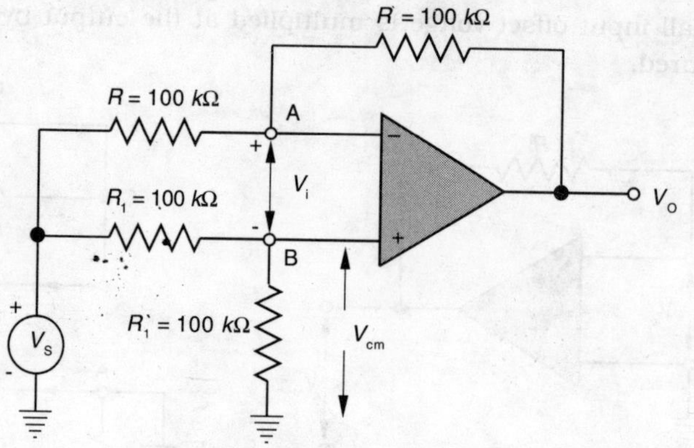


**Fig. 8.11 Measurement of common-mode rejection ratio**

Since $R_1 = R$ and $R'_1 = R'$, $\quad V_{cm} = \dfrac{R'}{R+R'} V_s$

As $R'_1 >> R_1$ and $R' >> R$, $\quad V_{cm} = V_s$

Voltage across $R$, $\quad V_R = V_i - \dfrac{V_s R_1}{R_1 + R_1'}$

and voltage across $R'$, $\quad V_{R'} = V_o - V_i - \dfrac{V_s R'}{(R_1 + R_1')}$

If $I_1$ and $I_2$ are the currents through $R$ and $R'$, we have

$$I_1 = \frac{V_R}{R} = \left(V_i - \frac{V_s R_1}{R_1 + R_1'}\right) \Big/ R$$

and

$$I_2 = \frac{V_{R'}}{R'} = \left(V_o - V_i - \frac{V_s R_1'}{R_1 + R_1'}\right) \Big/ R'$$

Equating the currents in $R$ and $R'$ and assuming $R = R_1$ and $R' = R_1'$, we have

$$\frac{V_i}{R} - \frac{V_s}{R+R'} = \frac{V_o}{R'} - \frac{V_i}{R'} - \frac{V_s}{R+R'}$$

or

$$V_i \left[\frac{R+R'}{RR'}\right] = \frac{V_o}{R'}$$

$$V_i = \left(\frac{R}{R+R'}\right) V_o \qquad \longrightarrow (2)$$

We know that

$$V_o = A_d V_i + A_{cm} V_{cm}$$

$$V_o = \frac{A_d R V_o}{R+R'} + \frac{A_{cm} R' V_s}{R+R'} \qquad \left(\text{because } V_{cm} = \frac{R' V_s}{R+R'}\right)$$

If $\frac{A_d R}{(R+R')} >> 1$, then $A_d R V_o + A_{cm} R' V_s \cong 0$

Common-mode rejection ratio, $\rho = \left|\frac{A_d}{A_{cm}}\right| = \frac{R'}{R}\left|\frac{V_o}{V_s}\right|$ ⟶ (3)

If $A_d$ = 50,000, $R'$ = 100 kΩ and $R$ = 100 Ω, then

$$\frac{A_d R}{R+R'} = \frac{50{,}000 \times 100}{100{,}000+100} = 50$$

It satisfies the inequality assumed above. The above measurement is to be made at the rated common-mode voltage swing.

## 7. Slew Rate

Slew rate is the maximum rate at which the output voltage can change. This rate $dV_o/dt$ can be measured using the non-inverting circuit of Fig. 8.4, with $R = \infty$ and $R_f = 0$, that represents the worst case. If the amplifier has a single-ended input, then the circuit of Fig. 8.2(a) is used with $Z = R = 1$ kΩ and $Z_f = R_f = 10$ kΩ. A high-frequency square wave input signal $V_s$ is applied. The slopes $dV_o/dt$ with respect to time of the leading and trailing edges of the output are measured. It is common to specify the slower of the two rates as the slew rate of the device.

## IMPORTANT POINTS TO REMEMBER

1. Difference amplifier amplifies the difference between two input signals.
2. An operational amplifier is a high gain differential amplifier directly coupled which has the property of amplifying a signal of frequency ranging from 0 to 5 *MHz*.
3. Operational amplifier has two input terminals namely (i) inverting inut terminal (–)(ii) noninverting input terminal (+) and one output terminal.
4. Virtual ground in an op-amp is a node that has zero voltage with respect to ground, but is not physically grounded.
5. Common-mode rejection ratio is defined as the ratio of differential voltage gain to common-mode voltage gain. Since it is very large, it is measured in *dB*.
6. The voltage which is applied to the input of an op-amp to make its output voltage zero is called input offset voltage.
7. The average of the currents that flow into the inverting and non-inverting input terminals of an op-amp is called input bias current.
8. The difference between the currents entering the inverting and non-inverting terminals of a differential amplifier is known as input offset current.
9. Input offset current drift is the ratio of the change of input offset current to the change of temperature.
10. Input offset voltage drift is the ratio of the change of input offset voltage to the change in temperature.
11. Output offset voltage is the difference between the *dc* voltages present at the output terminal and ground when the two input terminals are grounded.
12. Power supply rejection ratio is the ratio of the change in input offset voltage to the corresponding change in the power supply, when all other power supply voltages remain constant.

13. Slew rate is time rate of change of closed amplifier output voltage under large-signal conditions. It is the dynamic characteristic of an operational amplifier.
14. Open-loop voltage gain of an op-amp is the ratio of output signal voltage to the input differential signal voltage.
15. The input resistance of practical op-amp is very high and its output resistance is very low.

## IMPORTANT FORMULAE

1. Voltage gain of inverting op-amp, $A_{Vf} = \frac{-Z_f}{Z} = \frac{-R_f}{R}$
2. Voltage gain of non-inverting op-amp, $A_{Vf} = 1 + \frac{R_f}{R}$
3. Output voltage of an ideal differential amplifier, $v_o = A_d (v_1 - v_2)$
4. Common-mode rejection ratio, $CMRR = \frac{A_d}{A_{cm}}$ and $\rho = 20 \log_{10}$ CMRR $= 20 \log_{10} \left| \frac{A_d}{A_{cm}} \right|$
5. Output voltage of an op-amp in terms of CMRR, $v_o = A_d v_d \left[ 1 + \frac{1}{\rho} \frac{v_{cm}}{v_d} \right]$
6. Input bias current, $I_B = \frac{I_{B_1} + I_{B_2}}{2}$ when $V_o = o$
7. Input offset current, $I_{io} = I_{B_1} - I_{B_2}$ when $V_o = o$
8. Input offset current drift $= \frac{\Delta I_{io}}{\Delta T}$
9. Input offset voltage drift $= \frac{\Delta V_{io}}{\Delta T}$
10. Slew rate of an op-amp $= \frac{dV_o}{dt}_{max}$ $V/\mu s$
11. Open-loop voltage gain of an op-amp, $A_v = \frac{V_o}{V_{id}}$
12. Output resistance of an op-amp, $R_o = \left( \frac{A_v}{A_V} - 1 \right) R_L$
13. Input resistance of an op-amp, $R_i = 2R \left( \frac{V_o'}{V_o - V_o'} \right)$
14. Input offset voltage, $V_{io} = \left( \frac{R}{R + R'} \right) V_o$
15. Common-mode rejection ratio, $\rho = \frac{R'}{R} \left| \frac{V_o}{V_s} \right|$

## SOLVED PROBLEMS

**1. A certain op-amp has an open-loop voltage gain of 100,000 and a common-mode gain of 0.2. Determine the CMRR and express it in decibels.**

**Data:** $A_v$ = 100,000 and $A_{cm}$ = 0.2

**To find:** Common-mode rejection ratio, CMRR

**Solution:** Common mode rejection ratio = $\dfrac{A_d}{A_{cm}}$.

Here $A_d = A_v$ $\qquad \therefore \quad \text{CMRR} = \dfrac{100{,}000}{0.2} = \mathbf{500{,}000}$

CMRR in decibels, $= 20 \log_{10} (500{,}000) = \mathbf{114\ dB}$

**2. (a) For a differential amplifier, the first set of signals is $v_1$ = +40 μV and $v_2$ = – 40 μV and the second set is $v_1$ = 840 μV and $v_2$ = 760 μV. If the common-mode rejection ratio is 100, calculate the percentage difference in the output voltage obtained for the two sets of input signals. Repeat part (a) if ρ = 10,000.**

**Data:** In the first case $v_1$ = 40 μV and $v_2$ = – 40 μV. In the second case $v_1$ = 840 μV and $v_2$ = 760 μV, ρ = 100

**To find:** Percentage in the output voltage

**Solution:** (a) For the first case, $v_d = v_1 - v_2 = 40 + 40 = 80$ μV and

$$v_{cm} = \frac{1}{2}(v_1 + v_2) = \frac{1}{2}(40 - 40) = 0$$

Output voltage,
$$v_o = A_d v_d \left(1 + \frac{v_{cm}}{\rho v_d}\right)$$

$$v_o = A_d \times 80 \left(1 + \frac{0}{\rho \cdot v_d}\right) = \mathbf{80\ A_d\ \mu V}$$

For the second case,
$$v_d = v_1 - v_2 = 840 - 760 = 80\ \mu V$$

$$v_{cm} = \frac{1}{2}(840 + 760) = 800\ \mu V$$

Output voltage,
$$v_o = A_d \times 80 \left(1 + \frac{800}{\rho \cdot 80}\right) = 80\ A_d \left(1 + \frac{10}{100}\right) \mu V$$

**These two measurements differ by 10 per cent.**

(b) For ρ = 10,000, the second set of signals results in an output

$$v_o = 80 A_d \left(1 + \frac{10}{10{,}000}\right) = 80\ A_d\ (1 + 10^{-3})\ \mu V$$

whereas the first set of signals gives an output, $v_o$ = 80 $A_d$ μV.

Hence, **the two measurements differ by only 0.1 per cent.**

**3. The base currents in a differential amplifier are 30 μA and 32 μA. Calculate the value of input offset current and input bias current.**

**Data:** $I_{B_1} = 30\ \mu A$ and $I_{B_2} = 32\ \mu A$

**To find:** Input offset current, $I_{io}$ and input bias current, $I_B$

**Solution:** Input offset current, $I_{io} = I_{B_1} \sim I_{B_2} = 30 \sim 32 = \mathbf{2\ \mu A}$

Input bias current, $$I_B = \frac{I_{B_1} + I_{B_2}}{2}$$

$$= \frac{30 + 32}{2} = \mathbf{31\ \mu A}$$

**4. The data sheet of an op-amp gives $A_v$ = 100,000 and CMRR = 80 *dB*. What is the common-mode voltage gain?**

**Data:** $A_v$ = 100,000 and CMRR = 80 *dB*. Here $A_d = A_v$

**To find:** Common-mode voltage gain, $A_{cm}$

Common-mode rejection ratio, $\rho = 20 \log_{10}\left(\frac{A_d}{A_{cm}}\right)$

$$80 = 20 \log_{10}\left(\frac{100{,}000}{A_{cm}}\right)$$

$$4 = \log_{10}\left(\frac{100{,}000}{A_{cm}}\right)$$

or $$\log_{10}(10{,}000) = \log_{10}\left(\frac{100{,}000}{A_{cm}}\right)$$

$$10{,}000 = \frac{100{,}000}{A_{cm}}$$

or $$\mathbf{A_{cm} = 10}$$

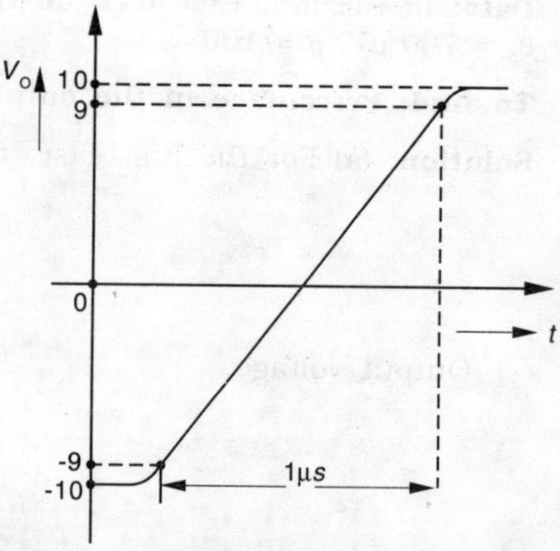


**Fig. 8.12**

**5. The output voltage of a certain op-amp appears as shown in Fig. 8-12 in response to a step input. Determine the slew rate.**

**Data:** From the figure $\Delta t$ = 1μs

**To find:** Slew rate

**Solution:** The output goes from the lower limit to the upper limit in 1μs. Since this response is not ideal, the limits are taken at the 90% points, as indicated. So, the upper limit is +9*V* and the lower limit is –9*V*.

$$\text{Slew rate} = \frac{\Delta V_o}{\Delta t} = \frac{9 - (-9)}{1 \times 10^{-6}} = \mathbf{18\ V/\mu s}$$

**6. When a pulse is applied to an op-amp, the output voltage goes from –8 *V* to +7 *V* in 0.75 μs. What is the slew rate?**

**Data:** $V_1$ = –8 *V*, $V_2$ = 7 *V* and *t* = 0.75 μs

**To find:** Slew rate

**Solution:** Change in output voltage, $\Delta V_0 = V_2 - V_1$

$$= 7 - (-8) = 15\ V$$

$$\text{Slew rate,} = \frac{\Delta V_0}{t} = \frac{15}{0.75 \times 10^{-6}} = \mathbf{20V/\ \mu s}$$

**7. An operational amplifier has a slew rate of 35 $V/\mu s$. How long will it take the output to change from 0 to 15 V.**

**Data:** Slew rate = 35 $V/\mu s$, $\Delta V$ = 0 to 15 $V$

**To find:** Time to take change in voltage, $\Delta t$

**Solution:** $$\text{Slew rate,} = \frac{\Delta V}{\Delta t}$$

or $$35\ V/\mu s = \frac{15V}{\Delta t}$$

or $$\Delta t = \mathbf{0.429\ \mu s}$$

**8. For the circuit of inverting amplifier, $R$ = 10 $k\Omega$, $R_f$ = 100 $k\Omega$ and $V_i$ = 1 $V$. Calculate (a) $I_i$ (b) $V_o$ and (c) $A_{vf}$.**

**Data:** $R$ = 10 $k\Omega$, $R_f$ = 100 $k\Omega$, and $V_i$ = 1 $V$

**To find:** Input current $I_i$, output voltage $V_o$ and closed loop voltage gain $A_{Vf}$

**Solution:** (a) Input current, $$I_1 = \frac{V_i}{R} = \frac{1V}{10 \times 10^3\ \Omega} = \mathbf{0.1\ mA}$$

(b) Output voltage, $$V_o = \frac{-R_f}{R} . V_i = \frac{-100\ k\Omega}{10 k\Omega} \times 1V = \mathbf{-10V}$$

(c) Closed-loop voltage gain, $$A_{Vf} = \frac{-R_f}{R} = \frac{-100}{10} = \mathbf{10}$$

**9. In the circuit of non-inverting amplifier, let $R$ = 5 $k\Omega$, $R_f$ = 20 $k\Omega$ and $V_i$ = 1 $V$. A load resistor of 10 $k\Omega$ is connected at the output. Calculate, (i) $V_o$, (ii) $A_{vf}$, (iii) the load current $I_L$ (iv) the output current $I_o$ indicating proper direction of flow.**

**Data:** $R$ =5 $k\Omega$, $R_f$ = 20 $k\Omega$, $V_i$ = 1 $V$ and $R_L$ = 10 $k\Omega$

**To find:** Output voltage $V_o$, closed-loop gain $A_{Vf}$, load current $I_L$ and output current $I_o$

**Solution:** (i) Output voltage, $$V_o = \left(1 + \frac{R_f}{R}\right) V_i = \left(1 + \frac{20}{5}\right) 1 = \mathbf{5V}$$

(ii) Closed-loop gain, $$A_{Vf} = \frac{V_o}{V_i} = \frac{5\ V}{1V} = \mathbf{5}$$

(iii) Load current, $$I_L = \frac{V_0}{R_L} = \frac{5V}{10k\Omega} = \mathbf{0.5mA}$$

(iv) Output current, $$I_0 = \frac{V_0 - V_i}{R_f} = \frac{5-1}{20} = \mathbf{0.2mA}$$

**10. A differential amplifier has an open-circuit voltage gain of 100. The input signals are 3.25 *V* and 3.15 *V*. Determine the output voltage.**

**Data:** $A_v = 100$, $v_1 = 3.25\ V$ and $v_2 = 3.15\ V$

**To find:** Output voltage, $v_0$

**Solution:** Output voltage, $v_0 = A_v\ (v_1 - v_2)$

$$= 100\ (3.25 - 3.15) = \mathbf{10\ V}$$

The 10 *V* output is therefore the amplified difference between the input signals. The common-mode signal is

$$V_{cm} = \frac{v_1 + v_2}{2} = \frac{3.25 + 3.15}{2} = \mathbf{3.20\,V}$$

It follows that the net input signal $v_1 = 3.25 - 3.20 = 0.05\ V$ and $v_2 = 3.15 - 3.20 = -0.05\ V$.

Using the net input signals,

Output voltage, $v_0 = A_v\,(v_1 - v_2) = 100[0.05 - (-0.05)] = \mathbf{10\ V}$

This shows that the common-mode signal is not amplified.

## QUESTIONS

1. Draw the schematic block diagram of the basic op-amp with inverting and non-inverting inputs. Draw its equivalent circuit.
2. Write six characteristics of the ideal op-amp.
3. Describe some of the characteristics of a practical op-amp.
4. Draw the schematic diagram of an ideal inverting op-amp with voltage shunt feedback impedances $Z$ and $Z_f$. Derive an expression for voltage gain with feedback.
5. What is the concept of virtual ground? Why is it different from ordinary ground?
6. What are the benefits of negative feedback in an op-amp circuit?
7. Why is it necessary to reduce the gain of an op-amp from its open-loop gain?
8. Draw the equivalent circuit of practical inverting op-amp with voltage shunt feedback impedances $z$ and $z_f$ using Miller's theorem. Derive an expression for closed-loop voltage gain in terms of admittances.
9. Draw the schematic diagram of an ideal non-inverting op-amp with voltage series feedback and derive an expression for the voltage gain.
10. What is an ideal differential amplifier?
11. Distinguish between difference input signal and common-mode input signal.
12. Define common-mode rejection ratio.
13. For a given value of open-loop voltage gain, does a higher CMRR result in a higher or lower common-mode gain?
14. Distinguish between open-loop voltage gain and closed-loop voltage gain of an op-amp.
15. Explain why the CMRR is infinite for an ideal op-amp.
16. List at least nine op-amp parameters.
17. Which two parameters, not including the frequency response, are frequency dependent?
18. Define (i) input bias current, (ii) input offset current, (iii) input offset voltage and (iv) output offset voltage of an op-amp.
19. Define (i) common-mode rejection ratio, (ii) power supply rejection ratio, and (iii) slew rate of an op-amp.
20. Show the balancing arrangement for (i) an inverting and (ii) a non-inverting op-amp.
21. Draw the circuit and explain how to measure (i) open-loop voltage (ii) input resistance and (iii) output resistance of an op-amp.

22. Draw the circuit and explain how to measure input offset voltage of an op-amp.
23. Draw the circuit and explain how to measure input offset current of an op-amp.
24. Draw the circuit and explain how to measure common-mode rejection ratio of an op-amp.
25. Draw the circuit and explain how to measure slew rate of an op-amp.
26. Describe the effect of input bias current.
27. Describe the effect of input offset voltage.

## EXERCISES

1. A certain op-amp has a CMRR of 250,000. Convert this to decibels. **[Ans: 120 *dB*]**
2. The open-loop gain of a certain op-amp is 175,000. Its common-mode gain is 0.18. Determine the CMRR in decibels. **[Ans: 119.75 *dB*]**
3. An op-amp data sheet specifies a CMRR of 300,000 and an $A_v$ of 90,000. What is the common-mode gain? **[Ans: 0.3]**
4. Determine the bias current given that the input currents to an op-amp are 8.3 $\mu A$ and 7.9 $\mu A$. **[Ans: 8.1 $\mu A$]**
5. How long does it take the output voltage of an op-amp to go from -10 $V$ to + 10 $V$, if the slew rate is 0.5$V/\mu s$. **[Ans: 40 $\mu s$]**
6. Calculate the closed-loop voltage gain of an op-amp inverting amplifier, given that $R$ = 8 $k\Omega$ and $R_f$ = 56 $k\Omega$. **[Ans: – 7]**
7. For the inverting amplifier $R$ = 2 $k\Omega$ and $R_f$ = 1 $M\Omega$. Determine the following values: (1) $A_v$ (ii) $R_i$ and (iii) $R_o$. **[Ans: (i) -1000; (ii) 1 $k\Omega$; (iii) 0 $\Omega$]**
8. Design an inverting amplifier with a gain of -5 and an input resistance of 10 $k\Omega$. **[Ans: $R$ = 10 $k\Omega$ and $R_f$ = 50 $k\Omega$]**
9. Calculate the output voltage of non-inverting amplifier if $R$ = 50 $k\Omega$, $R_f$ = 500 $k\Omega$ and $V_i$ = 0.4 $V$. **[Ans: 4.4 V]**
10. Calculate the voltage gain of non-inverting amplifier op-amp with $R_f$ = 10 $k\Omega$, $R$ = 2 $k\Omega$. Find the output voltage when input is 200 $mV$. **[Ans: 6 and 1.2 V]**
11. A certain inverting amplifier has a closed loop gain of 25. The op-amp has an open-loop gain of 100,000. If another op-amp having an open-loop gain of 200,000 is substituted in the configuration, what will be the closed loop gain? **[Ans: 25]**
12. A certain op-amp has bias current of 50 $\mu A$ and 49.3 $\mu A$. Calcuate the input offset current of an op-amp. **[Ans: 700 *nA*]**
13. If the output of a particular op-amp increases by 8 $V$ in 12 $\mu s$, what is its slew rate? **[Ans: 0.67 V/$\mu$s]**
14. A 100 $pF$ capacitor has a maximum charging current of 150 $\mu A$. What is the slew rate? **[Ans: 1.5 V/$\mu$s]**
15. (a) Find the closed-loop gain for a non-inverting amplifier with $R$ = 10 $k\Omega$ and $R_f$ = 20 $k\Omega$. (b) How does $A_{Vf}$ change if a third resistance $R'$ = 10 $k\Omega$ is connected in series with $R$? In parallel with $R$? **[Ans: (a) 3 (b) 2, 5]**

# 9 APPLICATIONS OF OPERATIONAL AMPLIFIERS

## Chapter Outline

- Instrumentation amplifier
- Active filters
- Comparators
- Schmitt Trigger
- Analog to Digital Converters
- Digital to Analog Converters
- Clippers and Clampers
- Small-Signal Half-Wave Rectifiers
- Positive and Negative Clampers
- Absolute-Value Output Circuit
- Peak Detector
- Sample and Hold Circuit
- Voltage Regulators

## INSTRUMENTATION AMPLIFIER

The instrumentation amplifier is one of the most useful, precise and versatile amplifiers available today. Instrumentation amplifiers are used in environments with high common-mode noise such as in data acquisition systems where remote sensing of input variables is required.

Generally, a **transducer** is used at the measuring site to obtain the required information. The transducer is a device that converts one form of energy into another. Figure 9.1 shows the block diagram of an instrumentation system. An instrumentation amplifier is used to measure the output signal produced by a transducer and often to control the physical signal producing it. The input stage of the system consists of a preamplifier and some sort of transducer, depending on the physical quantity to be measured. The output stage may consist of meters, oscilloscopes, charts or magnetic recorders.

The connecting lines between the blocks represent transmission lines. Transmission lines are used when the transducer is at a remote site monitoring hazardous conditions such as high temperatures or the liquid levels of flammable chemicals. These transmission lines send signals from one unit to another. The length of the transmission lines mainly depends on the physical quantities to be monitored and on system requirements. The output of the transducer is the signal source of the instrumentation amplifier.

The instrumentation amplifier must amplify the small signal from the remote sensor and reject the large common-mode voltage. In short, the instrumentation amplifier is intended for precise, low level signal amplification where low noise, low thermal and time drifts, high input resistance, low output offset, low output resistance and accurate closed-loop gain are required. Besides, low power consumption, high common-mode rejection ratio and high slew rate are the key characteristics for superior performance.

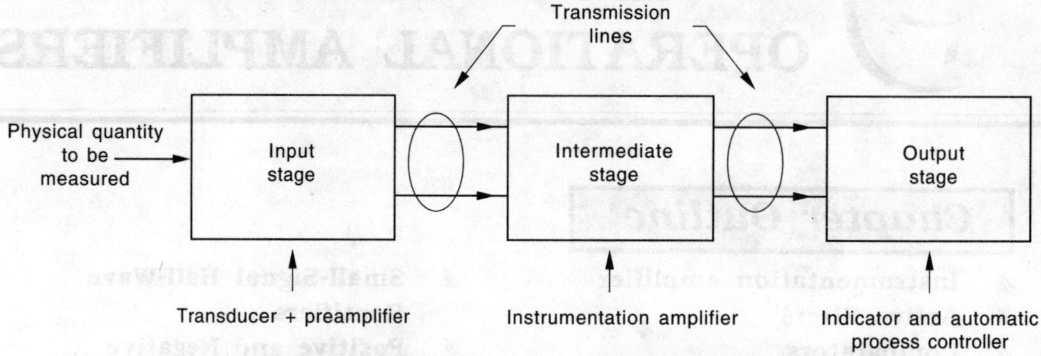


**Fig. 9.1 Block diagram of an instrumentation system**

There are many instrumentation operational amplifier ICs, such as the μA 725, ICL 7605, LH 0036, AD 622, AD 524 and AD 547 that make a circuit extremely stable and accurate. These ICs are relatively expensive, because they are very precise special purpose circuits in which most of the electrical parameters are optimized. Most of the instrumentation systems use a transducer in a bridge circuit.

## A Specific Instrumentation Amplifier

The schematic diagram of an instrumentation amplifier, the AD 622 is shown in Fig. 9.2 where IC pin numbers are given for reference.

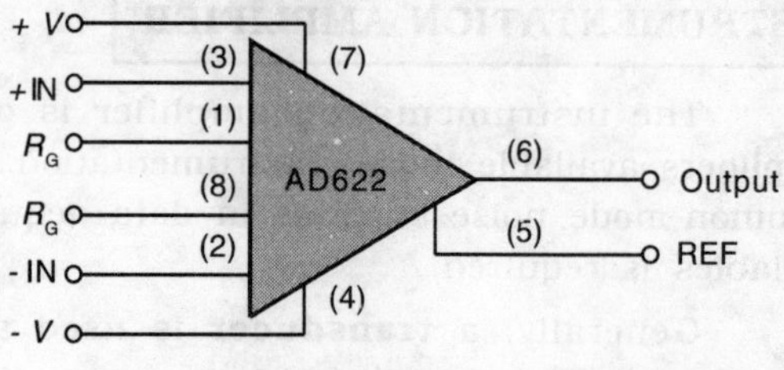


**Fig. 9.2 Schematic diagram of instrumentation amplifier AD 622**

The voltage gain of IC AD 622 can be adjusted from 2 to 1000 with an external resistor $R_G$. There is unity gain with no external resistor. The input impedance is 10 $G\Omega$. The CMRR has a minimum value of 66 *dB*. It has a bandwidth of 800 *kHz* at a gain of 10 and a slew rate of 1.2 $V/\mu s$.

## The Basic Instrumentation Amplifier

The basic instrumentation amplifier is made up of three operational amplifiers and several resistors. Op-amps $A_1$ and $A_2$ are non-inverting configurations which provide high input impedance and voltage gain. Op-amp $A_3$ is a unity-gain differential amplifier. $R_G$ is a gain setting resistor.

Op-amp $A_1$ receives the differential input signal $V_{i1}$ on its non-inverting input (+) and amplifies this signal with a voltage gain of

$$A_v = 1 + \frac{R_1}{R_G}$$

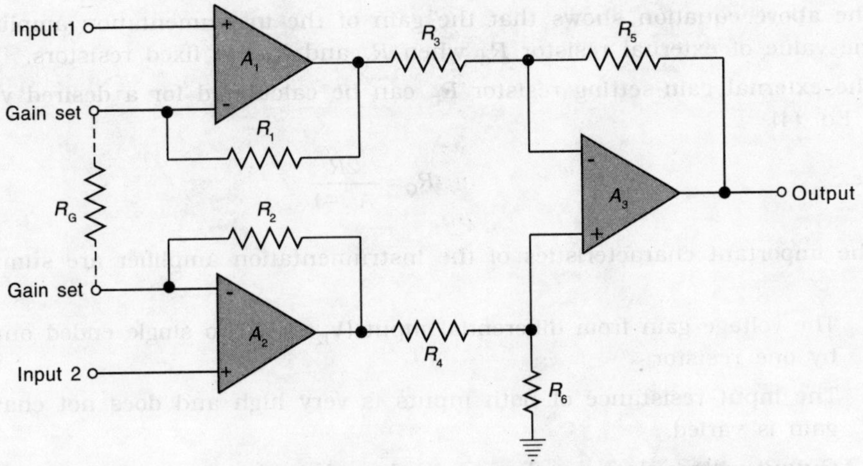


**Fig. 9.3 The basic instrumentation amplifier using three op-amps**

Op-amp $A_1$ also has $V_{i2}$ as an input signal to its inverting (–) input through op-amp $A_2$ and the path formed by $R_2$ and $R_G$. The input signal $V_{i2}$ is also amplified by op-amp $A_1$ with a voltage gain of

$$A_v = \frac{R_1}{R_G}$$

The common-mode voltage $V_{cm}$ on the non-inverting input is also amplified by the small common-mode gain of op-amp $A_1$. The total output voltage of op-amp $A_1$ is given by

$$V_{01} = \left(1 + \frac{R_1}{R_G}\right) V_{i1} - \left(\frac{R_1}{R_G}\right) V_{i2} + V_{cm} \qquad \longrightarrow (1)$$

Similarly, the total output voltage of op-amp $A_2$ is given by

$$V_{02} = \left(1 + \frac{R_2}{R_G}\right) V_{i2} - \left(\frac{R_2}{R_G}\right) V_{i1} + V_{cm} \qquad \longrightarrow (2)$$

Total output voltage of op-amp $A_3$ is the difference between $V_{01}$ and $V_{02}$. Hence

$$V_0 = V_{02} - V_{01}$$

$$V_0 = \left(1 + \frac{R_1}{R_G} + \frac{R_2}{R_G}\right) V_{i2} - \left(1 + \frac{R_1}{R_G} + \frac{R_2}{R_G}\right) V_{i1}$$

If $R_1 = R_2 = R$, then

$$V_0 = \left(1 + \frac{2R}{R_G}\right) V_{i2} - \left(1 + \frac{2R}{R_G}\right) V_{i1}$$

$$V_0 = \left(1 + \frac{2R}{R_G}\right) (V_{i2} - V_{i1}) \qquad \longrightarrow (3)$$

The closed-loop gain of instrumentation amplifier is given by

$$A_{vf} = \frac{V_0}{V_{i2} - V_{i1}}$$

or

$$\mathbf{A_{vf} = 1 + \frac{2R}{R_G}} \qquad \longrightarrow (4)$$

The above equation shows that the gain of the instrumentation amplifier can be set by the value of external resistor $R_G$ when $R_1$ and $R_2$ are fixed resistors.

The external gain-setting resistor $R_G$ can be calculated for a desired voltage gain by using Eq. (4).

i.e.,
$$R_G = \frac{2R}{A_{vf} - 1}$$

The important characteristics of the instrumentation amplifier are summarised as follows:

1. The voltage gain from differential input $(V_{i2} - V_{i1})$ to single ended output, is set by one resistor.
2. The input resistance of both inputs is very high and does not change as the gain is varied.
3. Output voltage does not depend on the voltage common to both $V_{i1}$ and $V_{i2}$, and depends only on their difference, as it provides very high CMRR.

## Instrumentation Amplifier Using Transducer Bridge

Figure 9.4 shows the circuit of a differential instrumentation amplifier with a transducer bridge. A resistive transducer whose resistance changes as a function of some physical energy is connected in one arm of the bridge and is denoted by $(R_T \pm \Delta R)$, where $R_T$ is the resistance of the transducer and $\Delta R$ the change in resistance.

When the bridge is balanced,
$$V_b = V_a$$

According to voltage-divider rule,
$$\frac{R_B V}{R_B + R_C} = \frac{R_A V}{R_A + R_T}$$

or
$$\frac{R_B}{R_B + R_C} = \frac{R_A}{R_A + R_T}$$

or
$$\frac{R_C}{R_B} = \frac{R_T}{R_A} \longrightarrow (1)$$

Generally, resistors $R_A$, $R_B$ and $R_C$ are selected so that they are equal in value to the transducer resistance $R_T$ at some **reference condition.** The reference condition is the specific value of some physical quantity under measurement at which the bridge is balanced. This specific value is normally formed by the designer and depends on the transducer's characteristics, the type of the physical quantity to be measured and the desired application.

Initially, the bridge is balanced at a desired reference condition. As the physical quantity to be measured changes, the resistance of the transducer also changes that causes the bridge to unbalance $(V_b \neq V_a)$.

Let the change in resistance of the transducer be $\Delta R$. Since $R_B$, $R_C$ and $V_b$ have constant values, the voltage $V_a$ only varies as a function of the change in transducer resistance. Applying voltage divider rule, we have

$$V_a = \frac{R_A V}{R_A + (R_T + \Delta R)}$$

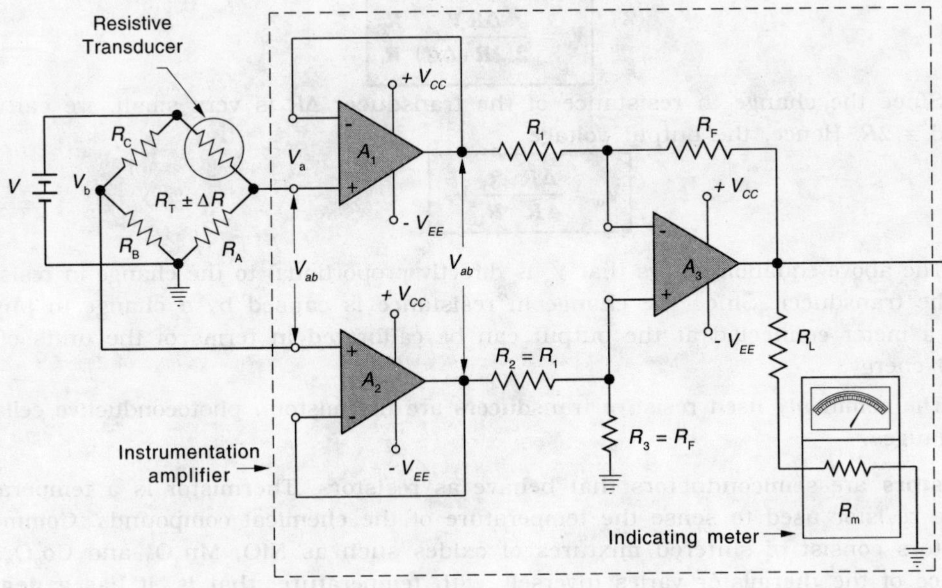


**Fig. 9.4 Differential instrumentation amplifier using a transducer bridge**

The differential voltage $V_{ab}$ across the output terminals of the bridge is given by

$$V_{ab} = V_a - V_b$$

$$= \frac{R_A V}{R_A + (R_T + \Delta R)} - \frac{R_B V}{R_B + R_C}$$

If $R_A = R_B = R_C = R_T = R$, (say), then

$$V_{ab} = \frac{RV}{2R + \Delta R} - \frac{RV}{2R}$$

$$V_{ab} = -\frac{\Delta R V}{2(2R + \Delta R)} \qquad \longrightarrow (2)$$

The negative(–) sign in the above equation shows that $V_a < V_b$ because of the increase in the value of $R_T$. It also shows that $V$ should be made large to maximize the bridge differential output voltage, $V_{ab}$.

The differential output voltage $V_{ab}$ of the bridge is then applied to the differential instrumentation amplifier. The voltage followers preceding the basic differential amplifier help to eliminate loading of the bridge circuit. Since the basic differential amplifier is an inverting amplifier, its voltage gain is $(-R_F/R_1)$. Therefore, the output voltage $V_0$ of the circuit is given by

$$V_0 = \left(-\frac{R_F}{R_1}\right) V_{ab}$$

$$= \frac{-R_F}{R_1} \times \left[\frac{-\Delta R\, V}{2(2R + \Delta R)}\right]$$

$$V_o = \frac{\Delta R\, V}{2(2R+\Delta R)} \cdot \frac{R_F}{R_1} \qquad \longrightarrow (3)$$

Since the change in resistance of the transducer $\Delta R$ is very small, we can write $(2R + \Delta R) \cong 2R$. Hence, the output voltage

$$V_o = \frac{\Delta R}{4R} \frac{R_F}{R_1} V \qquad \longrightarrow (4)$$

The above equation shows that $V_o$ is directly proportional to the change in resistance $\Delta R$ of the transducer. Since the change in resistance is caused by a change in physical energy, a meter connected at the output can be calibrated in terms of the units of that physical energy.

The commonly used resistive transducers are thermistors, photoconductive cells and strain gauges.

**Thermistors** are semiconductors that behave as resistors. Thermistor is a temperature-sensitive resistor used to sense the temperature of the chemical compounds. Commercial thermistors consist of sintered mixtures of oxides such as NiO, $Mn_2O_3$ and $Co_2O_3$. The resistance of the thermistor varies inversely with temperature; that is, it has a **negative temperature coefficient of resistance**, measured in ohm per unit change in degree celsius. Essentially, the thermistor converts temperature to resistance. The temperature of the chemical mixture determines the resistance of the thermistor. Thermistors with a high temperature coefficient of resistance are more sensitive to temperature change and are therefore used to measure and control the temperature. Thermistors are available in a wide variety of shapes and sizes.

**Photoconductive cell** belongs to the family of photodetectors whose resistance varies with an incident radiant energy or light. As the intensity of incident light increases, the resistance of the cell decreases. In the dark the resistance of the photoconductive cell is about $10\, M\Omega$ at room temperature for a 1.4 *cm* diameter cell. The intensity of light is expressed in metre candles (lux).

A common semiconductor for a photoconductive cell is cadmium sulphide (CdS), since it has a spectral response similar to that of the human eye. The primary advantages of CdS photoconductors are their high dissipation capability, their excellent sensitivity in the visible spectrum and their low resistance when stimulated by light. Hence a CdS photoconductor can operate a relay directly, without intermediate amplifier circuits.

Some photoconductive cells are very sensitive to light and hence can be used in UV and IR regions. The photoconductive cell is typically composed of a ceramic base, a layer of photoconductive material (CdS), a moisture-proof enclosure and metallic leads. They are also known as photocells or light-dependent resistors (LDRs).

A **Strain gauge** is a conducting wire whose resistance changes by a small amount when it is lengthened or shortened. The strain gauge is bonded to a structure so that the percent change in length of the strain gauge and structure are identical.

Strain gauges are made from metal alloys such as constantan, nichrome V, dynaloy, stabiloy or platinum alloy. For high temperature work they are made of wire. For moderate temperature, strain gauges are made by forming the metal alloy into very thin sheets by a photoetching process.

While mounting a strain gauge, the surface of the mounting beam must be cleaned, sanded and rinsed with alcohol, Freon or Methyl ethyl ketone. The gauge is then fastened permanently to the cleaned surface by Eastman 910, epoxy, polymide adhesive or ceramic cement.

Semiconductors strain gauges are more sensitive than the wire type and therefore provide better accuracy and resolution. The sensitivity of a strain gauge is defined as unit change in resistance per unit change in length. It is a dimensionless quantity.

1. **Temperature Indicator** The circuit of Fig. 9.4 can be used as a temperature indicator if the transducer in the bridge circuit is a thermistor and the output meter is calibrated in degrees Celsius. The bridge can be balanced at a desired reference condition. As the temperature varies from its reference value, the resistance of the thermistor changes and thereby the bridge becomes unbalanced. The meter then shows the deflection and it can be calibrated to read a desired temperature range by selecting an appropriate gain for the differential instrumentation amplifier. If $\Delta R$ is the change in the value of the thermistor resistance, then $\Delta R$ = (Temperature coefficient of resistance) × (Final temperature - Reference temperature).

2. **Temperature Controller** The circuit of temperature controller may be constructed by using a thermistor in the bridge circuit and by replacing a meter with a relay in the circuit of Fig. 9.4. The output of the differential instrumentation amplifier drives a relay that controls the current in the heat-generating circuit.

3. **Light-intensity meter** The circuit of Fig. 9.4 can be used as a light-intensity meter if a transducer in the bridge circuit is a photocell. The bridge can be balanced for darkness condition. When exposed to light, the bridge gets unbalanced and meter deflects. The meter can be calibrated in terms of lux to measure the change in light intensity.

4. **Analog weight scale** The circuit in Fig. 9.4 can be used as a weight scale with transducer as strain gauge.

In the analog weight scale, strain gauge elements are connected in all four arms of the bridge. The gauge elements are mounted on the base of the weight platform. When an external force or weight is applied to the platform, one pair of elements in opposite arms elongates, whereas the other pair of elements in opposite arms compress. In other words, when an object exerts a force, $R_{T_1}$ and $R_{T_3}$ both decrease in resistance and $R_{T_2}$ and $R_{T_4}$ both increase in resistance, or vice versa [Fig. 9.5].

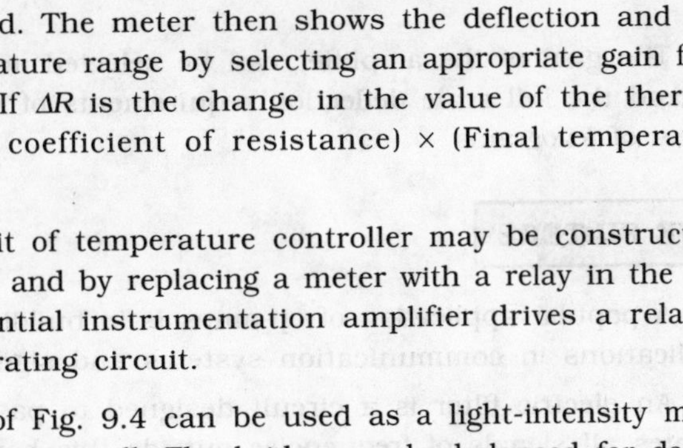

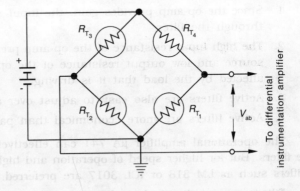

**Fig. 9.5 Strain gauge bridge circuit for analog weight scale**

When no weight is placed on the platform, the bridge is balanced, $R_{T_1} = R_{T_2} = R_{T_3} = R_{T_4} = R$ (say) and therefore output voltage of the weight scale is zero.

When a weight is placed on the scale platform, the bridge becomes unbalanced. If $\Delta R$ is increase or decrease in resistance by the same amount, then the resulting unbalanced voltage is given by

$$V_{ab} = -V \cdot \frac{\Delta R}{R} \qquad \longrightarrow (1)$$

Since $V_a < V_b$, the output voltage $V_{ab}$ is negative.

The voltage $V_{ab}$ is amplified by the differential instrumentation amplifier, which drives the meter. Since the gain of the amplifier is $(-R_F/R_1)$, the output voltage is given by

$$V_o = V\left(\frac{\Delta R}{R}\right)\left(\frac{R_F}{R_1}\right) \qquad \longrightarrow (2)$$

The gain of the amplifier can be selected according to the sensitivity of the strain gauge and the full scale deflection requirements of the meter. The meter can be calibrated in terms of kilograms.

## ACTIVE FILTERS

A popular application of op-amps is in building active filters. Filters find wide range of applications in communication systems and signal processing.

An electric filter is a circuit designed to pass a specified band of frequencies and attenuates all signals of frequencies outside this band. This property is called **selectivity.** Filter networks may be either active or passive. **Passive filter networks** contain only resistors, inductors and capacitors. Active filters use the transistors or op-amps combined with passive *RC*, *RL* or *RLC* circuits. The active devices provide voltage gain along with frequency selectivity whereas the passive circuits provide only frequency selectivity.

An active filter offers the following advantages over a passive filter:

1. Since the op-amp provides gain, the input signal is not attenuated as it passes through the filter.
2. The high input resistance of the op-amp prevents excessive loading of the driving source and low output resistance of the op-amp prevents the filter from being affected by the load that it is driving.
3. Active filters are also easy to adjust over a wide range of frequencies.
4. Active filters are more economical than passive filters.

The operational amplifier μA 741 can effectively serve as an active component in active filters. But for higher speed of operation and higher unity gain-bandwidth operational amplifiers such as LM 318 or ICL 3017 are preferred.

Filters can be classified as: low-pass, high-pass, band-pass, band-elimination, and all-pass filters.

The **passband** of a filter is the range of frequencies that are allowed to pass through the filter with minimum attenuation (usually defined as less than -3 *dB* of attenuation).

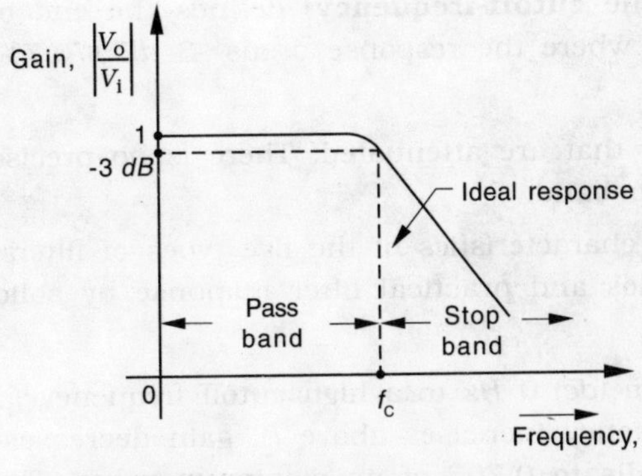


(a) Comparison of an ideal low-pass filter response with actual response

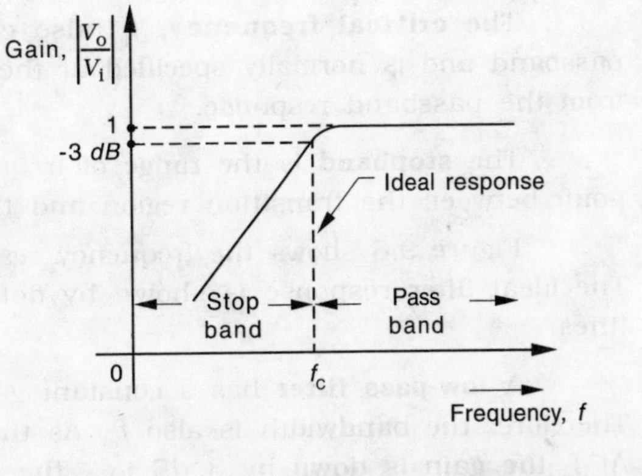


(b) Comparison of an ideal high-pass filter response with actual response

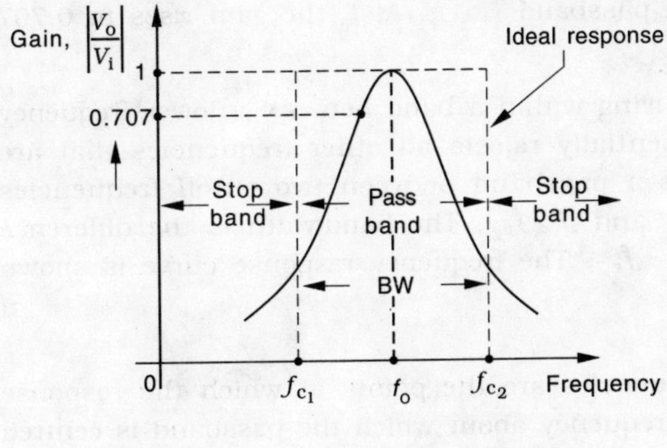


(c) Comparison of an ideal band-pass filter response with actual response

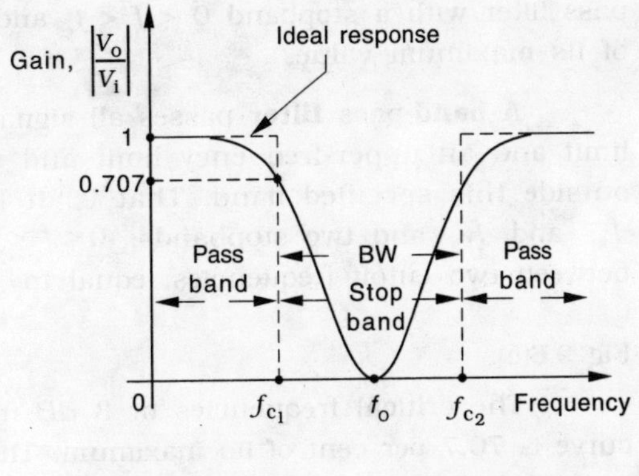


(d) Comparison of an ideal band-stop filter response with actual response

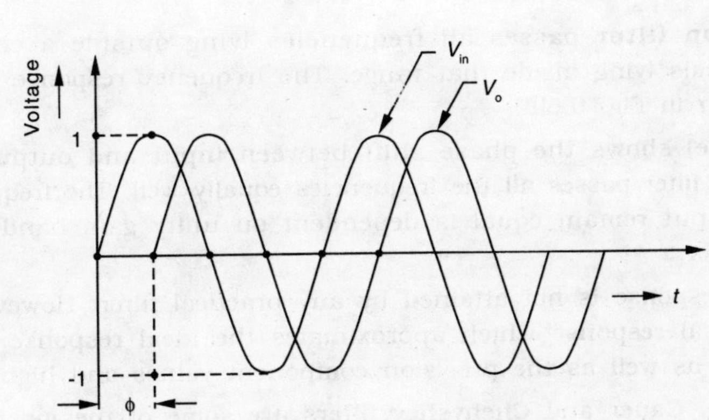


(e) Phase shift between input and output voltages of an all-pass filter

**Fig. 9.6 Frequency response of the major active filters**

The **critical frequency,** $f_c$ (also called the **cutoff-frequency**) defines the end of passband and is normally specified at the point where the response drops -3 *dB* (70.7%) from the passband response.

The **stopband** is the range of frequencies that are attenuated. There is no precise point between the transition region and the stop band.

Figure 9.6 shows the frequency response characteristics of the five types of filters. The ideal filter response is shown by dotted lines and practical filter response by solid lines.

A **low-pass filter** has a constant gain from (*dc*) 0 *Hz* to a high cutoff frequency $f_c$. Therefore, the bandwidth is also $f_c$. As the frequency increases above $f_c$, gain decreases. At $f_c$ the gain is down by 3 *dB* i.e., the gain falls to 0.707 of its maximum value. The frequency response of the low-pass filter is shown in Fig. 9.6 (a).

A **high-pass filter** attenuates the voltage gain for all frequencies below the cutoff frequency. Above $f_c$, the magnitude of the gain is constant. Figure 9.6(b) shows a high-pass filter with a stopband $0 < f < f_c$ and a passband $f > f_c$. At $f_c$ the gain rises to 0.707 of its maximum value.

A **band-pass filter** passes all signals lying within a band between a lower frequency limit and an upper-frequency limit and essentially rejects all other frequencies that are outside this specified band. That is, it has a passband between two cutoff frequencies $f_{c_1}$ and $f_{c_2}$ and two stopbands: $0 < f < f_{c_1}$ and $f > f_{c_2}$. The bandwidth is the difference between two cutoff frequencies, equal to $f_{c_2} - f_{c_1}$. The frequency response curve is shown in Fig 9.6(c).

The critical frequencies or 3 *dB* frequencies are the points at which the response curve is 70.7 per cent of its maximum. The frequency about which the passband is centred is called the **centre frequency**, $f_0$, defined as the geometric mean of the critical frequencies.

$$f_0 = \sqrt{f_1 f_2}$$

A **band-stop filter** passes all frequencies lying outside a certain range and attenuates all signals lying inside that range. The frequency response curve of a band-stop filter is shown in Fig. 9.6(d).

Figure 9.6(e) shows the phase shift between input and output voltages of an all-pass filter. This filter passes all the frequencies equally well. The frequency upto which the input and output remain equal is dependent on unity gain bandwidth product of operational amplifier.

The ideal response is not attained by any practical filter. However, it is possible to obtain a practical response which approximates the ideal response by using special design techniques, as well as the precision component values and high-speed op-amps.

Butterworth, Cauer and Chebyshev filters are some of the most commonly used practical filters that approximate the ideal response.

The actual response curves of the filters in the stopband either steadily decrease or increase or both with increase in frequency. The rate at which the gain of the filter changes in the stopband is determined by the order of the filter.

## Introduction to Butterworth Filter

In many low-pass filter applications, it is necessary for the closed-loop gain to be close to 1 within the passband. The **Butterworth filter** is best suited for this type of application. The Butterworth filter is also called **maximally flat** or **flat-flat filter.** The Butterworth characteristic provides a very flat amplitude response in the passband and a roll-off rate of –20 *dB*/decade. But, the phase response is not linear and the phase shift (or time delay) of signals passing through the filter varies nonlinearly with frequency. Therefore, a pulse applied to a filter with a Butterworth response will cause overshoots on the output because each frequency component of the pulse's rising and falling edges experiences a different time delay. **The rate of decrease in gain, below or above the critical frequency of the filter is known as roll-off.**

Figure 9.7 shows the ideal (dashed line) and the practical (solid lines) frequency response for three types of Butterworth filters. As the roll-offs become steeper, they approach the ideal filter.

Butterworth filters are not designed to keep a constant phase angle at the cutoff frequency. A basic low-pass filter of –20 *dB*/decade has a phase angle of –45° at $f_c$. A –40 *dB*/decade Butterworth filter has a phase angle of –90° at $f_c$, a –60 *dB*/decade filter has a phase angle of –135° at $f_c$. Therefore, for each increase of –20 *dB*/decade, the phase angle increases by –45° at $f_c$. So, a first-order filter has a roll-off rate of –20 *dB*/decade; a second-order filter has a roll-off rate of –40 *dB*/decade, a third-order filter has a roll-off rate of –60 *dB*/decade; and so on.

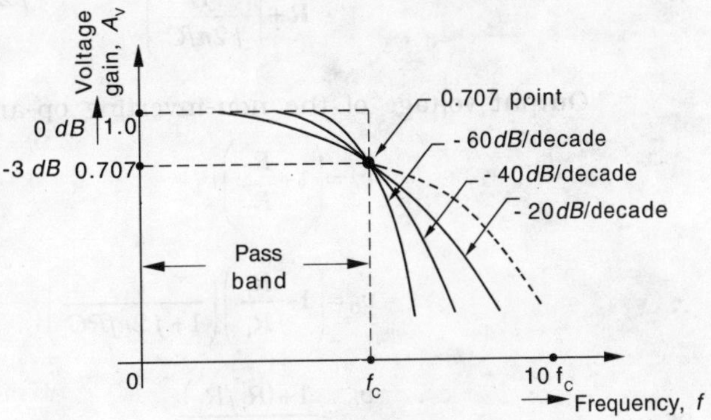


**Fig. 9.7 Frequency response curves for three types of low-pass Butterworth filters**

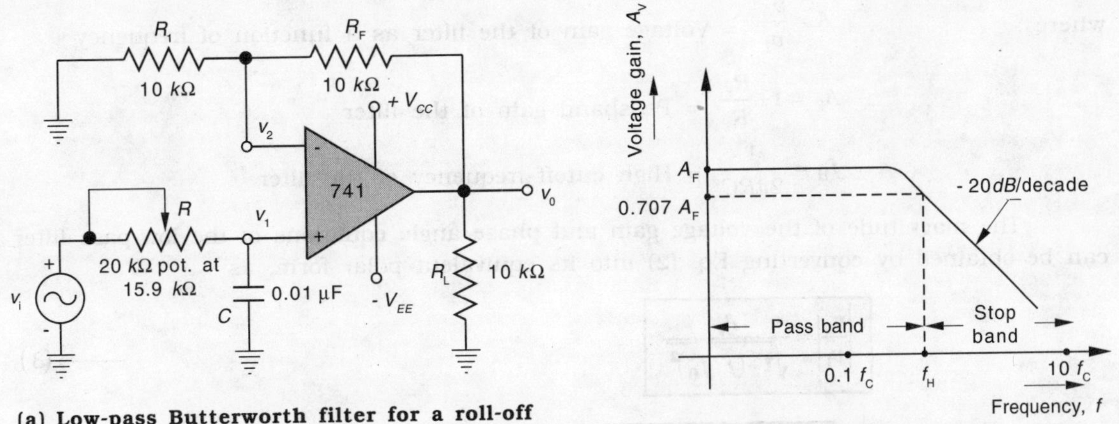


**(a) Low-pass Butterworth filter for a roll-off of -20 *dB*/decade**

**(b) Frequency response**

**Fig. 9.8 First-order low-pass Butterworth filter**

### First-Order Low-Pass Butterworth Filter

A first-order low-pass Butterworth filter that uses an *RC* network for filtering is shown in Fig. 9.8(a). The op-amp is used in the non-inverting configuration. Hence it does not load down the *RC* network. Resistors $R_1$ and $R_f$ control the gain to the desired level.

According to the voltage divider rule, the voltage at the non-inverting terminal i.e., across the capacitor *C* is given by

$$v_1 = \frac{-jX_C}{R - jX_C} \cdot v_i \qquad \longrightarrow (1)$$

where $j = \sqrt{-1}$ and $-jX_c = \frac{1}{j\,\omega_c} = \frac{1}{j\,2\pi fC}$

$$\therefore \quad v_1 = \frac{\frac{1}{j\,2\pi fC}}{R + \left(\frac{1}{j\,2\pi fC}\right)} \cdot v_i = \frac{v_i}{1 + j\,2\pi fRC}$$

Output voltage of the non-inverting op-amp is

$$v_O = \left(1 + \frac{R_f}{R_1}\right) v_1$$

$$\therefore \quad v_O = \left(1 + \frac{R_f}{R_1}\right)\left(\frac{v_i}{1 + j\,2\pi fRC}\right)$$

or

$$\frac{v_O}{v_i} = \frac{1 + (R_f/R_1)}{1 + j\,2\pi fRC}$$

or

$$A = \frac{v_O}{v_i} = \frac{A_F}{1 + j(f/f_H)} \qquad \longrightarrow (2)$$

where $A = \frac{v_O}{v_i}$ = Voltage gain of the filter as a function of frequency

$A_F = 1 + \frac{R_F}{R_1}$ = Passband gain of the filter

$f_H = \frac{1}{2\pi RC}$ = High cutoff frequency of the filter

The magnitude of the voltage gain and phase angle equations of the low-pass filter can be obtained by converting Eq. (2) into its equivalent polar form, as

$$\left|\frac{v_o}{v_i}\right| = \frac{A_F}{\sqrt{1 + (f/f_H)^2}} \qquad \longrightarrow (3)$$

and phase angle,

$$\phi = -\tan^{-1}(f/f_H) \qquad \longrightarrow (4)$$

**Special cases:**

1. At very low frequencies, that $f < f_H$, $\left|\frac{v_o}{v_i}\right| = A_F$

2. At $f = f_H$, $\left|\frac{v_o}{v_i}\right| = \frac{A_F}{\sqrt{2}} = 0.707\ A_F$

3. At $f > f_H$, $\left|\frac{v_o}{v_i}\right| < A_F$

Thus the low-pass filter has a constant voltage gain $A_F$ from 0 *Hz* to the high cutoff frequency $f_c$. At cutoff frequency $f_c$ the voltage gain is 0.707 $A_F$. After $f_c$ the gain decreases at a constant rate with increase in frequency. That is, when the frequency is increased ten-fold (10*f*), the voltage gain is divided by 10. In other words, the gain decreases by 20 *dB* (= 20 log 10) each time the frequency is increased by 10. Hence the rate at which the voltage gain rolls off after cutoff frequency $f_c$ is 20 *dB*/decade or 6 *dB*/octave, where octave signifies a two-fold increase in frequency. If the cutoff frequency is 1 *kHz*, then the gain decreases by 20 *dB* when the frequency changes from 10 *kHz* to 100 *kHz*. It changes another 20 *dB* when the frequency increases from 100 *kHz* to 1 *MHz*.

## Filter Design

1. Choose the cutoff frequency, $f_c$.
2. Select a value of capacitor *C* less than or equal to 1 μ*F*. Mylar or tantalum capacitors are recommended for better performance.
3. Calculate the value of *R* using $R = \frac{1}{2\pi f_c C}$
4. Finally, select the values of $R_1$ and $R_f$ dependent on the desired passband gain $A_F$ using $A_F = 1 + \frac{R_f}{R_1}$

## Frequency Scaling

Once a filter is designed, we may need to change its cutoff frequency. The procedure of converting an original cutoff frequency $f_c$ to a new cutoff frequency $f_c'$ is known as **frequency scaling.** The frequency scaling is done by multiplying either *C* or *R* by the ratio of the original cutoff frequency to the new cutoff frequency. Generally capacitors are kept constant but resistor values are changed by using potentiometer, i.e., $R' = R\left(\frac{f_c}{f_c'}\right)$.

## First-order High-Pass Butterworth Filter

In high-pass filters, the role of the capacitor and resistor are reversed. A high-pass active filter with a 20 *dB*/decade roll-off is shown in Fig. 9.9(a). Notice that the input circuit is a single high-pass *RC* circuit. The high-pass response curve is shown in Fig. 9.9(b). At cutoff frequency the magnitude of the gain is 0.707 times the passband value. All frequencies above $f_L$ are passband frequencies, with the highest frequency determined by the closed loop bandwidth of the op-amp.

Let $v_1$ and $v_2$ be the voltages at the non-inverting and inverting terminals of op-amp respectively.

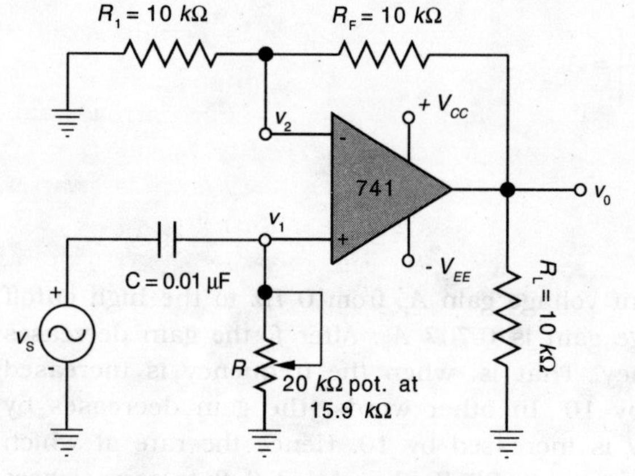


(a) **High-pass Butterworth filter for a roll of -20 *dB*/decade**

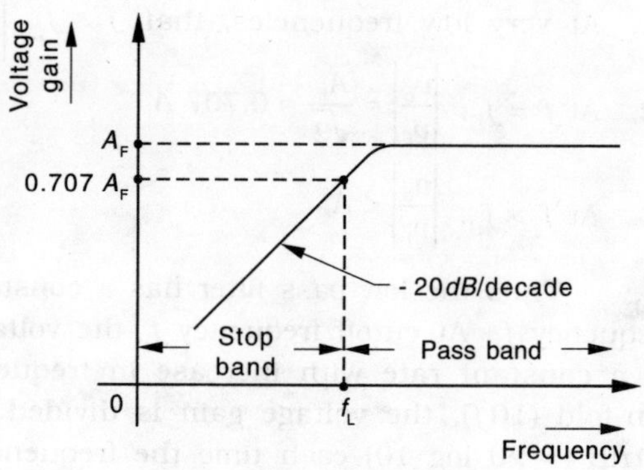


(b) **Frequency response**

**Fig. 9.9 First-order High-pass Butterworth filter**

According to the voltage divider rule, the voltage at the non-inverting terminal is given by

$$v_1 = \frac{v_i}{R - jX_c} R$$

$$v_1 = \frac{v_i}{R + \dfrac{1}{j\,2\pi RC}} R = \frac{j\,2\pi fRC}{1 + j\,2\pi RC} v_1$$

Output voltage of the non-inverting op-amp is

$$v_o = \left(1 + \frac{R_f}{R_1}\right) v_1$$

$$= \left(1 + \frac{R_f}{R_1}\right) \frac{j\,2\pi fRC}{1 + j\,2\pi RC} \cdot v_i$$

or

$$\frac{v_o}{v_i} = A_F \left[\frac{j\,2\pi fRC}{1 + j\,2\pi RC}\right]$$

$$A = \frac{v_o}{v_i} = A_F \left[\frac{j(f/f_L)}{1 + j(f/f_L)}\right] \longrightarrow (1)$$

where $A_F = 1 + \left(\dfrac{R_f}{R_1}\right)$ = Passband gain of the filter

$f$ = Frequency of the input signal

$f_L = \dfrac{1}{2\pi RC}$ = Low cutoff frequency

Magnitude of the voltage gain is

$$A = \left|\frac{v_o}{v_i}\right| = \frac{A_F (f/f_L)}{\sqrt{1 + (f/f_L)^2}} \longrightarrow (2)$$

## Band-Pass Filter

A band-pass filter is a circuit designed to pass signals only in a certain band of frequencies while rejecting all signals outside this band.

There are two types of band-pass filters: (i) wide band-pass and (ii) narrow band-pass. A wide band-pass filter is a circuit if its **figure of merit** or **quality factor** $Q < 10$. On the other hand, a band-pass filter, with its quality factor $Q > 10$ is known as narrow band-pass filter. Thus $Q$ is a measure of selectivity, meaning the higher the value of $Q$, the more selective is the filter or the narrower its bandwidth. The relation between quality factor $Q$, the 8 $dB$ bandwidth and the centre frequency $f_o$ is given by

$$Q = \frac{f_o}{BW} = \frac{f_o}{f_H - f_L} \longrightarrow (3)$$

### (i) Wide Band-Pass Filter

A wide band-pass filter is a cascaded arrangement of a high-pass filter and a low-pass filter. The order of band-pass filter depends on the order of the high-pass and low-pass filter sections. $Q$ values less than 10 are typical in this type of filter.

Figure 9.10 shows the ± 20 $dB$/decade wide band-pass filter which is a combination of first-order high-pass and first order low-pass filters. The critical frequency of each filter

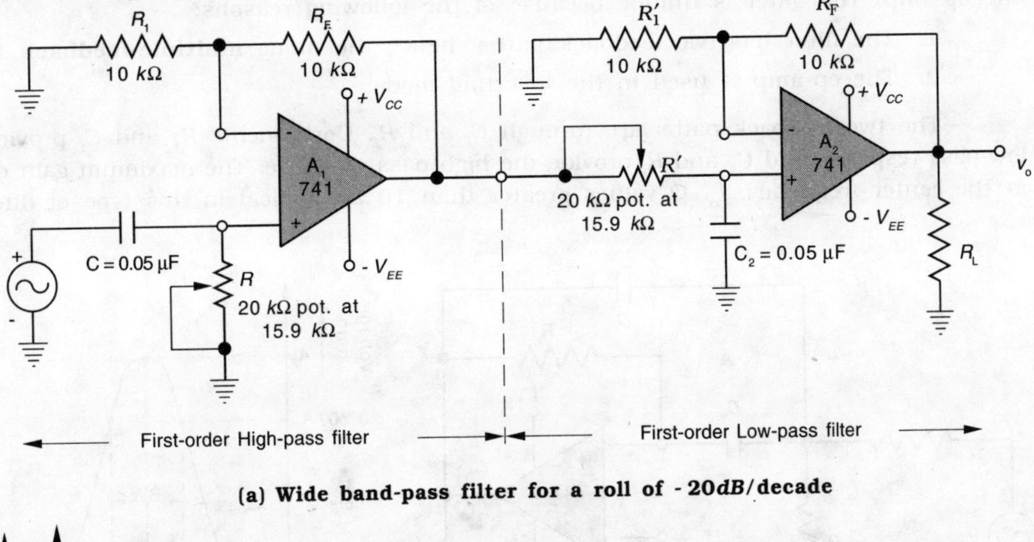


(a) Wide band-pass filter for a roll of - 20$dB$/decade

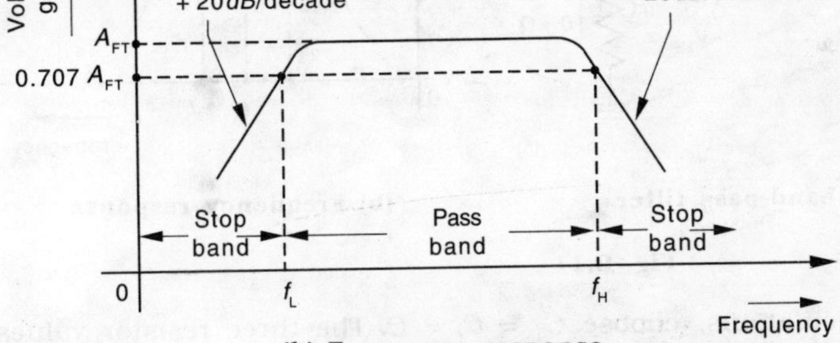


(b) Frequency response

**Fig. 9.10 First-order wide band-pass filter**

is chosen so that the response curves overlap sufficiently. To realise a band-pass response, $f_H$ must be larger than $f_L$.

The voltage gain magnitude of the band-pass filter is equal to the product of the voltage gain magnitudes of the high-pass and low-pass filters.

$$\left|\frac{v_o}{v_i}\right| = \frac{A_F}{\sqrt{1+(f/f_L)^2}} \times \frac{A_F(f/f_L)}{\sqrt{1+(f/f_H)^2}}$$

$$A = \left|\frac{v_o}{v_i}\right| = \frac{A_{FT}(f/f_L)}{\sqrt{\left[1+(f/f_L)^2\right]\left[1+(f/f_H)^2\right]}}$$

where
$A_{FT}$ = Total passband gain
$f$ = Frequency of the input signal
$f_L$ = Low cutoff frequency
$f_H$ = High cutoff frequency

**(ii) Narrow Band-Pass Filter**

Figure 9.11 shows the circuit of a multiple feedback band-pass filter using only one op-amp. This filter is unique because of the following reasons:

1. The filter has two feedback paths, hence the name **multiple-feedback filter.**
2. The op-amp is used in the inverting mode.

The two feedback paths are through $C_f$ and $R_f$. Components $R_1$ and $C_f$ provide the low-pass response and $C_1$ and $R_f$ provide the high-pass response. The maximum gain occurs at the center frequency, $f_o$. $Q$ values greater than 10 are typical in this type of filter.

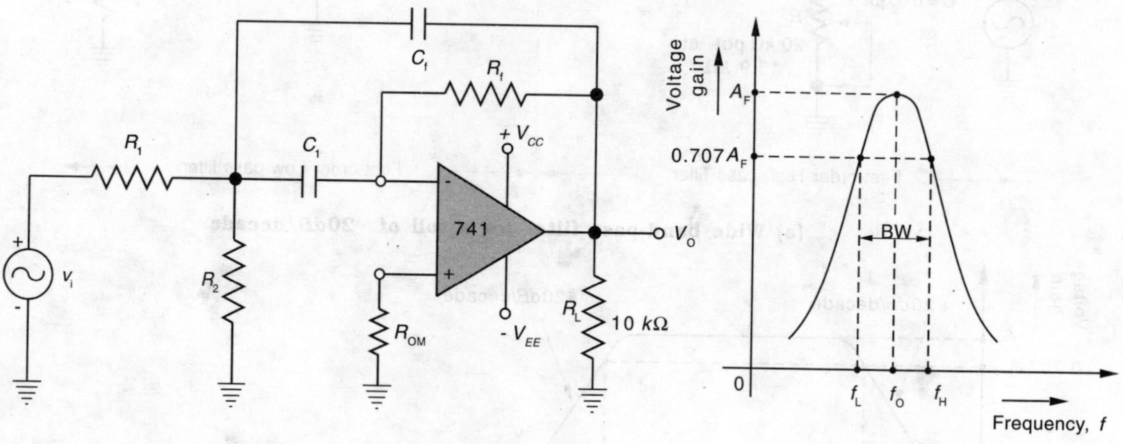


**(a) Multiple feedback narrow band-pass filter** **(b) Frequency response**

**Fig. 9.11**

To simplify the design calculations, choose $C_1 = C_f = C$. The three resistor values are calculated based on the desired value of $f_o$.

The quality factor $Q$ can be determined from the relation $Q = f_o/\text{BW}$. The resistor values can be found using the following formulae:

$$R_1 = \frac{Q}{2\pi f_o C A_F} \longrightarrow (1)$$

$$R_2 = \frac{Q}{2\pi f_o C\,(2Q^2 - A_F)} \longrightarrow (2)$$

$$R_f = \frac{Q}{\pi f_o C} \longrightarrow (3)$$

where $A_F$ is gain at center frequency $f_o$.

To develop a gain expression, solve for $Q$ in the $R_1$ and $R_f$ formulae as follows:

$$Q = 2\pi f_o A_F C R_1$$

$$Q = \pi f_o C R_f$$

Equating, $$2\pi f_o A_F C R_1 = \pi f_o C R_f$$

or $$2\,A_F R_1 = R_f$$

or $$\boldsymbol{A_F = \frac{R_f}{2\,R_1}} \longrightarrow (4)$$

In order for the denominator of the equation $R_2 = \dfrac{Q}{2\pi f_o C\,(2Q^2 - A_F)}$ to be positive, $A_F < 2Q^2$, which imposes a limitation on the gain.

Another advantage of the multiple feedback is that its center frequency $f_o$ can be changed to a new frequency $f_o'$ without changing the gain or bandwidth. This can be had by changing $R_2$ to $R_2'$ so that

$$R_2' = R_2\left(\frac{f_o}{f_o'}\right)^2 \longrightarrow (5)$$

The relation for center frequency is given by

$$f_o = \frac{1}{2\pi C\sqrt{(R_1 \parallel R_2)R_f}}$$

$$\boldsymbol{f_o = \frac{1}{2\pi C}\sqrt{\frac{R_1 + R_2}{R_1 R_2 R_f}}} \longrightarrow (6)$$

## Band-stop Filter

The band-stop filter is also called a **band-reject** or **band-elimination** filter. It rejects a specified band of frequencies and passes all others. The response is opposite to that of a band-pass filter. The band-stop filters can also be classified as (i) wide band-stop or (ii) narrow band-stop.

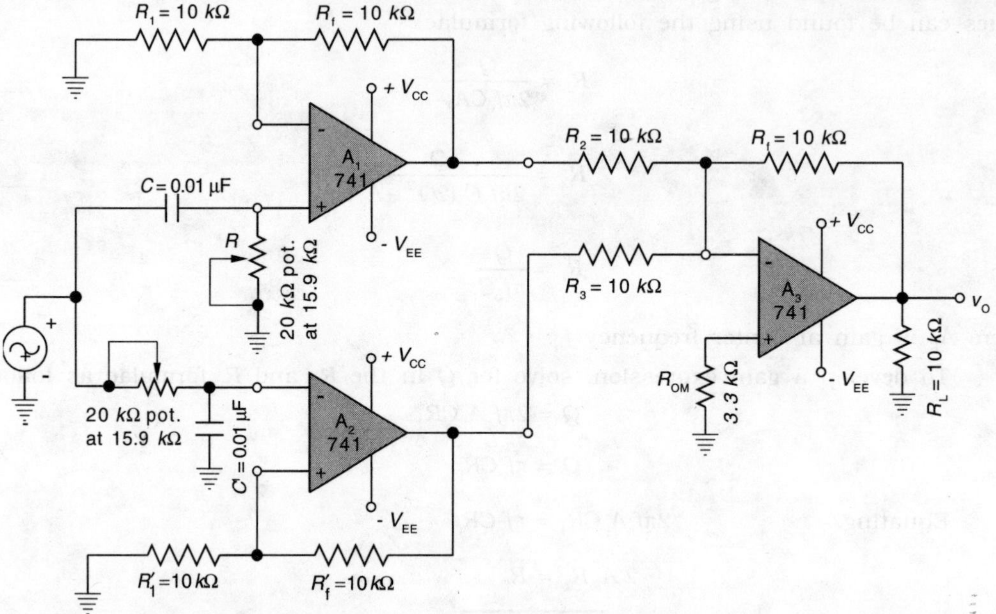


**(a) Circuit of a wide band-stop filter**

**Fig. 9.12 Wide band-stop filter**

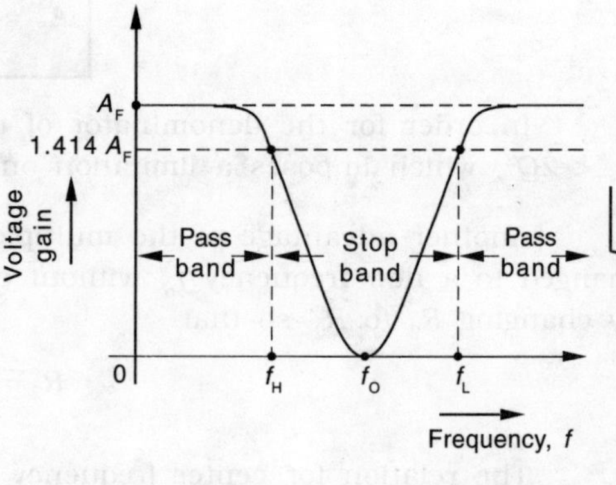


**(b) Frequency response**

## (i) Wide Band-Stop Filter

Figure 9.12 shows wide band-stop filter combining a low-pass filter, a high-pass filter and a summing amplifier. To realize a band-stop filter response, the low cutoff frequency $f_L$ of the high-pass filter must be larger than the high cutoff frequency $f_H$ of the low-pass filter. In this filter, the passband gain of both the high-pass and low-pass sections must be equal. The frequency response of the wide band-stop filter is shown in Fig. 9.12(b). The quality factor $Q$ is less than 10.

The voltage gain of wide band-stop filter changes at the rate of 20 $dB$/decade above $f_H$ and below $f_L$, with a maximum attenuation occurring at center frequency $f_o$.

## (ii) Narrow Band-Stop Filter

The narrow band-stop filter is also called the **notch filter**. It is used in minimizing the 50 $Hz$ "hum" in audio systems by setting the center frequency to 50 $Hz$. The most commonly used notch filter is the **twin - T** network shown in Fig. 9.13(a). This is a passive filter composed of two T-shaped networks. One T network is made up of two resistors and a capacitor and the other T network is made up of two capacitors and a resistor.

The **notch-out** frequency is the frequency at which maximum attenuation occurs and is given by

$$f_N = \frac{1}{2\pi RC}$$

The passive twin-T network has a relatively low figure of merit $Q$. However, the $Q$ of the twin-T network can be increased significantly if it is used with the voltage follower as shown in Fig. 9.13(b). In this circuit the output of the voltage follower is fed back to the junction $R/2$ and $2C$. The frequency response of the active notch filter is shown in Fig. 9.13(c). Because of its higher $Q\,(> 10)$, the bandwidth is much smaller than that of the wide band-stop filter.

Notch filters are used in communications and biomedical instruments for eliminating undesired frequencies.

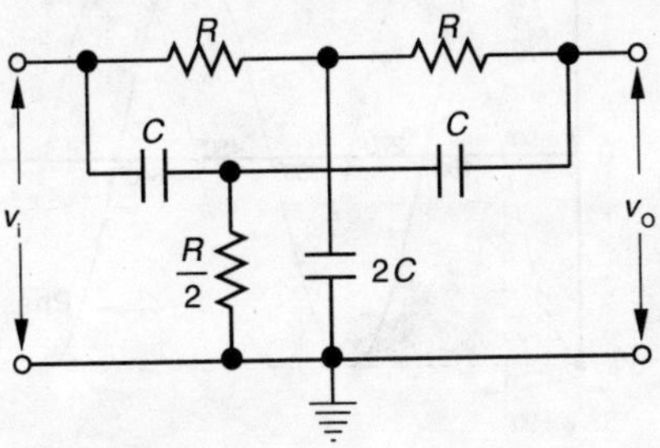


**(a) Twin-T notch filter**

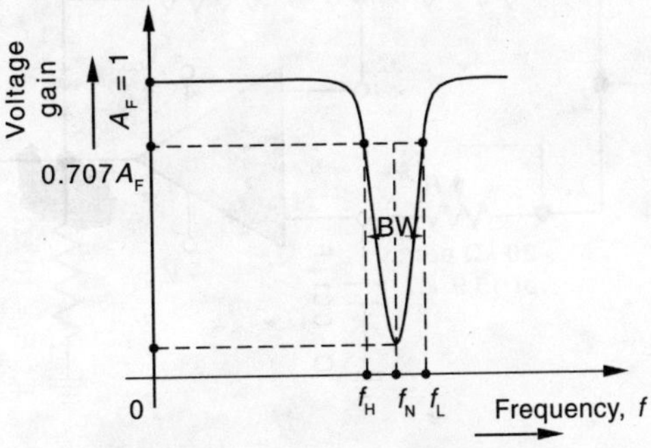


**(c) Frequency response**

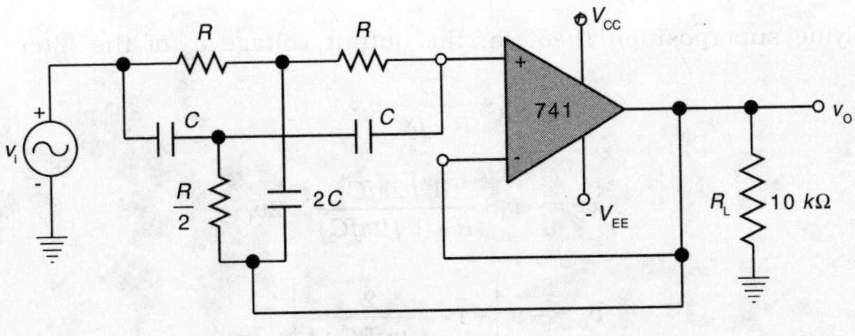


**(b) Circuit of active notch filter**

**Fig. 9.13 Narrow band-stop filter**

## Design Procedure

1. Choose the notch frequency, $f_N$.
2. Choose the value of capacitor, so that $C \le 1\ \mu F$.
3. Calculate the required value of $R$ from the relation, $f_N = \frac{1}{2\pi RC}$.

## All-Pass Filter

The all-pass filter passes all frequency components of the input signal without attenuation but provides predictable phase shifts for different frequencies of the input signal. The most important applications of all-pass filter are telephone wires. When signals are transmitted over transmission lines, such as telephone wires, they undergo change in phase.

In order to compensate such phase changes, all-pass filters are essential. The all-pass filters are also called **delay equalizers** or **phase correctors.**

The circuit of an all-pass filter using an op-amp is shown in Fig. 9.14. In this circuit, $R_f = R_1$.

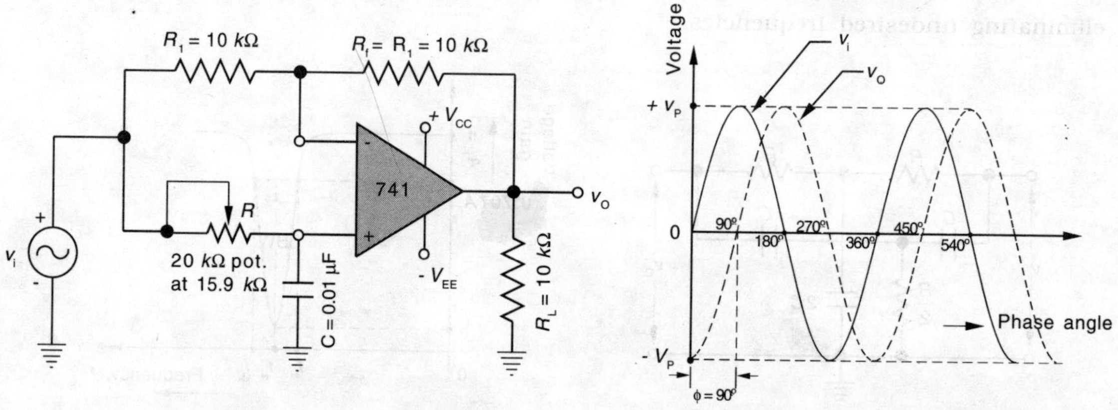


**(a) Circuit of all-pass filter**

**(b) Phase shift between input and output voltages**

**Fig. 9.14 All-pass filter**

By applying superposition theorem, the output voltage $v_o$ of the filter is given by

$$v_o = -v_i + \frac{-jX_c}{R - jX_c} v_i \times 2$$

$$v_o = -v_i + \frac{(1/j\,2\pi fC)}{R + (1/j\,2\pi fC)} \times 2v_i$$

$$v_o = -v_i \left[ -1 + \frac{2}{j\,2\pi fRC + 1} \right]$$

$$\boldsymbol{\frac{v_o}{v_i} = \left[ \frac{1 - j\,2\pi fRC}{1 + j\,2\pi fRC} \right]}$$

where $f$ is the frequency of the input signal.

The above equation shows that the amplitude of $v_o/v_i$ is unity; i.e., $|v_o| = |v_i|$ throughout the useful frequency range. Since the phase shift between $v_o$ and $v_i$ is a function of input frequency $f$ and is given by

$$\phi = -2\tan^{-1}\left(\frac{2\pi fRC}{1}\right)$$

From Fig. 9.14(b), it is seen that, output voltage $v_o$ lags input voltage $v_i$ by 90°. For fixed values of $R$ and $C$, the phase angle $\phi$ changes from 0 to −180° as the frequency $f$ is varied from 0 to ∞. If the positions of $R$ and $C$ are interchanged, the phase shift between input and output voltages becomes positive i.e., output voltage $v_o$ leads input voltage $v_i$.

## COMPARATORS

Operational amplifiers are often used as comparators to compare the amplitude of one voltage with another. A comparator is an open-loop op-amp, with two analog inputs and a digital output.

A comparator compares a signal voltage with a reference voltage. When the non-inverting voltage is larger than the inverting voltage the comparator produces a high output voltage. When the non-inverting voltage is less than the inverting voltage, the comparator produces a low output voltage. Hence it is used to indicate if a voltage is greater or lesser than a given reference voltage.

Comparators are used in circuits such as digital interfacing, Schmitt triggers, discriminators, level detectors and oscillators.

### Basic Comparator

The simplest way to build a comparator is to connect an op-amp without feedback resistors, as shown in Fig. 9.15(a). A fixed reference voltage (*dc*) $V_{ref}$ is applied to the inverting (−) input and an *ac* signal (sinusoidal) voltage $v_{in}$ is applied to the non-inverting input of op-amp. This circuit is called the **non-inverting comparator**. Semiconductor diodes $D_1$ and $D_2$ protect the op-amp from damage due to excessive input voltage $v_{in}$. If these diodes are made of silicon, the difference input voltage $v_{id}$ of the op-amp is clamped to either 0.7 *V* or −0.7 *V*; hence these diodes are called **clamp diodes**. The resistor *R* connected in series with input voltage source is used to control the current through $D_1$ and $D_2$. Another resistance $R_{DM}$ = *R* is connected between inverting input and *dc* source $V_{ref}$.

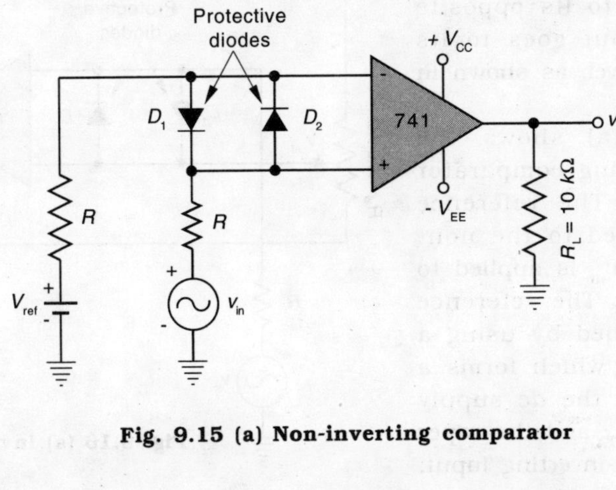


**Fig. 9.15 (a) Non-inverting comparator**

When $v_{in}$ is less than $V_{ref}$, the output voltage $v_o$ is at $-V_{sat}$ $(= -V_{EE})$ because the voltage at the inverting input is higher than at the non-inverting input. On the other hand, when $v_i$ is greater than $V_{ref}$, the non-inverting input becomes positive with respect to the

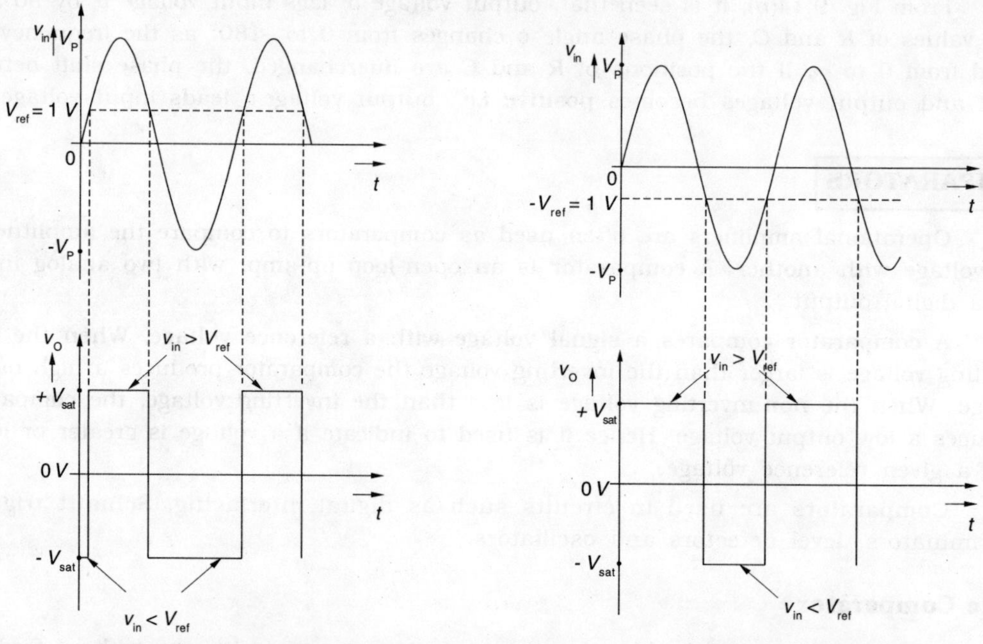


**(b) For $V_{ref}$ positive** **(c) For $V_{ref}$ negative**

**Fig. 9.15 Non-inverting comparator and its input and output waveforms**

inverting input and thereby output voltage $v_o$ becomes $+V_{sat}$ $(= +V_{cc})$. Thus the output voltage $v_o$ changes from one saturation level to another whenever $v_{in} = V_{ref}$, as shown in Fig. 9.15(b).

When the sine wave is positive, the output is at its maximum positive level. For $v_{in} < V_{ref}$ the amplifier is driven to its opposite state and the output goes to its maximum negative level, as shown in Fig. 9.15(c).

Figure 9.16(a) shows the circuit of an inverting comparator using an op-amp. The reference voltage $V_{ref}$ is applied to the non-inverting input and $v_{in}$ is applied to the inverting input. The reference voltage $V_{ref}$ is obtained by using a 10 $k\Omega$ potentiometer which forms a voltage divider with the *dc* supply voltages $+V_{cc}$ and $-V_{EE}$ and the wiper connected to the non-inverting input.

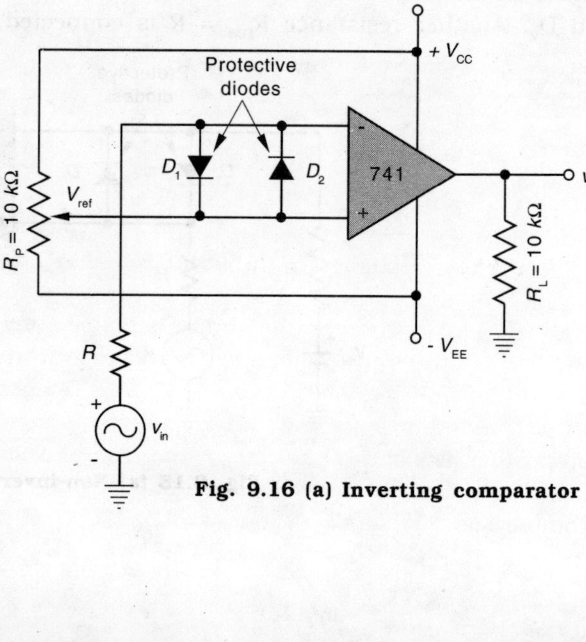


**Fig. 9.16 (a) Inverting comparator**

If the wiper is moved towards $-V_{EE}$, $V_{ref}$ becomes more negative while if it is moved towards $+V_{cc}$, $V_{ref}$ becomes more positive. Thus a reference voltage $V_{ref}$ of a desired amplitude and polarity can be obtained by simply adjusting the 10 $k\Omega$ potentiometer. The input and output waveforms of inverting comparator are shown in Fig. 9.16(b) and (c).

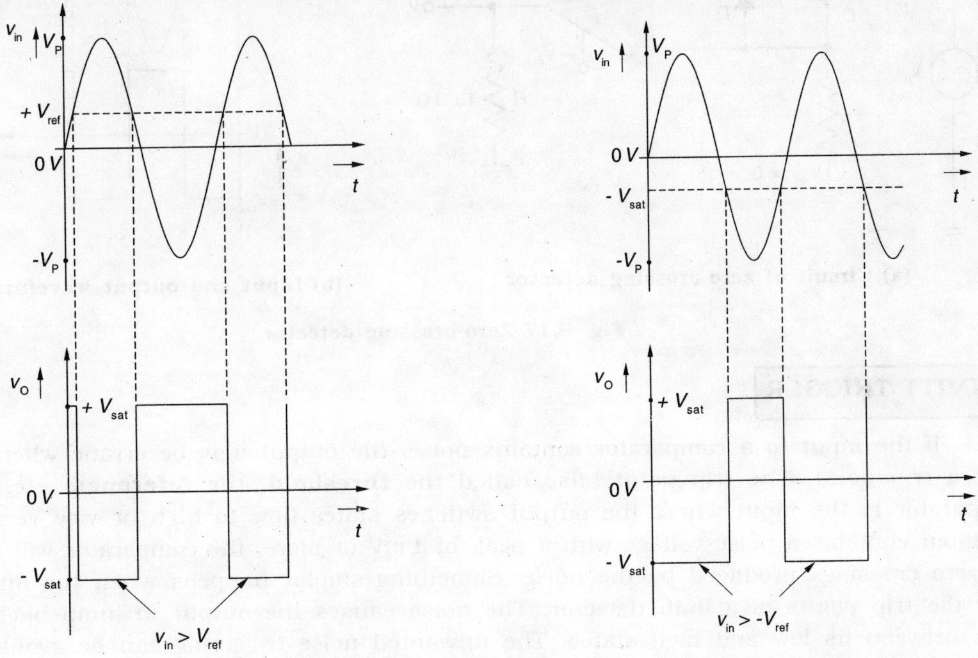


**(b) If $V_{ref}$ is positive** **(c) If $V_{ref}$ is negative**

**Fig. 9.16 Inverting comparator with input and output waveforms**

## Zero-Crossing Detector

An important application of the comparator is the **zero-crossing detector** or **sine wave-to-square wave converter**. The basic comparator of Fig. 9.15(a) or Fig. 9.16(a) can be used as the zero-crossing detector provided that reference voltage $V_{ref}$ is set to 0 $V$. The inverting comparator used as a zero-crossing detector is shown in Fig. 9.17(a). The output voltage $v_o$ is driven into negative saturation when input sinusoidal signal $v_{in}$ passes through zero in the positive direction. Conversely, when $v_{in}$ passes through zero in the negative direction, the output voltage $v_o$ switches and saturates positively, shown in Fig. 9.17(b).

If the frequency of the input signal is low, input voltage $v_i$ takes more time to cross 0 $V$. Then, the output voltage $v_o$ may not switch quickly from one state to the other. If the input to a comparator contains noise, the noise voltage becomes superimposed on the sinusoidal input voltage, then the output voltage $v_o$ may fluctuate between two saturation voltages $+V_{sat}$ and $-V_{sat}$, detecting zero reference crossings for noise voltages as well as input voltage $v_{in}$. These problems can be eliminated with the use of regenerative or positive feedback that causes the output voltage $v_o$ to change faster and eliminate any false output transitions due to noise voltages at the input of op-amp, at the point of comparison.

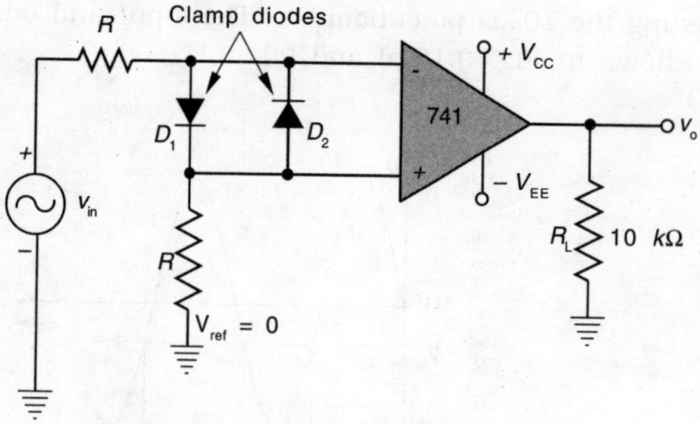


(a) Circuit of zero-crossing detector

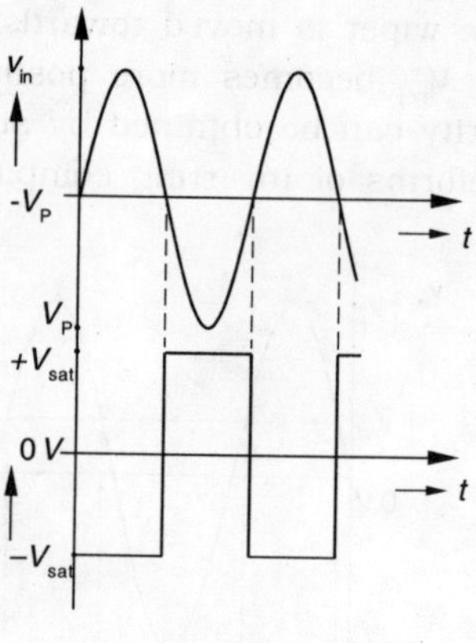


(b) Input and output waveforms

**Fig. 9.17 Zero-crossing detector**

## SCHMITT TRIGGER

If the input to a comparator contains noise, the output may be erratic when $v_i$ is near a trip point. The trip point (also called the **threshold**, the **reference**, etc.) of a comparator is the input where the output switches states (low to high or vice versa). If the input contains a noise voltage with a peak of 1 *mV* or more, the comparator will detect the zero crossings produced by the noise. Something similar happens when the input is near the trip points of a limit detector. The noise causes the output to jump back and forth between its low and high states. The unwanted noise triggering can be avoided by using a **Schmitt trigger**, with hysteresis.

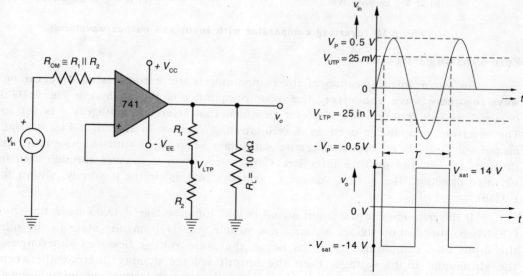

(a) Inverting comparator as Schmitt trigger

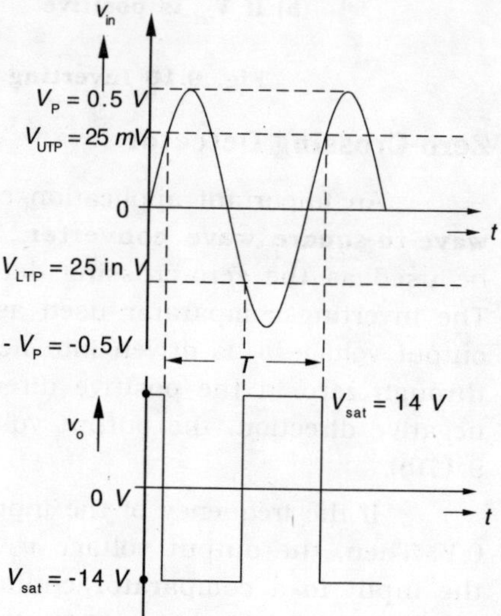


(b) Input and output waveforms

**Fig. 9.18 Schmitt trigger**

The circuit which converts any irregular shaped waveform to a square wave or pulse is called **Schmitt trigger** or **squaring circuit.** The circuit uses positive feedback. The input voltage $v_{in}$ triggers (changes the state of) the output voltage $v_o$ every time it exceeds certain voltage levels known as the **upper threshold voltage** $V_{UTP}$ and **lower threshold voltage** $V_{LTP}$, as shown in Fig. 9.18(b). The voltage divider feeds back a voltage to the non-inverting input, which acts as the comparison level for the input voltage $v_i$. Hence threshold voltages are obtained by using the voltage divider $R_1 - R_2$. The voltage across $R_2$ is a variable reference threshold voltage which depends on the value and polarity of the output voltage $v_o$. When $v_o = +V_{sat}$, the voltage across $R_2$ is called the **upper threshold voltage,** $V_{UTP}$. As long as $v_{in} < V_{UTP}$, $v_o$ is at $+V_{sat}$. The voltage fed back to the non-inverting input is given by

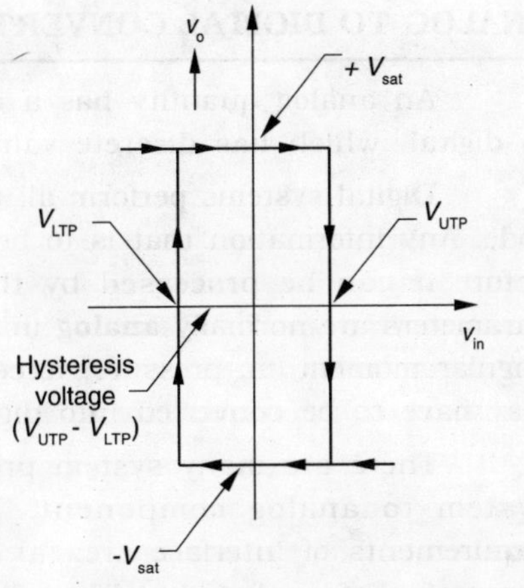


**(c) The input and output characteristics of the Schmitt trigger**

**Fig. 9.18 Schmitt trigger**

$$V_{UTP} = \frac{R_2}{R_1 + R_2}(+V_{sat}) \longrightarrow (1)$$

When $v_{in}$ exceeds $V_{UTP}$, the output voltage drops to its negative maximum, $-V_{sat}$. The voltage across $R_2$ is known as **lower threshold voltage**, $V_{LTP}$, i.e., voltage fed back to the non-inverting input is given by

$$V_{LTP} = \frac{R_2}{R_1 + R_2}(-V_{sat}) \longrightarrow (2)$$

Thus, if the threshold voltages $V_{LTP}$ and $V_{UTP}$ are made larger than the input noise voltages, the positive feedback eliminates the false output transitions. Because of regenerative action, positive feedback makes output voltage $v_o$ switch faster between $+V_{sat}$ and $-V_{sat}$. The resistance $R_{OM} = R_1 || R_2$ in the inverting input also minimizes the offset problems.

Note that the two transitions given by Eqs. (1) and (2) do not occur at the same values of $v_i$. A plot of input-output characteristics of the Schmitt trigger is shown in Fig. 9.18(c). The curve is called a **hysteresis** curve since it has the form of a magnetic hysteresis loop. The amount of hysteresis is defined by the difference of the two trigger levels.

$$V_{hys} = V_{UTP} - V_{LTP}$$

$$= \frac{R_2}{R_1 + R_2}[+V_{sat} - (-V_{sat})]$$

$$\boldsymbol{V_{hys} = \frac{2 R_2 V_{sat}}{R_1 + R_2}} \longrightarrow (3)$$

## ANALOG TO DIGITAL CONVERTERS

An analog quantity has a continuous set of values over a given range in contrast to digital, which has discrete values.

Digital systems perform all their internal operations in binary or some type of binary code. Any information that is to be input to a digital system must be put into binary form before it can be processed by the digital circuits. On the input side, measurements parameters are normally **analog** in nature. The physical quantities like velocity, acceleration, angular momentum, pressure, force, etc. are all analog in nature. These physical quantities may have to be converted into digital form.

There are many system problems that require connecting a digital portion of the system to analog component. The connecting circuits are called **interface**. The requirements of interface are: (a) the conversion of digital signals to analog, and (b) the conversion of analog signals to digital.

**The process of conversion from an analog signal to digital signal is referred to as analog-to-digital conversion.**

Figure 9.19 shows a circuit diagram consisting of five blocks involved when a computer is monitoring and controlling a physical variable that is analog in nature.

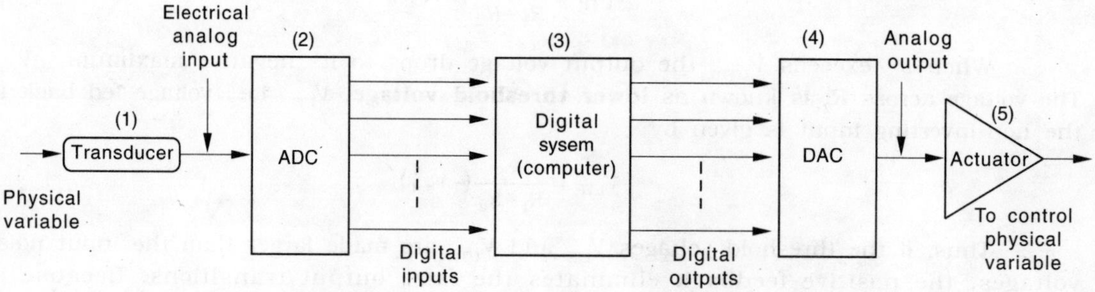


**Fig. 9.19 ADC and DAC are used to interface a computer to the analog world so that the computer can monitor and control a physical variable**

There are many types of A/D converters: single-ramp integrating, dual-ramp integrating, signal counter, tracking and successive approximation.

### Successive Approximation A/D Converter

Figure 9.20 shows the block diagram of a successive approximation A/D converter. It consists of an 8-bit successive approximation register (SAR), whose output is applied to an 8-bit D/A converter. The analog output ($V_a$) of the D/A converter is then compared with an analog input signal ($V_{in}$) by the comparator. The output of the comparator is a serial data input to the SAR. The SAR then adjusts its digital output (8-bits) until it is equivalent to analog input $V_{in}$. The 8-bit latch at the end of the conversion holds the resultant digital data. The A/D converter circuit works as follows:

At the start of a conversion cycle the SAR is reset by holding the start signal (S) HIGH. On the arrival of first clock pulse LOW-to-HIGH transition, the most significant bit (MSB) $Q_7$ of the SAR is set. The D/A converter then generates an analog signal equivalent to the $Q_7$ bit, which is compared with the analog input voltage $V_{in}$. If the comparator output is LOW, the D/A output > $V_{in}$ and the SAR clears its MSB $Q_7$. On the other hand, if the comparator output is HIGH, the D/A output < $V_{in}$ and the SAR keeps the MSB $Q_7$ set. Thus, on the arrival of next clock pulse LOW-to-HIGH transition, the SAR sets the next MSB $Q_6$. Depending on the output of the comparator, the SAR will then either keep or reset the bit $Q_6$. This process is repeated till the SAR tries all the bits. As soon as the LSB $Q_0$ is obtained, the SAR forces the conversion complete (CC) signal HIGH to indicate the parallel output lines contain valid data. The CC signal in turn enables the 8-bit latch, and thereby digital data appear at the output of the latch. Digital data are also available serially as the SAR determines each bit. To cycle the converter continuously the CC signal may be connected to the start conversion input as shown in Fig. 9.20(a).

The complete approximation procedure is illustrated for a 4-bit successive approximation A/D converter with the help of operation diagram shown in Fig. 9.20(b).

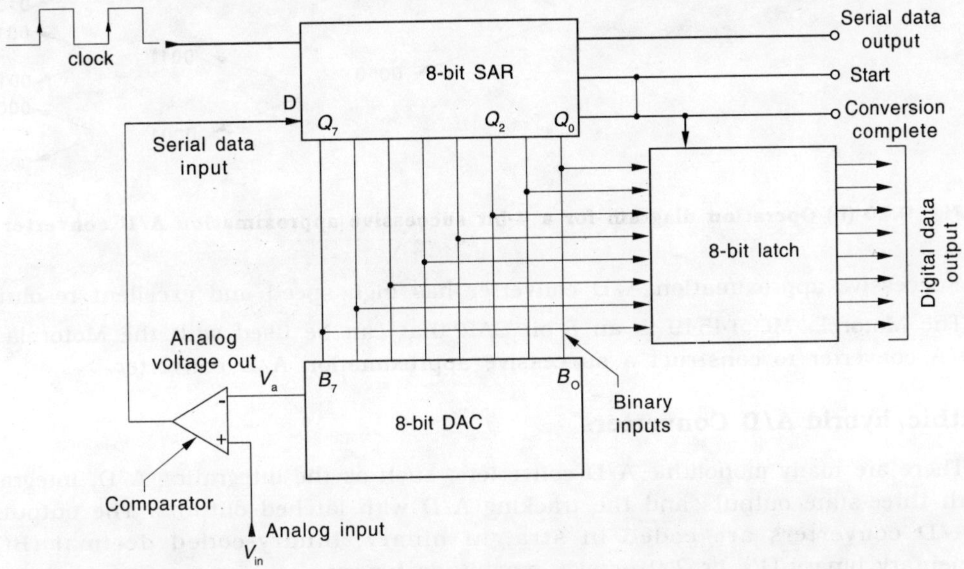


**Fig. 9.20 (a) Successive approximation-type A/D converter**

Suppose we apply 15 $V$ to the analog input. The successive approximation A/D converter first makes a "guess" at the analog input voltage. This guess is made by setting the MSB to 1. Then the digital signal reaching the latch is 1000 0000. The voltage produced by the latch is now compared with the analog input signal. If the reference signal is smaller than the analog input, the comparator output gives a signal to the control logic which now switches the output of $Q_6$ to 1. The digital signal now becomes 1100 0000. Once again its equivalent analog voltage is compared with the input signal. If reference voltage is greater than the analog input, then a signal is given to the control logic to reduce the reference voltage accordingly $Q_7$ is reset to 0 and $Q_6$ is set to 1. The digital signal becomes 0100 0000. This process is repeated until the analog input and reference voltage are equal.

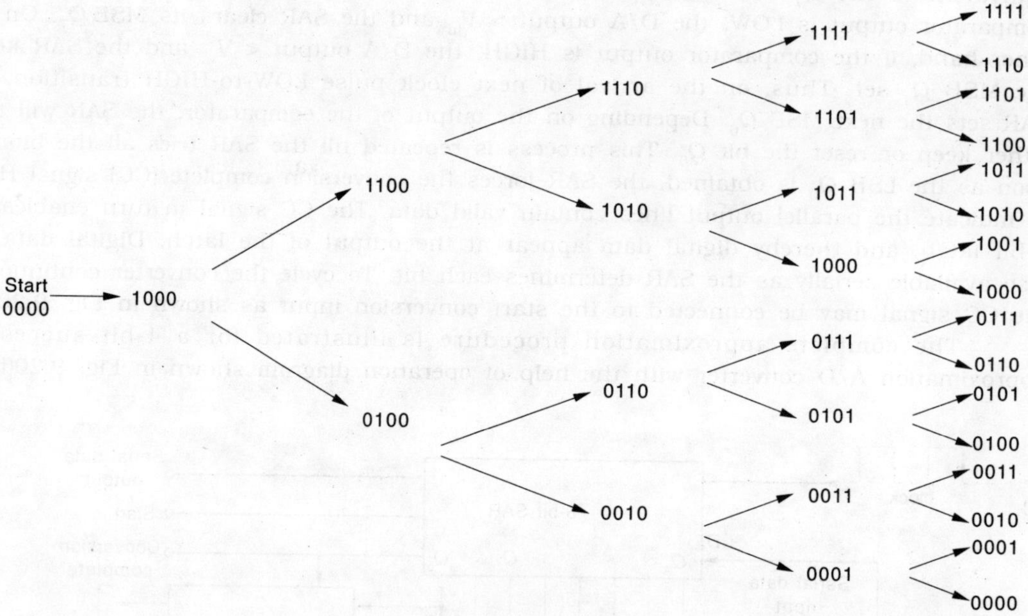


**Fig. 9.20 (b) Operation diagram for a 4-bit successive approximation A/D converter**

Successive approximation A/D converter has high speed and excellent resolution.

The Motorola MC 14549 is an 8-bit SAR that can be used with the Motorola MC 1408 D/A converter to construct a successive approximation A/D converter.

## Monolithic/hybrid A/D Converters

There are many monolithic A/D converters, such as the integrating A/D, integrating A/D with three-state outputs and the tracking A/D with latched outputs. The outputs of these A/D converters are coded in straight binary, binary-coded decimal (BCD), complementary binary (1's or 2's) or sign-magnitude binary.

Teledyne 8703 is an 8-bit monolithic CMOS A/D converter with three state output.

As in the case of monolithics, there are many hybrid A/D converters, such as the successive-approximation A/D with input buffer amplifier, low-power CMOS A/D, the fast A/D with sample-and-hold and the ultra fast A/D with input buffer amplifier. Datel Intersil's hybrid ADC-815MC is a very high speed 8-bit successive approximation A/D converter. It is capable of 8-bit resolution in only 600 *ns* and has a 6 *V* analog-input voltage range with parallel and serial outputs.

## Specifications of A/D converters

The manufacturers of A/D converters usually specify the following specifications:

1. Range of input voltage (5 *V*)
2. Conversion time (0.05 to 100,000 μs)
3. Accuracy (+ ½ LSB to ± 2 LSB)

4. Format of digital output
5. Input impedance.
6. Maximum power dissipation (15 to 8600 *mW*)
7. Power supply voltage (+ 5 *V*)

## DIGITAL TO ANALOG CONVERTERS

A simple D/A converter that uses an op-amp and binary weighted resistors is shown in Fig. 9.21. The op-amp is connected in the inverting mode. Since the number of binary inputs is four, the converter is called a 4-bit converter. As there are 16 ($=2^4$) combinations of binary inputs analog output will have 16 different values. There are four electronic switches. They are used to simulate binary inputs. Each switch is controlled by one bit line. When the bit corresponding to switch is a 1, then the switch is "thrown" to the up position. If the bit is a 0, then the switch is "thrown" down.

When the switch is up, the corresponding input becomes *V*. When a switch is down,

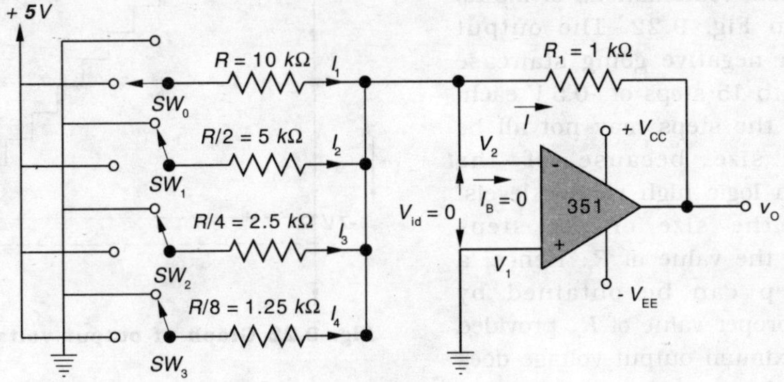


**Fig. 9.21 D/A converter with binary weighted resistors**

that input becomes zero. Now suppose that the binary number is $b_3b_2b_1b_0$.

If $b_j = 1$, then switch j is up; if $b_j = 0$, then switch j is down. If switch $sw_0$ is closed, the voltage across *R* is 5*V* because $V_1 = V_2 = 0\,V$. The current therefore through $R_f$ is $5V/10\,k\Omega = 0.5\,mA$. Since, the input bias current $I_B$ is negligible, the current through feedback resistor $R_f$ is also 0.5 *mA*, which in turn produces an output voltage, $V_O = -R_f \times 0.5\,mA = -1k\Omega \times 0.5\,mA = -0.5\,V$. Note that the op-amp is working as a current-to-voltage converter.

If switch $sw_1$ is closed and $sw_0$ is opened, then

$$V_O = -\left(\frac{5V}{5k\,\Omega}\right)(1k\,\Omega) = -1V$$

Similarly, if both switches $sw_0$ and $sw_1$ are closed, then output voltage

$$V_O = -V R_f\left[\frac{b_0}{R} + \frac{b_1}{R/2}\right] = -5\times 1k\Omega\left[\frac{1}{10k\Omega} + \frac{1}{5k\Omega}\right]$$

$$= -\left(\frac{1}{2} + 1\right) = -1.5V$$

When all the switches are closed, the output will be maximum. The output voltage is given by

$$V_o = -IR_f = -[I_1 + I_2 + I_3 + I_4]R_f = -\left[\frac{b_o V}{R} + \frac{b_1}{R/2} \cdot V + \frac{b_2}{R/4} \cdot V + \frac{b_3}{R/8} \cdot V\right] R_f$$

$$V_0 = -V R_f \left[\frac{b_0}{R} + \frac{b_1}{R/2} + \frac{b_2}{R/4} + \frac{b_3}{R/8}\right]$$

$$\mathbf{V_0 = -\frac{V R_f}{R}[8b_3 + 4b_2 + 2b_1 + b_0]} \quad \longrightarrow (1)$$

By adjusting $VR_f/R$ we can adjust the level of the output voltage. Thus, the output voltage $V_o$ will be proportional to the binary number. The constant of proportionality is $VR_f/R$.

The graph of output voltage versus possible combinations of inputs is shown in Fig. 9.22. The output voltage is a negative going staircase waveform with 15 steps of –0.5 *V* each. In practice, the steps may not all be the same size because of the variations in logic high voltage levels. Note that the size of the steps depends on the value of $R_f$. Hence, a desired step can be obtained by selecting a proper value of $R_f$, provided that the maximum output voltage does not exceed the saturation level of an op-amp.

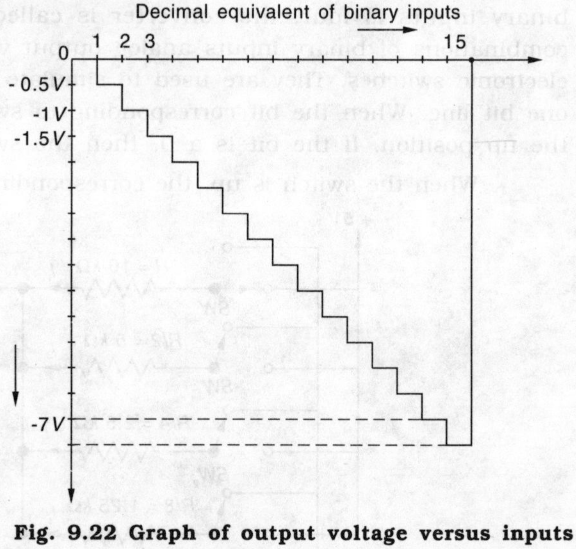


**Fig. 9.22 Graph of output voltage versus inputs**

## D/A Converter with *R* and 2*R* Resistors

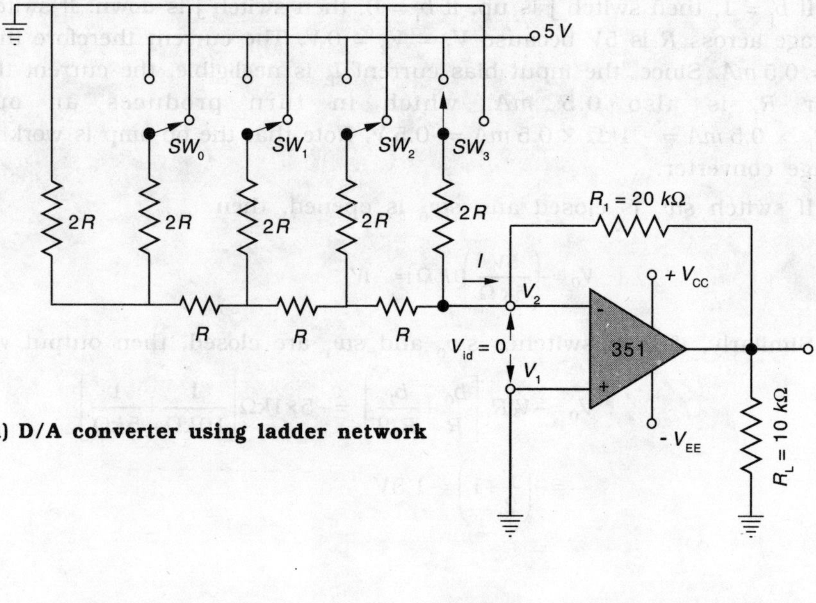


**(a) D/A converter using ladder network**

The $R-2R$ ladder D/A converter is made of only two resistor values wired as shown in Fig. 9.23(a). The binary inputs are simulated by four switches $sw_0$ to $sw_3$. Each bit of the binary word connects the corresponding switch either to ground or to the inverting terminal of the op-amp. Assume that the most significant bit (MSB) switch $sw_3$ $(\pm b_3)$ is connected to $+5V$ and other switches are connected to ground. Thevenizing the circuit to the left of switch $sw_3$, we get

$$\text{Thevenin resistance, } R_{TH} = [\{[(2R \| 2R + R) \| 2R] + R\} \| 2R] + R$$

$$= [\{[(R + R) \| 2R] + R\} \| 2R] + R$$

$$R_{TH} = [\{R + R\} \| 2R] + R = R + R = 2R$$

The equivalent circuit (when switch $sw_3$ is closed) is shown in Fig. 9.23(b). Since the input is at virtual ground ($V_2 \cong 0$ V), no current flows through Thevenin resistance $R_{TH}$. As the resistor $2R$ is connected to $+5V$, the current through it is $I = \frac{V}{2R}\left( = \frac{5V}{20k\Omega} = 0.25\ mA \right)$. The same current flows through the feedback resistor $R_f$ and

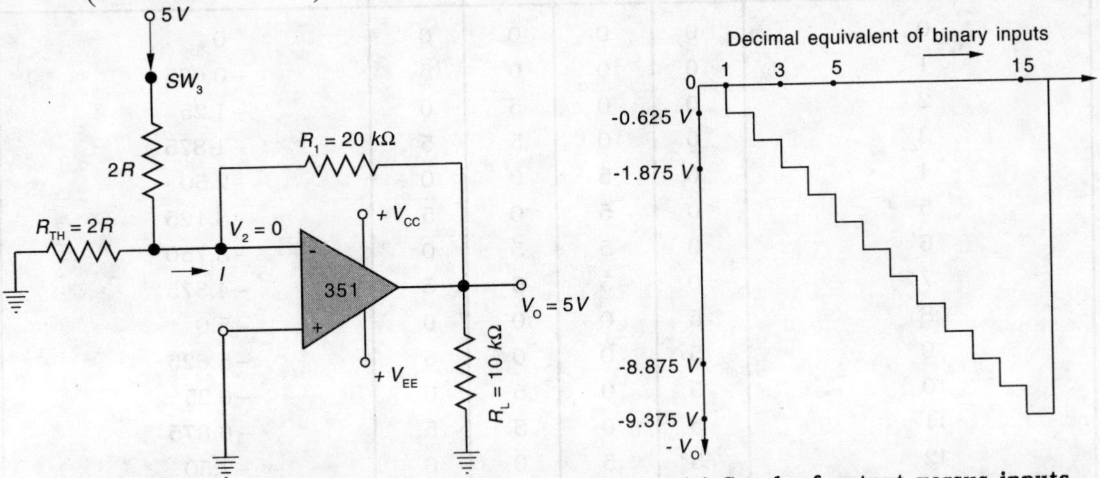


(b) **Equivalent circuit when $b_3$ is HIGH**

(c) **Graph of output versus inputs**

**Fig. 9.23 D/A converter with *R* and 2*R* resistors**

it produces the output voltage.

$\therefore$ Output voltage, $V_0 = -IR_f$

$$= -0.25\ mA \times 20\ k\Omega = -5\ V$$

The output voltage equation can be written as

$$V_o = -R_f(I_1 + I_2 + I_3 + I_4)$$

$$= -R_f\left( b_3 \frac{V}{2R} + b_2 \frac{V}{4R} + b_1 \frac{V}{8R} + b_0 \frac{V}{16R} \right)$$

$$\mathbf{V_0 = -R_f\left[\frac{b_3}{2R} + \frac{b_2}{4R} + \frac{b_1}{8R} + \frac{b_0}{16R}\right] V}$$

or $$V_0 = \frac{-R_f V}{R}\left[b_3 2^{-1} + b_2 2^{-2} + b_1 2^{-3} + b_0 2^{-4}\right]$$

If all the inputs are HIGH, the maximum output voltage is given by

$$V_0 = \frac{-20k\Omega}{10k\Omega}\left[\frac{1}{2}+\frac{1}{4}+\frac{1}{8}+\frac{1}{16}\right]5V$$

$$= -10V\left[\frac{15}{16}\right] = \mathbf{-9.375V}$$

The output voltage corresponding to all possible combinations of binary inputs are calculated and is given in Table 9.1.

**Table 9.1**

| Decimal equivalent of binary inputs | Input (V) | | | | Output voltage ($V_0$) |
|---|---|---|---|---|---|
| | $b_3$ | $b_2$ | $b_1$ | $b_0$ | |
| 0 | 0 | 0 | 0 | 0 | 0 |
| 1 | 0 | 0 | 0 | 5 | −0.625 |
| 2 | 0 | 0 | 5 | 0 | −1.25 |
| 3 | 0 | 0 | 5 | 5 | −1.875 |
| 4 | 0 | 5 | 0 | 0 | −2.50 |
| 5 | 0 | 5 | 0 | 5 | −3.125 |
| 6 | 0 | 5 | 5 | 0 | −3.750 |
| 7 | 0 | 5 | 5 | 5 | −4.375 |
| 8 | 5 | 0 | 0 | 0 | −5.0 |
| 9 | 5 | 0 | 0 | 5 | −5.625 |
| 10 | 5 | 0 | 5 | 0 | −6.25 |
| 11 | 5 | 0 | 5 | 5 | −6.875 |
| 12 | 5 | 5 | 0 | 0 | −7.50 |
| 13 | 5 | 5 | 0 | 5 | −8.125 |
| 14 | 5 | 5 | 5 | 0 | −8.875 |
| 15 | 5 | 5 | 5 | 5 | −9.375 |

The $R-2R$ ladder-type D/A converter has some advantages over the more basic unit. Since only two values of resistors are required $R$ and $2R$, the circuit is well suited for IC fabrication. As only two types of resistors are involved it is less susceptible to drift.

The circuit can also be used as analog multiplier.

## Ratings of D/A Converters

**Voltage levels:** Arbitrary levels for the voltage $V$ and the input voltage are limited and are specified by the manufacturer. If a D/A converter is built using discrete components then the resistance ratio can be varied. In an IC device, these are chosen by the manufacturer and cannot be varied by the user. Hence, the variation of the voltages is the only way that the output level can be varied.

**Resolution (Step Size):** Resolution is the difference between two consecutive output voltages and is determined by the number of bits that the D/A converter can accommodate.

For instance, the D/A converter accommodates 4-bits. Then there are $2^4 = 16$ possible levels. The resolution is one part in 16 or 6.25%.

**Sources of Errors:** There are mainly three errors in DACs: Linearity, offset and gain errors.

**Linearity Error:** The linearity error is defined as the amount by which the actual output differs from the ideal straight line output. The error is normally expressed as a percentage of the full scale range.

**Gain error:** It is defined as the difference between the calculated gain of the current to voltage converter and the actual gain achieved. It is due to the errors in the feedback resistor of the current to voltage converter that uses op-amp.

**Offset error:** It is defined as the nonzero level of the output voltage when all inputs are zero.

It is due to the presence of offset voltage in op-amp and leakage currents in the current switches.

**Settling Time:** The time required for the output of the DAC to settle to within ± ½ LSB of the final value for a given digital input, is known as settling time.

**Conversion Time:** The time required by the ADC circuitry to convert an analog input into an equivalent digital output is called the conversion time.

**Accuracy:** The actual output voltage of a D/A converter departs from the ideal value. The factors that contribute to lack of linearity also contribute to a lack of accuracy. Practical operational amplifiers have problems associated with them which affect accuracy. The accuracy is also affected by operating conditions.

## CLIPPERS AND CLAMPERS

Waveshaping circuits are used in communications systems and digital computers. Waveshaping techniques include limiting, clipping and clamping.

**The process of deriving a desired waveform from a given waveform is known as waveshaping.**

The circuit elements that are used to alter the waveform are diodes, op-amps, resistors, inductors and capacitors. Shaping or waveform shaping is correcting the effects of distortion by appropriate amplitude and phase equalization or correction.

### Positive and Negative Clippers

**The circuit with which the waveform is shaped by removing (or clipping) a particular portion of the input signal above or below a certain level is known as clipping circuit.**

A circuit which removes positive parts of the input signal is known as a **positive clipper.**

Figure 9.24 is a positive clipper that clips the positive parts of the input voltage. It consists of an op-amp and a rectifier diode D. The op-amp is used as a voltage follower with a diode in the feedback path. The clipping level is determined by the reference voltage

$V_{ref}$. The reference voltage $V_{ref}$ should be less than the input voltage range of the op-amp. As shown in the figure, $V_{ref}$ is derived from the positive supply voltage $+V_{CC}$.

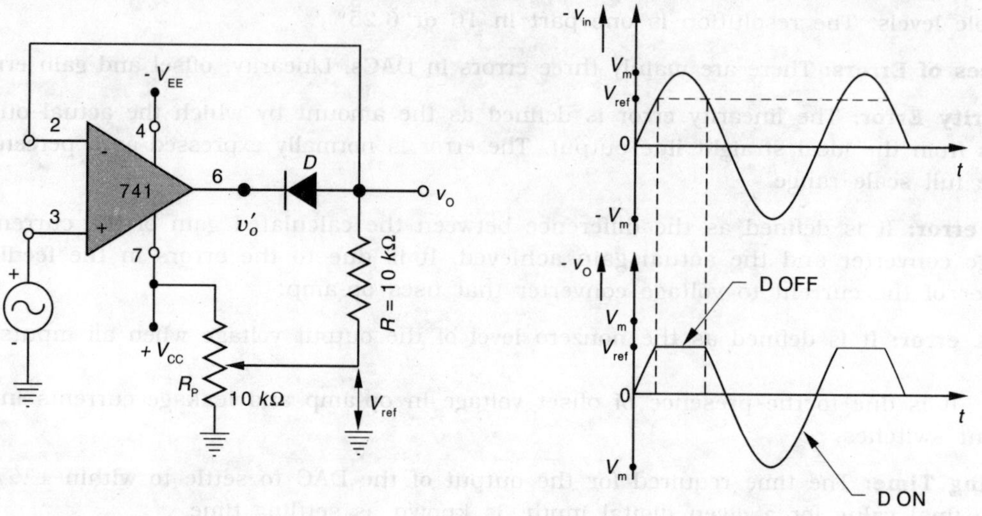


**(a) Positive clipper circuit** **(b) Input and output waveforms for $+V_{ref}$**

**Fig. 9.24 Positive clipper**

Let a sinusoidal signal be applied to the circuit. During the positive half-cycle of the input, the diode D conducts only until $v_{in} = V_{ref}$. This happens because, when $v_{in} < V_{ref}$, the voltage $V_{ref}$ at the inverting input is higher than at the noninverting input; hence the output voltage $v_o'$ of the op-amp becomes sufficiently negative to drive diode D into conduction. When diode conducts, it closes the feedback loop and thereby, the op-amp functions as a voltage follower, that is, output voltage $v_o$ follows the input voltage $v_{in}$ until $v_{in} = V_{ref}$. If $v_{in}$ is slightly greater than $V_{ref}$, the output voltage $v_o'$ of the op-amp becomes sufficiently positive to drive diode D into cutoff. When diode does not conduct it opens the feedback loop and the op-amp operates open-loop. Therefore, the op-amp drives its output $v_o'$ toward positive saturation ($+V_{cc}$).

When diode D is reverse-biased, the output voltage $v_o = V_{ref}$. Thus, when $v_{in} > V_{ref}$, $v_o' \cong + V_{cc}$ and $v_o = V_{ref}$ [Fig. 9.24 (b)].

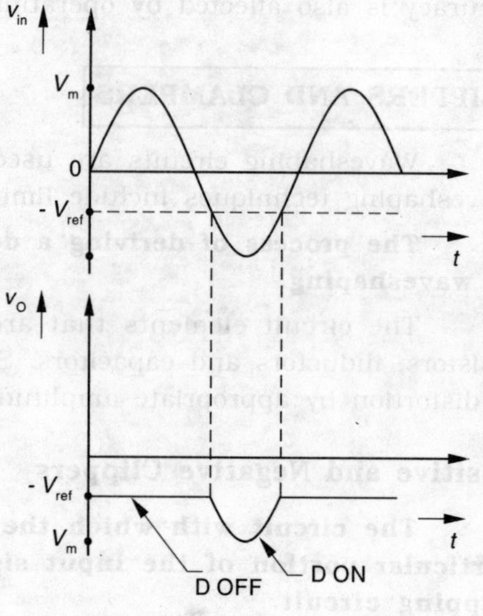


**(c) Input and output waveforms for $-V_{ref}$**

**Fig. 9.24 Positive clipper**

Thus diode D is ON for $v_{in} < V_{ref}$ and OFF for $v_{in} > V_{ref}$. The op-amp alternates between open-loop and closed loop operations as the diode D is turned OFF and ON, respectively. For this reason high speed op-amps such as HA 2500, LM 310 and μA 318 must be used and compensated for unity gain.

If potentiometer $R_p$ is connected to the negative supply $-V_{EE}$ instead of $+V_{cc}$, the reference voltage $V_{ref}$ will be negative. This causes the entire output waveform above $-V_{ref}$ to be clipped off, as shown in Fig. 9.24(c). Thus, the output follows the input only when $v_{in} < -V_{ref}$.

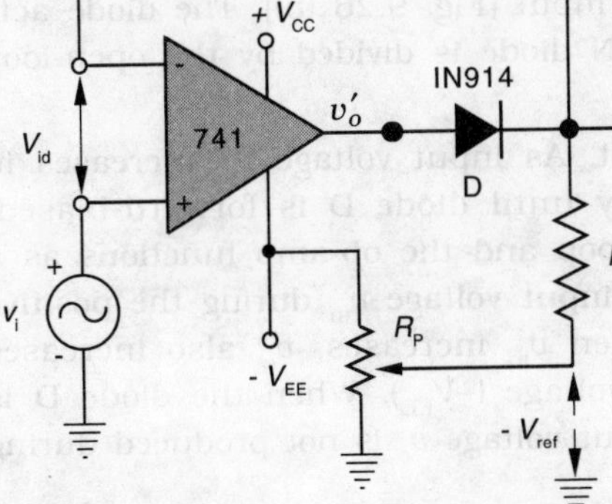


**(a) Negative clipper circuit**

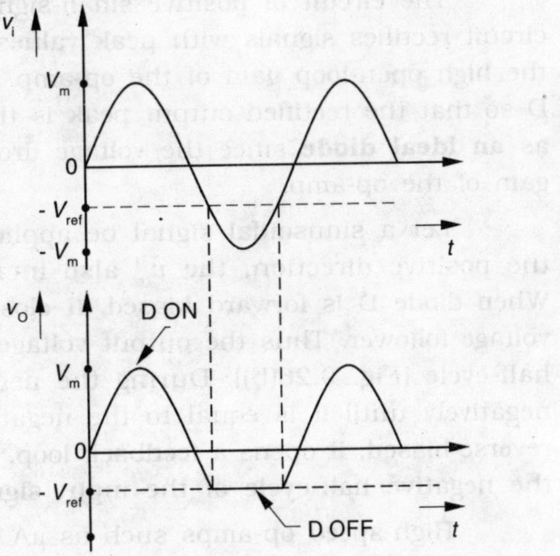


**(b) Input and output waveforms for $-V_{ref}$**

**Fig. 9.25 Negative clipper**

**A circuit which removes negative parts of the input signal is known as a negative clipper.**

The positive clipper can be converted into a negative clipper by simply reversing diode D and changing the polarity of reference voltage $V_{ref}$. Figure 9.25(a) shows a negative clipper that clips off the negative parts of the input signal below the reference voltage [Fig. 9.25(b)].

When $v_{in} > V_{ref}$, diode D conducts. The output voltage $v_o$ then follows the input voltage $v_{in}$ during this period. When $v_{in} < -V_{ref}$, diode D becomes reverse-biased and thereby the negative portion of the output voltage below $-V_{ref}$ is clipped off.

If potentiometer $R_p$ is connected to the positive supply $+V_{CC}$, the reference voltage $V_{ref}$ will be positive. The output voltage $v_o$ below $V_{ref}$ will be clipped off, as shown in Fig. 9.25(c). Thus the diode D is ON for $v_{in} > +V_{ref}$ and OFF for $v_{in} < +V_{ref}$.

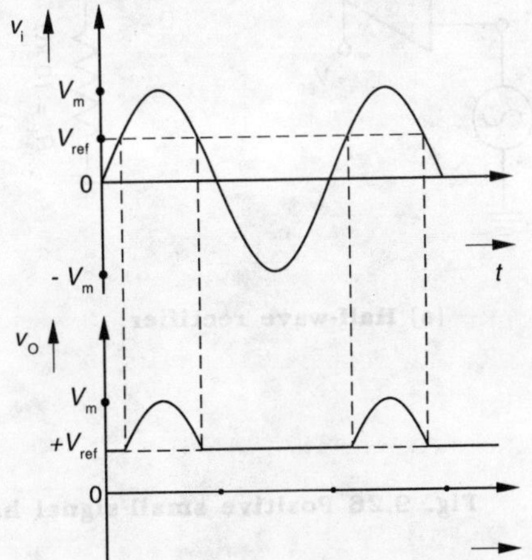


**(c) Input and output waveforms for $+V_{ref}$**

**Fig. 9.25 Negative clipper**

## SMALL-SIGNAL HALF-WAVE RECTIFIERS (PEAK RECTIFIERS)

### Positive Small-signal Half-wave Rectifier

The circuit of positive small-signal half-wave rectifier is shown in Fig. 9.26(a). This circuit rectifies signals with peak values down to a few millivolts. This is possible because the high open-loop gain of the op-amp automatically adjusts the voltage drive to the diode D so that the rectified output peak is the same as the input [Fig. 9.26 (b)]. The diode acts as **an ideal diode** since the voltage drop across the ON diode is divided by the open-loop gain of the op-amp.

Let a sinusoidal signal be applied to the circuit. As input voltage $v_{in}$ increases in the positive direction, the $v_o'$ also increases positively until diode D is forward-biased. When diode D is forward biased, it closes a feedback loop and the op-amp functions as a voltage follower. Thus the output voltage $v_o$ follows the input voltage $v_{in}$ during the positive half-cycle [Fig. 9.26(b)]. During the negative cycle, when $v_{in}$ increases, $v_o'$ also increases negatively until it is equal to the negative saturation voltage ($-V_{EE}$). When the diode D is reverse-biased, it opens a feedback loop. Hence the output voltage $v_o$ is not produced during the negative half-cycle of the input signal.

High-speed op-amps such as μA 318, HA 2500 and LM310 must be used since op-amp alternates between open-loop and closed-loop operations.

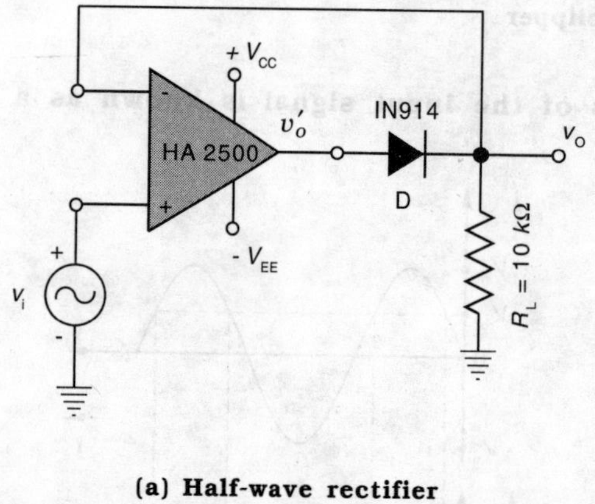


(a) Half-wave rectifier

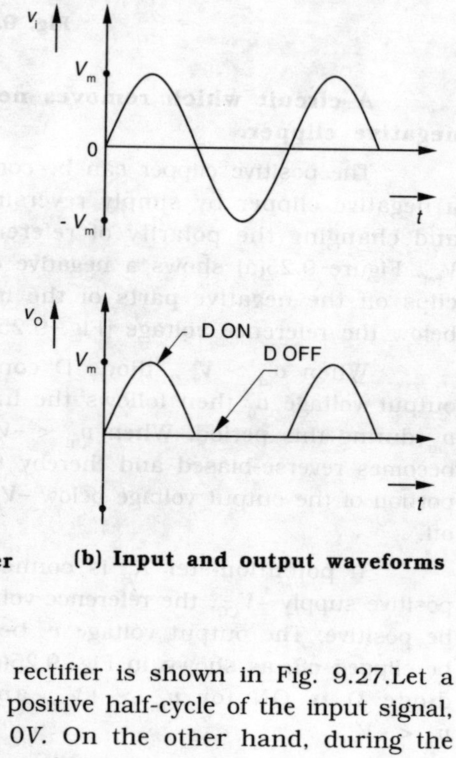


**Fig. 9.26 Positive small-signal half-wave rectifier** **(b) Input and output waveforms**

### Negative Small-signal Half-wave Rectifier

The circuit of negative small-signal half-wave rectifier is shown in Fig. 9.27.Let a sinusoidal signal be applied to the circuit. During the positive half-cycle of the input signal, diode D becomes reverse-biased and therefore $v_o = 0V$. On the other hand, during the negative half-cycle, diode D is forward-biased; hence $v_o$ follows $v_{in}$.

The circuit of negative half-wave rectifier using two diodes and an op-amp is shown in Fig. 9.28. Since two diodes are used the output $v_o'$ of the op-amp does not saturate.

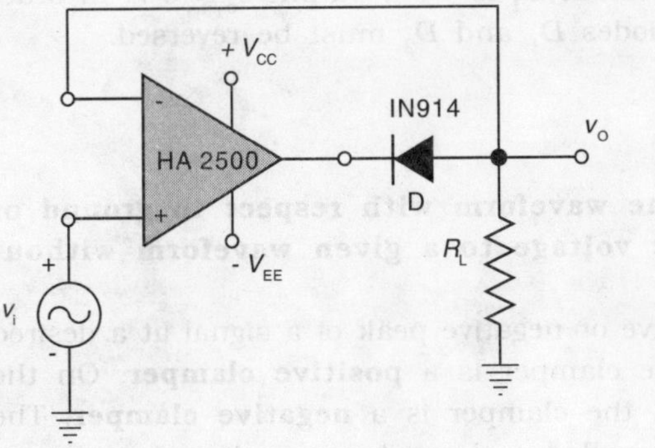


**(a) Half-wave rectifier.**

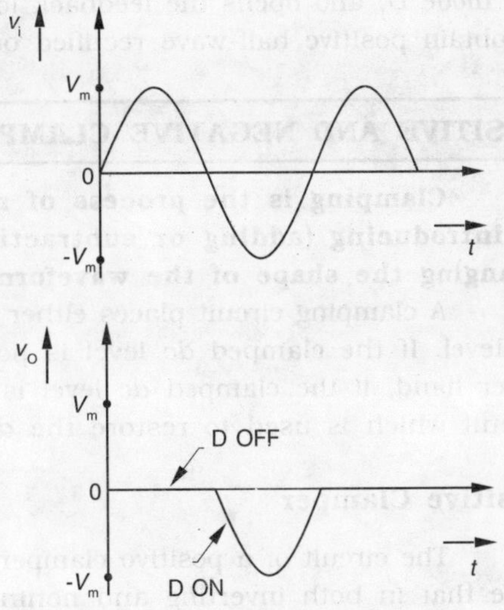


**(b) Input and output waveforms**

**Fig. 9.27 Negative small-signal half-wave rectifier**

This minimises the response time and increases the operating frequency range of the op-amp. The output $v_o$ is measured at the anode of diode $D_1$ with respect to ground. The output impedance is low when $D_1$ is ON and high ($=R_f$) when $D_1$ is OFF. This problem can be solved by connecting a voltage follower stage at the output.

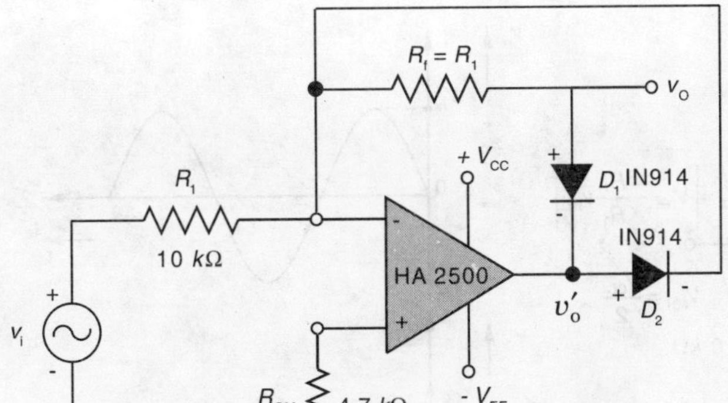


**(a) Negative half-wave rectifier circuit**

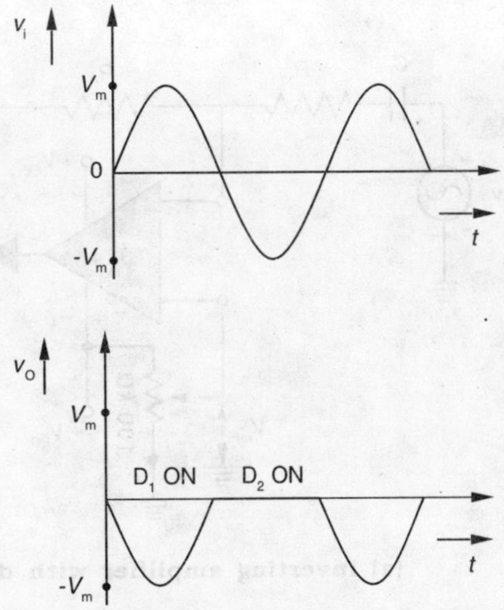


**(b) Input and output waveforms**

**Fig. 9.28 Negative half-wave rectifier**

During the positive half-cycle of $v_{in}$, output $v_o'$ of op-amp is negative, that forward biases diode $D_1$ and closes the feeback loop through $R_f$. Since $R_1 = R_f$, $v_o = (R_f/R)v_i = v_i$. During the negative half-cycle of $v_{in}$, output $v_o'$ of op-amp is positive, that reverse biases the diode $D_1$ and opens the feedback loop. Since diode $D_1$ is OFF, output $v_o = 0$ V. In order to obtain positive half-wave rectified outputs, diodes $D_1$ and $D_2$ must be reversed.

## POSITIVE AND NEGATIVE CLAMPERS

**Clamping is the process of raising the waveform with respect to ground or of introducing (adding or subtracting) a *dc* voltage to a given waveform without changing the shape of the waveform.**

A clamping circuit places either the positive or negative peak of a signal at a desired *dc* level. If the clamped *dc* level is positive, the clamper is a **positive clamper**. On the other hand, if the clamped *dc* level is negative, the clamper is a **negative clamper**. The circuit which is used to restore the *dc* level is called a *dc* **restorer** or **inserter**.

### Positive Clamper

The circuit of a positive clamper using the 124 series op-amp is shown in Fig. 9.29. Note that in both inverting and noninverting amplifier circuits the *dc* level added to the output voltage is approximately equal to $V_{cc}/2$ to obtain a maximum undistorted symmetrical sine wave.

A clamper with a variable positive *dc* level is shown in Fig. 9.29(a). The output voltage $v_o$ is sum of *ac* and *dc* input voltages applied to (–) and (+) input terminals. Let us first consider the input voltage at the (+) input. Since this voltage is positive, the output $v_o'$ of op-amp is also positive, that forward biases the diode *D*. This closes the feedback loop and the op-amp functions as a voltage follower. This is possible because the capacitor $C_1$ is an open circuit for *dc* voltage. Hence, output voltage $v_o = V_{ref}$.

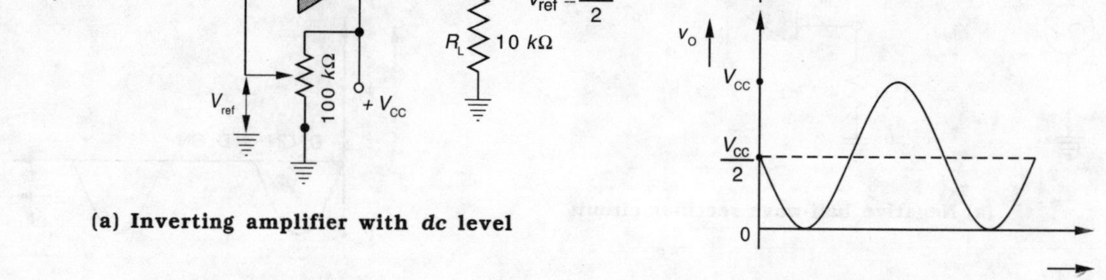


(a) Inverting amplifier with dc level

(b) Input and output waveforms

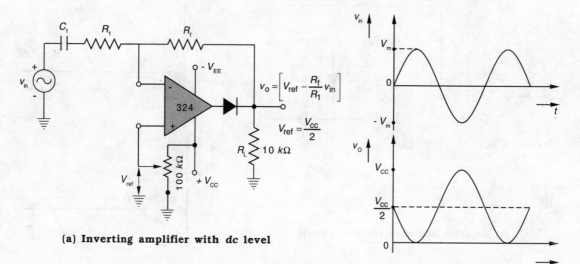

**Fig. 9.29 Positive clamper**

Next consider only the input voltage at the inverting (–) input. During the negative half-cycle, diode $D_1$ becomes forward-biased and thereby the capacitor $C_1$ gets charged to a voltage equal to $-V_m$ (negative peak value). During the positive half-cycle, diode $D_1$ becomes reverse-biased. Hence the voltage $V_m$ across the capacitor acquired during the negative half-cycle is retained. Since this voltage is in series with the positive peak voltage $V_m$, the output voltage, $v_o = 2V_m$. Thus the net output voltage $v_o = V_{ref} + 2V_m$. For precision clamping,

$$C_1 r_d << \frac{T}{2}$$

where $r_d$ = forward resistance of the diode (≅ 100Ω)

$T$ = time period of the input waveform

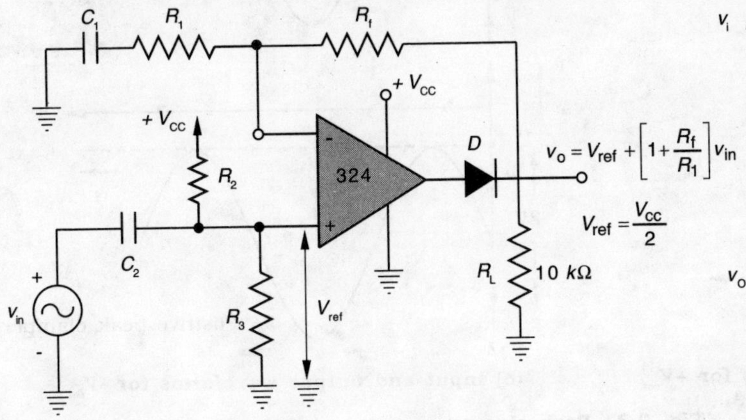


(a) Noninverting amplifier with *dc* level

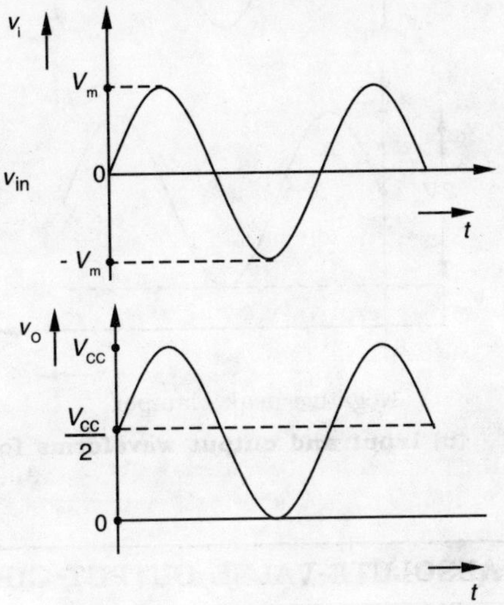


(b) Input and output waveforms

**Fig. 9.30 Positive clamper**

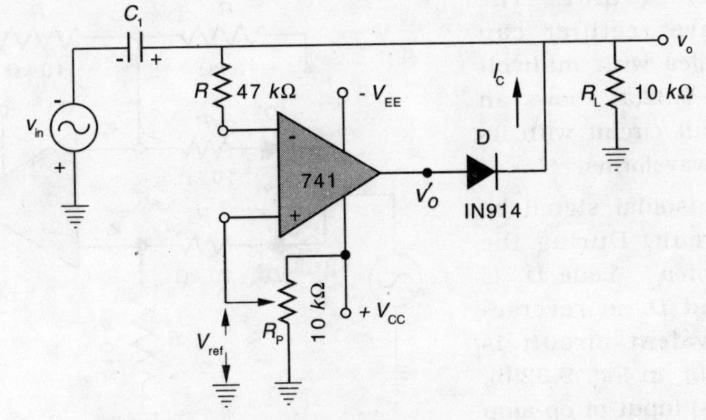


(a) Peak clamper circuit

**Fig. 9.31 Peak clamper**

In Fig. 9.31(a), resistor $R$ is used to protect the op-amp against excessive discharge currents from capacitor C, especially when the *dc* supply voltages are switched off. A positive peak clamping can be accomplished by reversing diode D and using negative reference voltage ($-V_{ref}$).

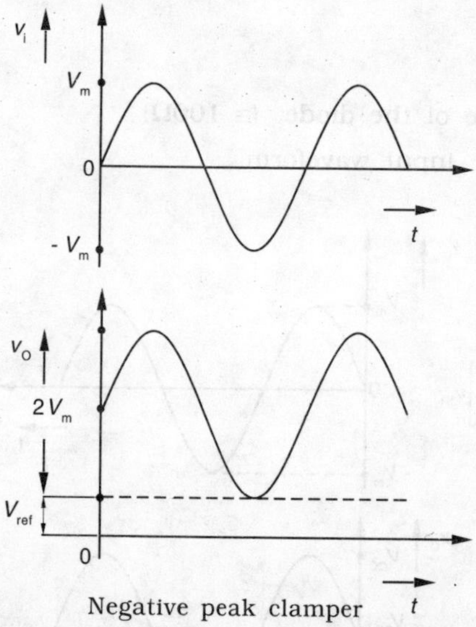


**(b) Input and output waveforms for $+V_{ref}$**

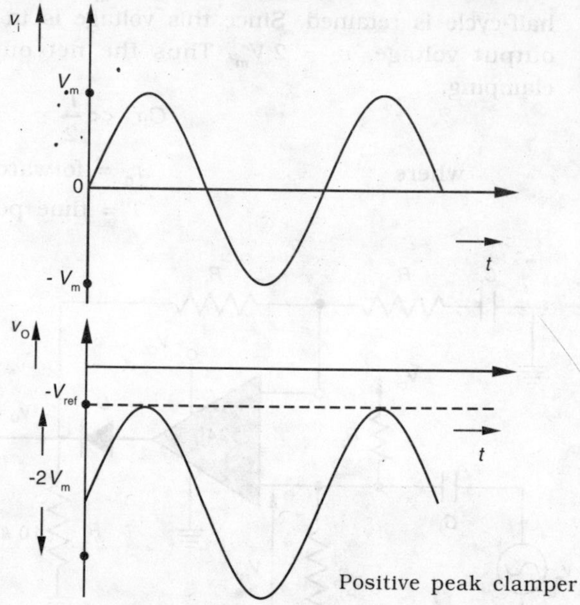


**(c) Input and output waveforms for $-V_{ref}$**

**Fig. 9.31 Peak clamper**

## ABSOLUTE-VALUE OUTPUT CIRCUIT

The precision full-wave rectifier is also called an absolute-value circuit and transmits one polarity of the input signal and inverts the other. Thus both half-cycles of an *ac* voltage are transmitted but are converted to a single polarity output. The precision full wave rectifier can rectify input voltages with millivolt amplitudes. Figure 9.32(a) shows an absolute-value output circuit with its input and output waveforms.

Let the sinusoidal signal be applied to the circuit. During the positive half-cycle of $v_{in}$, diode $D_1$ is forward biased and $D_2$ is reverse-biased. The equivalent circuit is shown in Fig. 9.32(b). In Fig. 9.32(b), the voltage at the (+) input of op-amp is given by

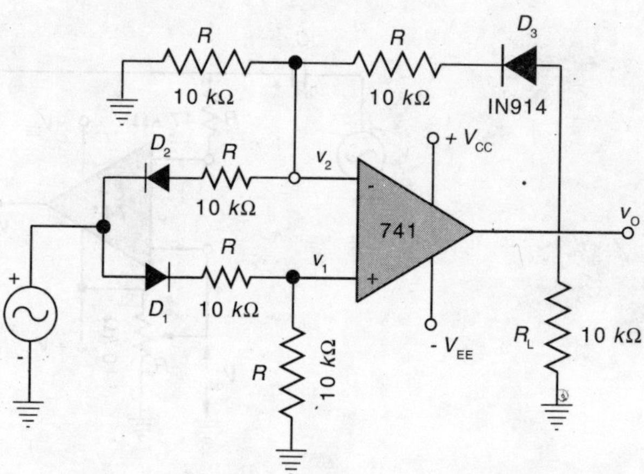


**(a) Absolute-value output circuit**

**Fig. 9.32 Absolute-value output circuit**

$$v_1 = \frac{V_m - V_{D_1}}{2}$$

where $V_{D_1}$ is the voltage drop across $D_1$ (= 0.7 $V$).

Similarly, voltage at the (+) input of op-amp is given by

$$v_2 = \frac{v_o(+) - V_{D_3}}{2}$$

where $v_o(+)$ = output voltage during the positive half-cycle

$V_{D_3}$ = voltage drop across diode $D_3$ = 0.7 $V$

Since $v_{id} \cong 0$, $v_1 = v_2$.

$$\therefore \quad \frac{V_m - V_{D_1}}{2} = \frac{v_o(+) - V_{D_3}}{2}$$

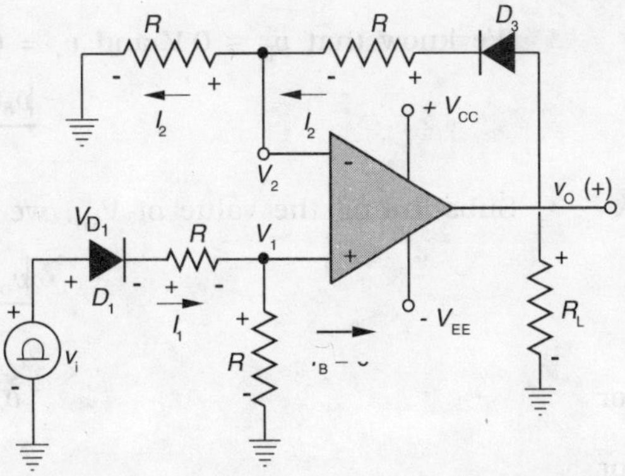


**(b) Equivalent circuit during positive half-cycle**

or $\quad v_o(+) = V_m \qquad \longrightarrow (1)$

During the negative half-cycle of $v_{in}$, diode $D_1$ is reverse-biased and diode $D_2$ is forward-biased. The equivalent circuit is shown in Fig. 9.32(c). This circuit can further be simplified by Thevenizing to the left of (–) input of op-amp. The corresponding Thevenin's voltage and resistance are

$$V_{TH} = -\left(\frac{V_m - V_{D_2}}{2}\right)$$

and

$$R_{TH} \cong \frac{R}{2}$$

Applying KCL at node $v_2$,

$$I_1 = I_2 + I_{B_2}$$

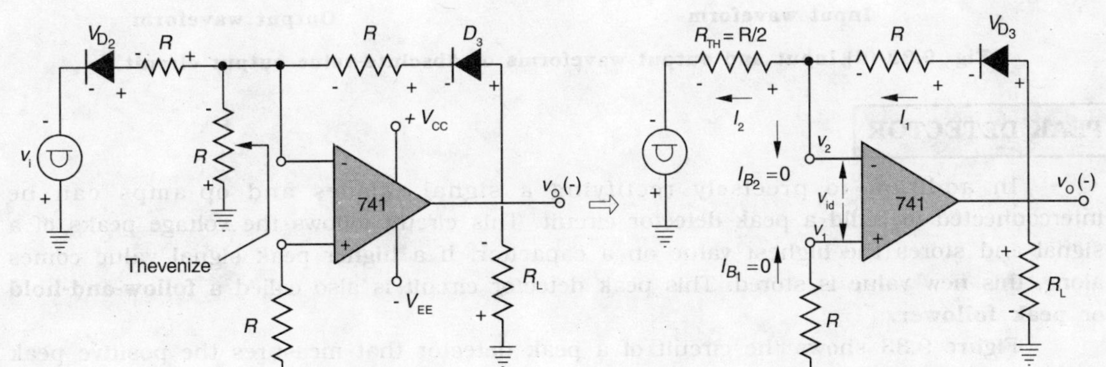


**(c) Equivalent circuit during negative half-cycle**

**Fig. 9.32 Absolute-value output circuit**

Since $I_{B_2} = 0$,
$$I_1 \cong I_2$$
$$\frac{|v_o(-) - V_{D_3}| - v_2}{R} = \frac{v_2 - V_{TH}}{R/2}$$

We know that $v_{id} \cong 0\,V$ and $v_1 = 0$. Therefore, $v_2 = 0$ and
$$\frac{|v_o(-) - V_{D_3}|}{R} = \frac{-V_{TH}}{R/2}$$

Substituting the value of $V_{TH}$, we have
$$\frac{|v_o(-) - V_{D_3}|}{R} = \frac{+(V_m - V_{D_2})/2}{R/2}$$

or
$$v_o(-) - V_{D_3} = V_m - V_{D_2}$$

or
$$v_o(-) = V_m \longrightarrow (2)$$

From Eqs. (1) and (2), it is seen that, regardless of the polarity of the input signal, output is always positive going, hence the name absolute-value output circuit. Because of the nature of its output waveform, the circuit may be used as a full-wave rectifier.

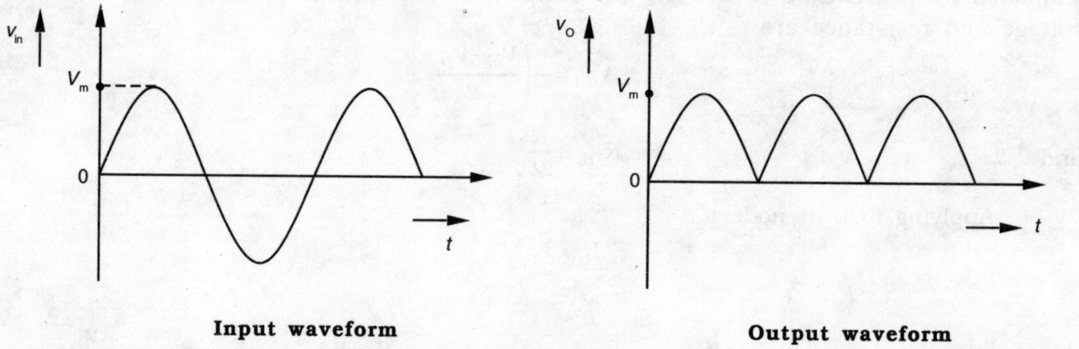


**Fig. 9.32 (d) Input and output waveforms of Absolute-value output circuit**

## PEAK DETECTOR

In addition to precisely rectifying a signal, diodes and op-amps can be interconnected to build a peak detector circuit. This circuit follows the voltage peaks of a signal and stores the highest value on a capacitor. If a higher peak signal value comes along, this new value is stored. This peak detector circuit is also called a **follow-and-hold** or **peak follower.**

Figure 9.33 shows the circuit of a peak detector that measures the positive peak values of the square-wave input. During the positive half-cycle of $v_i$, the diode $D_1$ becomes forward-biased and the capacitor $C$ gets charged to the positive peak value $V_m$ of the input voltage. In this situation, the op-amp functions as a voltage follower. During the negative half-cycle of $v_i$, diode $D_1$ is reverse-biased and the voltage across $C$ is retained. Since the circuit, charging time constant $Cr_d$ and discharging time $CR_L$ must satisfy the following

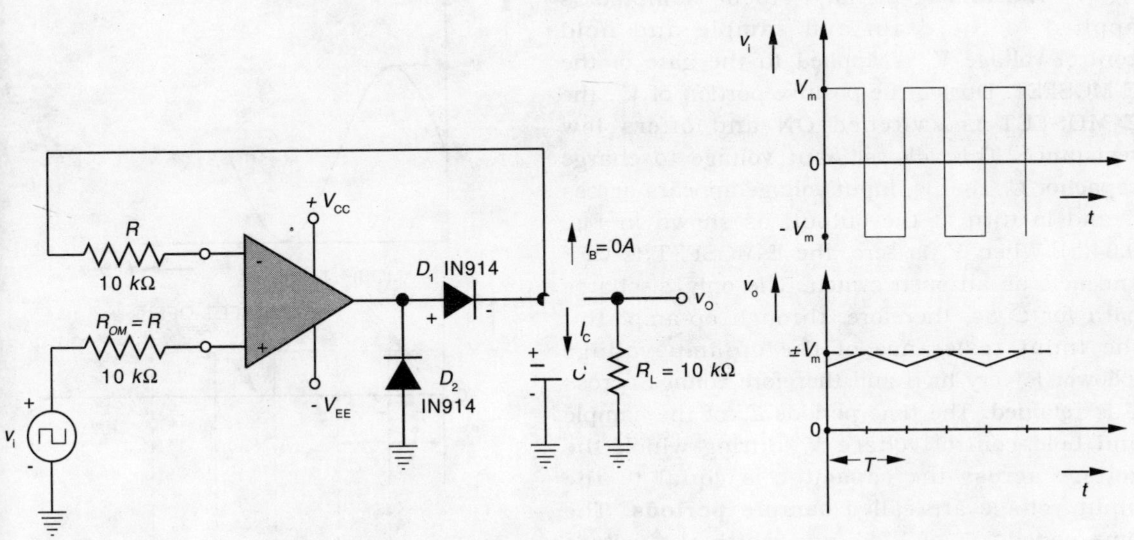


(a) **Peak detector circuit** (b) **Input and output waveforms**

**Fig. 9.33 Peak detector**

conditions:

$$Cr_d \leq \frac{T}{10}$$

where $r_d$ = forward resistance of the diode (≅ 100 Ω)

$T$ = time period of the input waveform

and $CR_L \geq 10T$

where $R_L$ = load resistance

The resistor $R$ is used to protect the op-amp against excessive discharge currents, especially when the power supply is switched off. The resistor $R_{OM}$ minimizes the offset problems caused by input currents. The purpose of diode $D_2$ is that it conducts during the negative half-cycle of $v_i$ and hence prevents the op-amp from going into negative saturation. This in turn reduces the recovery time of the op-amp.

Negative peaks of input signal $v_i$ can be detected simply by reversing diodes $D_1$ and $D_2$.

## SAMPLE AND HOLD CIRCUIT

A circuit that samples an input signal and holds on to its last sampled value until the input is sampled again, is called a **sample and hold circuit.**

A sample and hold circuit consists of an E-MOSFET and an op-amp, shown in Fig. 9.34. The E-MOSFET works as a switch that is controlled by the sample-and-hold control voltage $V_S$. The capacitor $C$ serves as a storage element.

The analog signal $v_i$ to be sampled is applied to the drain and sample and hold control voltage $V_S$ is applied to the gate of the E-MOSFET. During the positive portion of $V_S$, the E-MOSFET is switched ON and offers low resistance. This allows input voltage to charge capacitor $C$. That is, input voltage appears across $C$ and in turn at the output, as shown in Fig. 9.34(b). When $V_S$ is zero, the E-MOSFET is OFF and acts as an open switch. The only discharge path for $C$, is, therefore, through op-amp. But the input resistance of the op-amp voltage follower is very high and therefore voltage across $C$ is retained. The time periods $T_S$ of the sample and hold control voltage $V_S$ during which the voltage across the capacitor is equal to the input voltage are called **sample periods**. The time periods $T_H$ of $V_S$ during which the voltage across the capacitor is constant are called **hold periods**. The output of the op-amp is usually processed by a digital computer and observed during hold periods. To get close approximation of the input waveform, the frequency of $V_S$ must be significantly higher than that of the input.

A specially designed sample-and-hold IC such as the LF 398 or LM 102 provides improved performance.

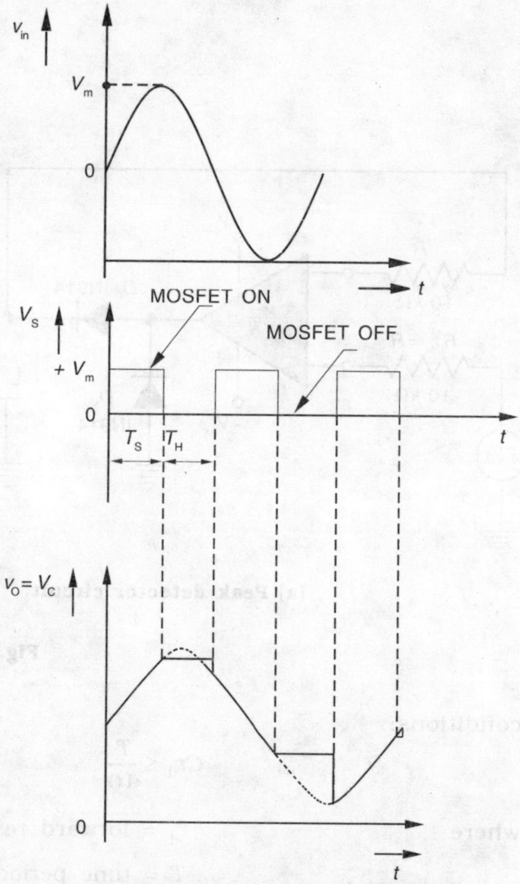


**(b) Input and output waveforms**

The sample-and-hold circuit is used in digital interfacing and communications such as analog-to-digital and pulse modulation systems.

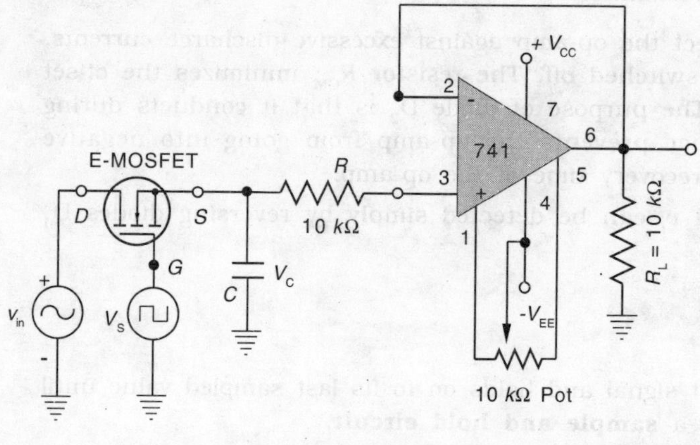


**(a) Sample and hold circuit**

**Fig. 9.34 Sample and hold circuit**

## VOLTAGE REGULATORS

A **voltage regulator** is a circuit which provides a constant voltage regardless of changes in input voltage, load current and temperature. The voltage regulator is one part of a power supply. An excellent *dc* voltage regulator can be built from an op-amp, Zener diode, two resistors and one potentiometer. IC voltage regulators are versatile and relatively inexpensive. They are available with features such as a programmable output, current/voltage boosting, internal short-circuit current limiting, thermal shutdown, and floating operation for high voltage applications. Most voltage regulators fall into two broad categories: linear regulators and switching regulators. Many types of IC regulators are available. The most popular types of IC voltage regulators are:

1. Fixed output voltage regulators: positive and/or negative output voltage
2. Adjustable output voltage regulators: positive or negative output voltage
3. Switching regulators
4. Special regulators

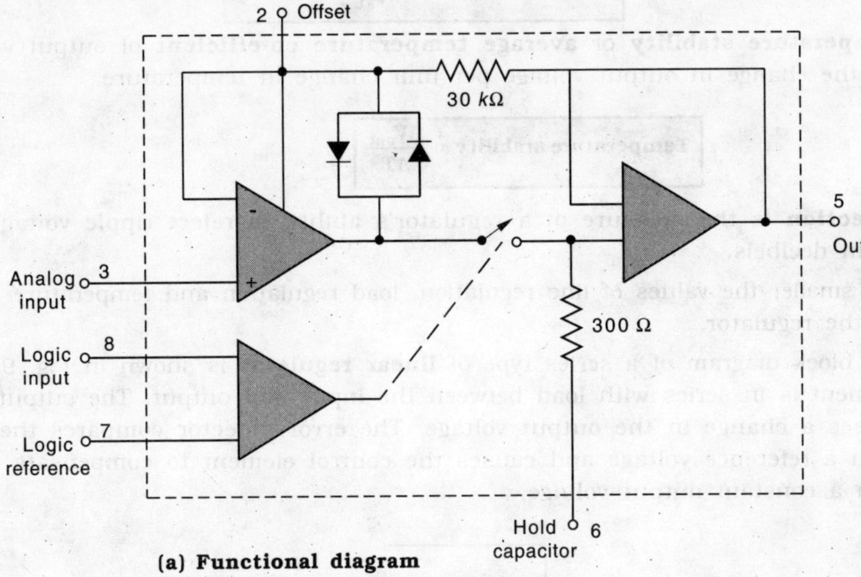


**(a) Functional diagram**

All types of regulators except switching regulators are called **linear regulators**. Most practical regulators are of the linear type.

The latest generation of IC voltage regulators has devices with only three pins: one for the unregulated input voltage, one for the regulated output voltage and one for ground. The new devices can supply load current from 100 $mA$ to more than 5 $A$. The new IC regulators are available in plastic or metal packages.

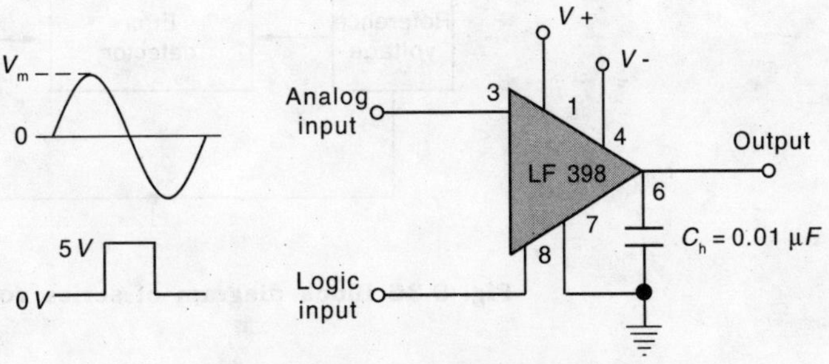


**(b) Typical connection diagram**

**Fig. 9.35 LF 398 sample and hold**

## Voltage Regulation

Two basic categories of voltage regulation are **line regulation** and **load regulation**. The purpose of line regulation is to maintain a nearly constant output voltage when the input voltage varies. The purpose of load regulation is to maintain a nearly constant output voltage when the load varies.

**Line regulation** is defined as the percentage change in output voltage for a given change in the input voltage.

$$\textbf{Line regulation} = \frac{\Delta V_{out}}{\Delta V_{in}} \times 100\%$$

**Load regulation** is defined as the percentage change in output voltage for a given change in load current. $V_{NL}$ is no-load voltage and $V_{FL}$ is full-load voltage.

$$\textbf{Load regulation} = \frac{V_{NL} - V_{FL}}{V_{FL}} \times 100\%$$

**Temperature stability** or **average temperature co-efficient** of output voltage is defined as the change in output voltage per unit change in temperature.

$$\textbf{Temperature stability} = \frac{\Delta V_{out}}{\Delta T}$$

**Ripple rejection** is the measure of a regulator's ability to reject ripple voltages. It is expressed in decibels.

The smaller the values of line regulation, load regulation and temperature stability, the better the regulator.

The block diagram of a series type of **linear regulator** is shown in Fig. 9.36. The control element is in series with load between the input and output. The output sample circuit senses a change in the output voltage. The error detector compares the sample voltage with a reference voltage and causes the control element to compensate in order to maintain a constant output voltage.

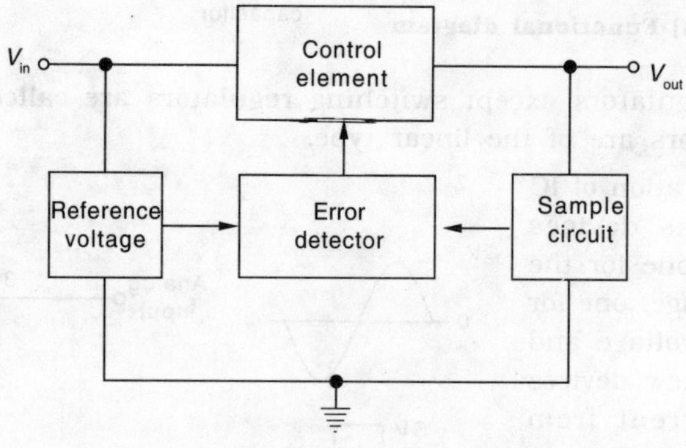


**Fig. 9.36 Block diagram of series voltage regulator**

## 1. Fixed Voltage Regulators

### (a) Positive Voltage Regulator Series with Seven Voltage Options

The 7800 series consists of three-terminal positive voltage regulators with seven voltage options. These ICs are designed as fixed voltage regulators and with adequate heat sinking can deliver output currents in excess of 1 *A*. The 7800 regulators can also be used as current sources.

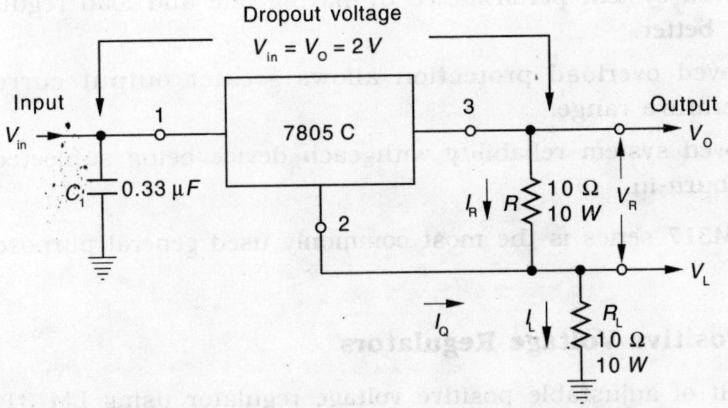


**Fig. 9.37 The 7805C as a 0.5A current source**

A typical connection diagram of the IC 7805 *C* as a 0.5 *A* current source is shown in Fig. 9.37. From the diagram, the current supplied to the load is given by

$$I_L = \frac{V_R}{R} + I_Q \longrightarrow (1)$$

where $I_Q$ = quiescent current (≅ 4.3 *mA*)

Since $V_R = V_{23} = 5\,V$ and $R = 10\,\Omega$,

$$I_L = \frac{5}{10} = 0.5\,A$$

The output voltage $V_o$ with respect to ground is given by

$$V_o = V_R + V_L \longrightarrow (2)$$

where $V_L = I_L R_L$

Since $R_L = 10\,\Omega$, $V_L = 0.5 \times 10 = 5\,V$

∴ $V_o = 5\,V + 5\,V = 10\,V$

Since the dropout voltage for the 7805 *C* is 2 *V*, the minimum input voltage required is given by the relation,

$$V_{in} = V_o + \text{dropout voltage}$$
$$= 10 + 2 = 12\,V$$

Hence, a current source circuit using a voltage regulator can be designed for a desired value of load current $I_L$ simply by selecting an appropriate value for *R*.

### (b) Negative Voltage Regulator Series with Nine Voltage Options

The 7900 series typical of three terminal IC regulators provide a fixed negative output voltage, with nine voltage options. The two extra voltage options are –2 *V* and –5.2 *V*.

## 2. Adjustable Voltage Regulators

Adjustable voltage regulators have the following performance and reliability advantages over the fixed types:

(i) Improved system performance by having line and load regulation of a factor of 10 or better.

(ii) Improved overload protection allows greater output current over operating temperature range.

(iii) Improved system reliability with each device being subjected to 100% thermal limit burn-in.

The IC LM317 series is the most commonly used general purpose adjustable voltage regulators.

## a) Adjustable Positive Voltage Regulators

The circuit of adjustable positive voltage regulator using LM 317 is shown in Fig. 9.38. There are two external resistors used in the circuit to set the output voltage. The three terminals of LM 317 are $V_{in}$, $V_{out}$ and Adjustment (ADJ). It is operated as a "floating" regulator because the adjustment terminal is not connected to ground, but it floats to whatever voltage is across $R_2$. When input voltage is applied, the LM 317 develops a nominal 1.25 $V$, referred to as the reference voltage $V_{ref}$, between the output and adjustment terminal. This constant reference voltage produces a constant current $I_1$ through $R_1$, regardless of the value of $R_2$. As resistor $R_1$ sets constant current $I_1$, it is called the **current set** or **program resistor**. The sum of $I_1$ and $I_{Adj}$ flows through the output set resistor $R_2$. Referring to Fig. 9.38, the output voltage $V_o$ is given by

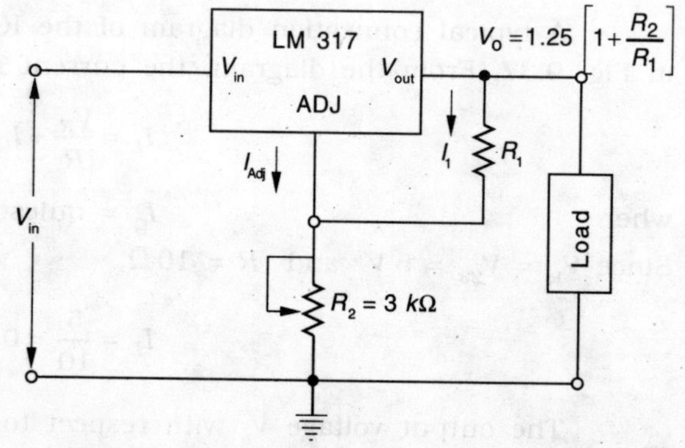


**Fig. 9.38 Adjustable positive voltage regulator**

$$V_o = R_1 I_1 + R_2 (I_1 + I_{Adj}) \qquad \longrightarrow (1)$$

where $$I_1 = \frac{V_{ref}}{R_1}$$

$R_1$ = Current ($I_1$) set resistor

$R_2$ = Output ($V_o$) set resistor

$I_{Adj}$ = Adjustment pin current

$$\therefore \quad V_o = R_1 \frac{V_{ref}}{R_1} + R_2 \left( \frac{V_{ref}}{R_1} + I_{Adj} \right)$$

$$= V_{ref}\left(1+\frac{R_2}{R_1}\right)+I_{Adj}R_2 \qquad \longrightarrow (2)$$

where $V_{ref}$ = 1.25 $V$ = reference voltage between the output and adjustment terminals.

Since the current $I_{Adj}$ is very small and constant, the voltage drop across $R_2$ due to $I_{Adj}$ is also very small and can be neglected.

$$\therefore \qquad \mathbf{V_o = 1.25\left(1+\frac{R_2}{R_1}\right)} \qquad \longrightarrow (3)$$

The above equation shows that the output voltage $V_o$ is a function of $R_2$ for a given value of $R_1$ and can be varied by adjusting the value of $R_2$.

The circuit diagram of adjustable positive voltage regulator using IC LM317 with capacitors and protective diodes is shown in Fig. 9.39. The disc tantalum capacitor $C_1$ (= 0.1 μ$F$) is used for input bypassing for almost all applications. An aluminium or tantalum capacitor $C_2$ in the range of 1 to 1000 μ$F$ is used to provide improved impedance and rejection of transients. In addition, the adjustment terminal is bypassed with $C_3$ to obtain very high ripple rejection ratios. Protective diodes $D_1$ and $D_2$ are used with the LM317 to prevent the capacitors from discharging through low current points into the regulator and for use with outputs greater than 25 $V$.

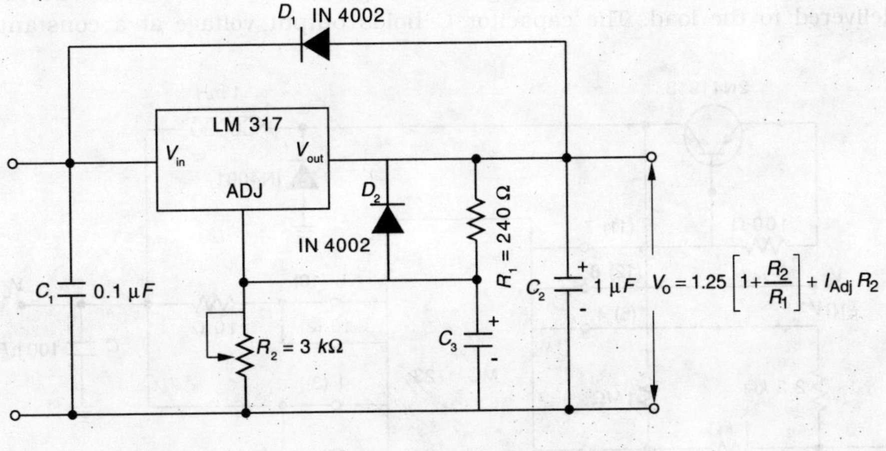


**Fig. 9.39 LM317 with capacitors and protective diodes**

## (b) Adjustable Negative Voltage Regulator

The LM 337 series of adjustable negative voltage regulators is a complement to the LM 317 series devices. They are available in the same voltage and current options as the LM 317 series devices. Like the LM 317, the LM 337 requires two external resistors for output voltage adjustment as shown in Fig. 9.40. The output voltage can be adjusted from –1.2 $V$ to 37 $V$, depending on the external resistor values. The capacitors are for decoupling and do not affect the *dc* operation.

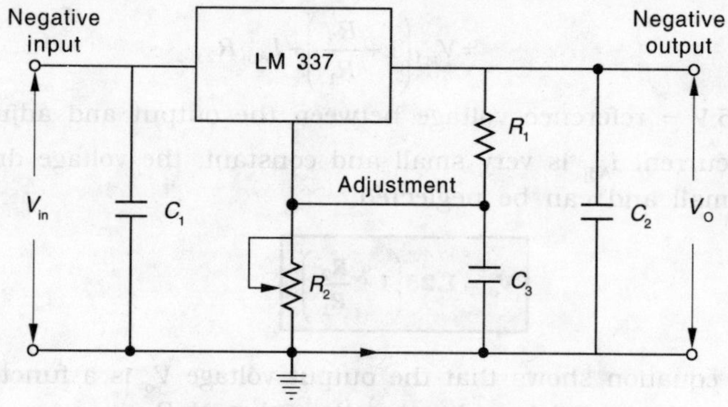


**Fig. 9.40 Adjustable negative voltage regulator**

## Switching Regulators

Switching regulators can provide greater load currents at low voltage than linear regulators and also consume less power. Ex: Motorola's MC 1723. It can be used in different ways, for example, as a fixed positive or negative output voltage regulator, variable output voltage regulator or switching regulator.

Figure 9.41 shows the connections for switching regulator MC1723 for positive output. It includes a power transistor, a choke *L*, a diode D and capacitors. The transistor minimizes the power dissipation during switching and choke smooths out the current pulses delivered to the load. The capacitor C holds output voltage at a constant *dc* level.

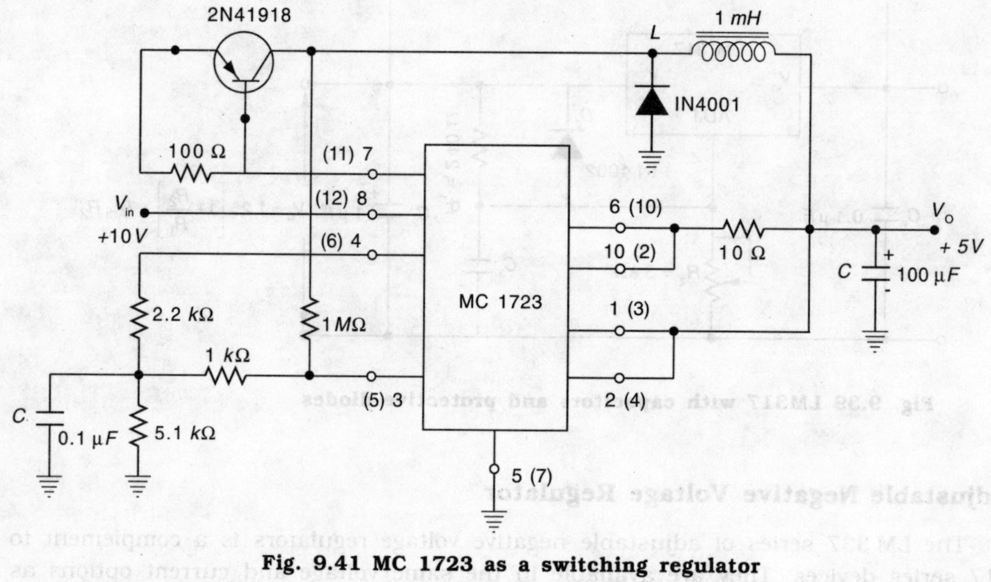


**Fig. 9.41 MC 1723 as a switching regulator**

To improve the efficiency of the regulator, the series-pass transistor is used as a switch rather as a variable resistor as in the linear mode. The series-pass transistor switches between cutoff and saturation at a high frequency and produces a pulse-width modulated (PWM) square wave output. This output is filtered through a low-pass *LC* filter to produce an average *dc* output voltage. Hence the output voltage is proportional to the pulse width and frequency.

## Theory of Switching Regulators

A basic switching regulator consists of the following components:

(i) Voltage source $V_{in}$

(ii) Switch $S_1$

(iii) Pulse generator $V_{pulse}$

(iv) Filter $F_1$

The block diagram of basic switching regulator is shown in Fig. 9.42.

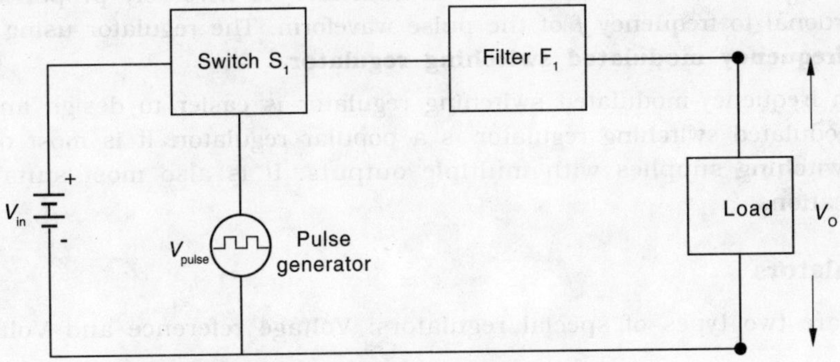


**Fig. 9.42 Basic switching regulator**

(i) **Voltage source ($V_{in}$)** may be a battery or an unregulated or a regulated voltage. The voltage source must satisfy the following requirements:

The voltage source must supply the required output power and the losses associated with the switching regulator.

It must be large enough to supply sufficient dynamic range for line and load variations.

It must be sufficiently high to meet the minimum requirement of the regulator system to be designed.

It must store energy for a specified amount of time during power failures.

**(ii) Switch $S_1$** is typically a transistor or a thyristor connected as a power switch. It is operated between cutoff and saturated modes. The pulse generator output alternately operates the switch ON and OFF.

**(iii) Pulse generator $V_{pulse}$** produces an asymmetrical square wave varying in either frequency or pulse width called frequency modulation or pulse width modulation, respectively. The most effective range is around 20 *kHz*.

The duty cycle of the pulse waveform determines the relationship between the input and output voltages. The duty cycle is defined as the ratio of the on-time $t_{on}$ to the period $T$ of the pulse waveform.

$$\text{Duty cycle} = \frac{t_{on}}{t_{on} + t_{off}} = \frac{t_{on}}{T} = t_{on}\, f \qquad \longrightarrow (1)$$

where $t_{on}$ = on-time of the pulse waveform

$t_{off}$ = off-time of the pulse waveform

Switching regulators operate frequencies in the range of 10 to 50 *kHz*. High operating frequencies reduce the ripple voltage at the expense of decreased efficiency and increased radiated electrical noise.

**(iv) Filter $F_1$** converts the pulse from the output of the switch into a *dc* voltage. The output voltage $V_o$ of the switching regulator is a function of duty cycle and the input voltage $V_{in}$. The output voltage is given by the relation,

$$V_o = \frac{t_{on}}{T} \times V_{in} \qquad \longrightarrow (2)$$

If time period $T$ is constant, then, $V_o$ is directly proportional to the on-time $t_{on}$ for a given value of $V_{in}$. Similarly, if $t_{on}$ is constant, then $V_o$ is inversely proportional to $T$ or directly proportional to frequency $f$ of the pulse waveform. The regulator using this method is known as **frequency modulated switching regulator.**

Though frequency-modulated switching regulator is easier to design and build, but pulse-width modulated switching regulator is a popular regulator. It is most often used in commercial switching supplies with multiple outputs. It is also most suitable in high-current applications.

## Special Regulators

There are two types of special regulators: Voltage reference and Voltage inverter types.

**(i) Voltage References:** Voltage references are a special type of voltage regulators which are used for reference purposes as reference voltages. Voltage references find applications in D/A and A/D converters that do not have internal references, amplifier biasing, low-temperature Zener replacements, high-stability current references, comparator circuits and voltmeter system references. ICL 8069 (a 1.2 *V* temperature-compensated), 9495 (a Teledyne 5-*V*) and MC 1403 (a Motorola 2.5-*V*) are examples of voltage references.

Figure 9.43 shows a D/A converter using the MC 1403 voltage reference. A stable current reference of 2 *mA* required for the MC 1408 is obtained from the MC 1403 with the addition of series resistors $R_1$ and $R_2$. The resistor $R_3$ improves temperature performance and capacitor $C$ decouples any noise present on the reference line.

**(ii) Voltage Inverters:** Voltage inverter produces an output voltage that is opposite in polarity to the input. Datel's VI-7660 is a monolithic CMOS voltage inverter which provides

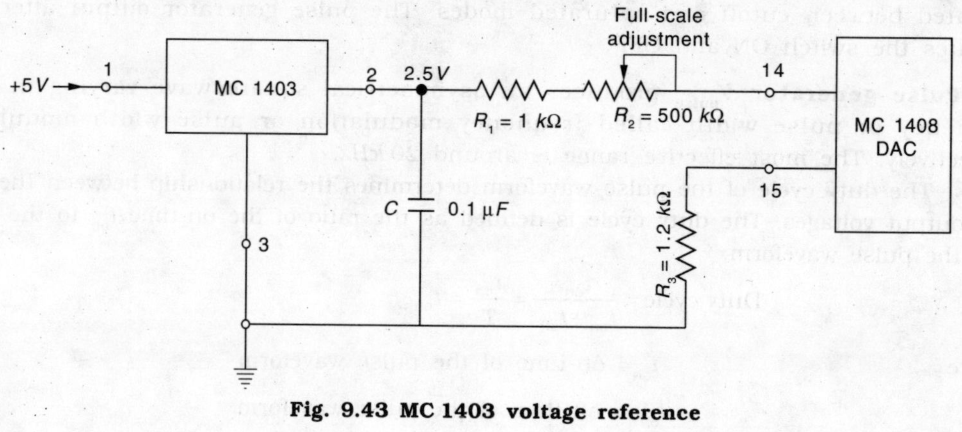


**Fig. 9.43 MC 1403 voltage reference**

–1.5 to –10 *V* from +1.5 to 10 *V* voltage sources with the addition of only two noncritical external capacitors. The block diagram of the VI-7660 voltage inverter is shown in Fig. 9.44. It consists of a *dc*-voltage regulator, *RC* oscillator, voltage-level translator, four output-power MOS switches and a logic network. The logic network senses the most negative voltage in the device and ensures that the output *n*-channel switches are not forward-biased. This assures latch-up free operation. When unloaded, the oscillator oscillates at a frequency of 10 *kHz* for an input supply voltage of 5 *V*. To maximise the conversion efficiency, it is necessary to lower the oscillator frequency. This can be achieved by connecting an additional capacitor (≅ 100 *pF*) between pins 7 and 8.

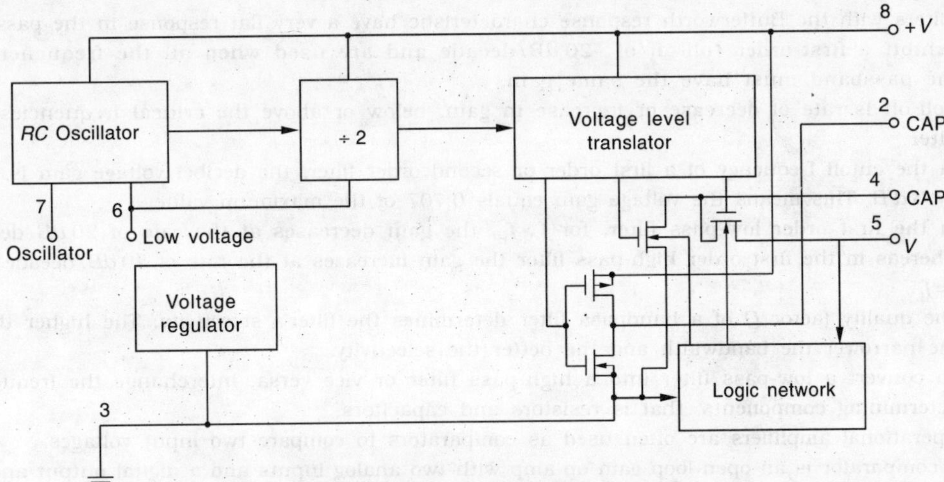


**Fig. 9.44 Block diagram of VI-7660 voltage inverter**

VI-7660 CMOS voltage inverter finds applications in data acquisition and microprocessor based systems.

## IMPORTANT POINTS TO REMEMBER

1. The instrumentation amplifier is intended for precise, low-level signal amplification where low noise, low thermal and time drifts and high input resistance are required.
2. An instrumentation amplifier has high CMRR, low output offset voltage and low output impedance.
3. The voltage gain of a basic instrumentation amplifier is set by a single external resistor.
4. An instrumentation amplifier is useful in applications where small signals are embedded in large common-mode noise.
5. A basic instrumentation amplifier is formed by three op-amps and seven resistors, including the gain-setting resistor.
6. Various differential amplifier configurations can be used as instrumentation amplifiers depending on the application requirements.
7. A filter is a circuit that passes certain frequencies and attenuates or rejects all other frequencies.
8. Filters may be classified in a number of ways: analog or digital, passive or active, audio or radio frequency.
9. A low-pass filter transmits signals below a certain frequency while rejecting higher frequencies.
10. A high-pass filter transmits signals above a certain frequency while rejecting lower frequencies.

11. Active filters use op-amps and reactive elements.
12. One advantage of active filters is that they eliminate inductors which are bulky and expensive at low frequencies.
13. A band-pass filter has a passband between two cutoff frequencies $f_H$ and $f_L$ such that $f_H > f_L$.
14. A band stop filter rejects a range of frequencies lying between a certain lower frequency $f_L$ and a certain higher frequency $f_H$.
15. The all pass filter has input and output amplitudes equal at all frequencies, however, the phase shift between the two is a function of frequency.
16. The order of the filter indicates the rate at which the gain changes while the input frequency is approaching or exceeding the cutoff frequency of the filter.
17. Filters with the Butterworth response characteristic have a very flat response in the passband exhibit a first-order roll-off of –20 *dB*/decade and are used when all the frequencies in the passband must have the same gain.
18. Roll-off is rate of decrease or increase in gain, below or above the critical frequencies of a filter.
19. At the cutoff frequency of a first order or second order filter, the decibel voltage gain is down by 3 *dB*. This means the voltage gain equals 0.707 of the maximum value.
20. In the first order low-pass filter, for $f > f_H$, the gain decreases at the rate of 20 *dB*/decade, whereas in the first order high-pass filter the gain increases at the rate of 20 *dB*/decade until $f = f_L$.
21. The quality factor $Q$ of a bandpass filter determines the filter's selectivity. The higher the $Q$, the narrower the bandwidth and the better the selectivity.
22. To convert a low-pass filter into a high-pass filter or vice versa, interchange the frequency-determining components, that is resistors and capacitors.
23. Operational amplifiers are often used as comparators to compare two input voltages.
24. A comparator is an open-loop gain op-amp with two analog inputs and a digital output and the output is either at $+V_{sat}$ or $-V_{sat}$, depending on which input is the larger.
25. A comparator switches to one state when the input reaches the upper trigger point (UTP) and back to the other state when the input drops below the lower trigger point LTP.
26. Comparator is used in digital interfacing, Schmitt trigger, analog-to-digital converters, oscillators and others.
27. The difference between the UTP and the LTP is the hysteresis voltage.
28. Hysteresis is the characteristic of a circuit in which two different trigger levels create an offset or lag in the switching action.
29. An immediate application of the comparator is the zero-crossing detector, in which the reference voltage $V_{ref} = 0\ V$.
30. The Schmitt trigger is a comparator with positive feedback that converts an irregular analog waveform to a square or pulse waveform.
31. A/D conversion is a common interfacing process often used when a linear analog system must provide inputs to a digital system.
32. Digital to analog (D/A) converter changes a digital input into an analog output.
33. The D/A converter can be formed by using an op-amp and either binary weighted resistors or an $R$ and $2R$ ladder network.
34. The A/D converter can be a single ramp or double-ramp integrating type, single counter type, tracking type or successive-approximation type.
35. As with D/A converters, there are many monolithic/hybrid A/D converters.
36. Typical applications of A/D converters include micro-processor interfacing, digital voltmeters and LED/LCD displays.
37. An ideal peak detector produces a *dc* output voltage that equals the peak of the input waveform.
38. A clipper is a circuit that removes certain parts of the input waveform.

39. The clipper is formed by using an op-amp and rectifier diode.
40. An active clamper can clamp low-level signals.
41. The sample and hold circuit samples an input signal and holds on to its last sampled value until the input is sampled again. It can be constructed by using an op-amp and an E-MOSFET (=switch)
42. Voltage regulators keep a constant *dc* output voltage when the input or load varies within limits.
43. A basic voltage regulator consists of a reference voltage source, an error detector, a sampling element and a control device.
44. Two basic categories of voltage regulators are linear and switching.
45. Switching regulators are more efficient than linear regulators and are particularly useful in low voltage, high-current applications. It provides low power dissipation.
46. Monolithic voltage regulators are available in a variety of different output voltage ratings and are also quicker and easier to use.
47. The 7800 series are three-terminal IC regulators with fixed positive output voltage.
48. The 7900 series are three-terminal IC regulators with fixed negative output voltage.
49. The LM 317 is a three-terminal IC regulator with a positive variable output voltage.
50. The LM 337 is a three-terminal IC regulator with a negative variable output voltage.
51. In a switching regulator, to improve its efficiency the series-pass transistor is used as a switch rather than as a variable resistor as in the linear mode.

## KEY FORMULAE

1. In instrumentation amplifier, the output voltage $V_o = \frac{R_f}{R_1}\frac{\Delta R}{4R}V_{dc}$
2. Closed-loop gain, $A_f = 1 + \frac{2R}{R_G}$
3. The gain magnitude and phase angle equations of the low-pass filter are
$$\left|\frac{v_o}{v_i}\right| = \frac{A_f}{\sqrt{1+(f/f_H)^2}} \quad \text{and} \quad \phi = \tan^{-1}\left(\frac{f}{f_H}\right)$$
4. Gain magnitude of first order high-pass filter is given by
$$\left|\frac{v_o}{v_i}\right| = \frac{A_f(f/f_L)}{\sqrt{1+(f/f_L)^2}} \quad \text{where } f_L = \frac{1}{2\pi RC}$$
5. The voltage gain magnitude of bandpass filter is $\left|\frac{v_o}{v_i}\right| = \frac{A_f(f/f_L)}{\sqrt{\left[1+(f/f_L)^2\right]\left[1+(f/f_H)^2\right]}}$
6. Quality factor $= \frac{f_o}{BW} = \frac{f_o}{f_H - f_L}$ and $f_o = \sqrt{f_H f_L}$
7. The gain magnitude and phase angle equations of all-pass filter are given by
$$\frac{v_o}{v_i} = \frac{1 - j2\pi RCf}{1 + j2\pi RCf} \quad \text{and} \quad \phi = -2\tan^{-1}\left(\frac{2\pi RCf}{1}\right)$$
8. In Schmitt trigger, $V_{UTP} = \frac{R_1}{R_1+R_2}(+V_{sat})$ and $V_{LTP} = \frac{R_1}{R_1+R_2}(-V_{sat})$
9. Hysteresis voltage, $V_{Hy} = V_{UTP} - V_{LTP} = \frac{2R_1}{R_1+R_2}.V_{sat}$

10. In D/A converter with binary-weighted resistor, output voltage is

$$V_o = -R_f\left[\frac{b_o}{R}+\frac{b_1}{R/2}+\frac{b_2}{R/4}+\frac{b_3}{R/8}\right]V$$

11. In D/A converter with $R$ and $2R$ resistors, output voltage is

$$V_o = -R_f\left[\frac{b_3}{2R}+\frac{b_2}{4R}+\frac{b_1}{8R}+\frac{b_0}{16R}\right]V$$

12. In absolute-value output circuit, $v_O(-) = V_m$
13. In fixed positive voltage regulator, $V_O = V_R + V_L$
14. In adjustable positive voltage regulator, $V_o = 1.25\left(1+\frac{R_2}{R_1}\right)$
15. In switching regulator, $V_o = \frac{t_{on}}{T} \times V_{in}$

## SOLVED PROBLEMS

**1. Determine the value of the external gain-setting resistor for a certain IC instrumentation amplifier with $R_1 = R_2 = 25\ k\Omega$. The closed-loop voltage gain is to be 500.**

**Data:** $R_1 = R_2 = 25\ k\Omega$ and $A_f = 500$

**To find:** External gain-setting resistor, $R_G$

**Solution:** Closed-loop voltage gain is given by

$$A_f = 1+\frac{2R}{R_G}$$

$$500 = 1+\frac{2\times 25\times 10^3}{R_G}$$

or $$R_G = \frac{50\times 10^3}{500-1} = \mathbf{100\ \Omega}$$

**2. In instrumentation amplifier, $R_1 = 2\ k\Omega$, $R_f = 10\ k\Omega$, $R_A = R_B = R_C = 100\ k\Omega$, $V_{dc} = +5\ V$, and op-amp supply voltages = ± 15 V. The transducer is a thermistor with the following specifications: $R_T = 100\ k\Omega$ at a reference temperature of 25° C; temperature coefficient of resistance = -1kΩ/°C or 1% /°C. Determine the output voltage at 0° C and at 100° C.**

**Data:** $R_1 = 2\ k\Omega$, $R_f = 10\ k\Omega$, $R_A = R_B = R_c = 100\ k\Omega$, $V_{dc} = +5\ V$, $R_T = 100\ k\Omega$, $t_{ref} = 25°\ C$, temperature co-efficient of resistance, $\alpha = -1\ k\Omega/°\ C$, $t_1 = 0°\ C$ and $t_2 = 100°\ C$

**To find:** Output voltage at 0° C and 100° C

**Solution:** Change in resistance of the thermistor is given by

$$\Delta R = \alpha\,(t_1 - t_2)$$

$$= \frac{-1k\Omega}{^\circ C}(0^\circ C - 25^\circ C) = 25k\Omega$$

Output voltage at 0° C, $$V_o = \frac{\Delta R V_{dc}}{2(2R+\Delta R)} \cdot \frac{R_f}{R_1}$$

$$= \frac{25\times5\times10}{4\times100\times2} = \mathbf{1.56\,V}$$

Similarly, $V_o$ at 100° C, $$\Delta R = \frac{-1k\Omega}{^\circ C}(100^\circ C - 25^\circ C) = -75k\Omega$$

$$V_o = \frac{(-75)(5)(10)}{4\times100\times2} = \mathbf{-4.68\,V}$$

**3. Design a low-pass filter at a cutoff frequency of 2 *kHz* with a passband gain of 2.**

**Data:** $f_H$ = 2 *kHz* and $A_f$ = 2

**To find:** Capacitance *C*, resistance *R* and feedback resistance $R_f$

**Solution:** We know $$R = \frac{1}{2\pi fC}$$

Choose *C* = 0.01 μ*F*, $$R = \frac{1}{2\times3.14\times2\times10^3\times10^{-8}} = \mathbf{7.95\Omega}$$

Use 10 *k*Ω potentiometer.

Since the passband gain is 2, $R_1$ and $R_f$ must be equal $\left(\text{because } A_t = 1 + \frac{R_f}{R_1}\right)$

Therefore, let $R_1$ = $R_f$ = **5 *k*Ω.**

**4. Let *R* = 10 *k*Ω and *C* = 0.001 μ*F* in the circuit of the low pass filter. What is the cutoff frequency?**

**Data:** *R* = 10 *k*Ω, *C* = 0.001 μ*F*

**To find:** Cutoff frequency, $f_C$

**Solution:** Cutoff frequency, $$f_C = \frac{1}{2\pi RC}$$

$$= \frac{1}{2\times3.14\times10\times10^3\times10^{-9}} = \mathbf{15.9\,kHz}$$

**5. Using the frequency scaling technique, convert the 2 *kHz* cutoff frequency of the low-pass filter of Example 3 to a cutoff frequency 3.2 *kHz*.**

**Data:** $f_C$ = 2 *kHz*, new cutoff frequency = 3.2 *kHz*

**To find:** Conversion of $f_C$ = 2 *kHz*, to $f'_C$ = 3.2 *kHz* using the frequency scaling.

**Solution:** $$\frac{\text{Original cutoff frequency}}{\text{New cutoff frequency}} = \frac{2\,kHz}{3.2\,kHz} = 0.625$$

∴ New resistor, $R'$ = 7.95 *k*Ω × 0.625 = **4.97 *k*Ω**

Use $R$ = 5 $k\Omega$ or 10 $k\Omega$ potentiometer and adjust it to 4.97 $k\Omega$.

New cutoff frequency, $f'_C = \dfrac{1}{2\pi RC}$

$$= \frac{1}{2\times3.14\times4.97\times10^3\times10^{-8}} = \mathbf{3.2\ kHz}$$

**6. Design a high-pass filter at a cutoff frequency of 3 *kHz* with a passband gain of 2.**

**Data:** $f_L = 3\ kHz,\ A_f = 2$

**To find:** Resistance $R$; capacitance $C$ and feedback resistance $R_f$

**Solution:** Let $C = 0.01\mu F$, then

$$R = \frac{1}{2\pi fC} = \frac{1}{2\pi\times3\times10^3\times10^{-8}} = 5.3\,k\Omega$$

$$A_f = 1 + \frac{R_f}{R_1} = 2, \text{ passband gain}$$

$\therefore$ $R_1 = R_f$

Similarly, use $R_1 = R_f = \mathbf{10\ k\Omega}$

**7. Design a wide band pass filter with $f_L$ = 400 *Hz*, $f_H$ = 2 *kHz* and a passband gain of 4. Calculate the value of $Q$ for the filter.**

**Data:** $f_L = 400\ Hz, f_H = 2\ kHz$ and $A_f = 4$

**To find:** Resistance, $R$, feedback resistance $R_f$ and capacitance $C$

**Solution:** Let $C' = 0.01\mu F$ Then $R' = \dfrac{1}{2\pi fC}$

$$\therefore \quad R' = \frac{1}{2\times3.14\times2\times10^3\times10^{-8}} = \mathbf{7.95\ k\Omega}$$

To design a high-pass filter, let $C$ = 0.05 $\mu F$.

$$R = \frac{1}{2\pi f_L C} = \frac{1}{2\times3.14\times400\times0.05\times10^{-6}} = \mathbf{7.95\,k\Omega}$$

Since the band-pass gain is 4, the gain of the high-pass as well as low-pass sections could be set equal to 2. That is, input and feedback resistors must be equal, say 10 $k\Omega$ each.

Centre cutoff frequency, $f_o = \sqrt{f_L f_H} = \sqrt{2000\times400} = \mathbf{894.4Hz}$

Quality factor, $Q = \dfrac{f_o}{f_H - f_L} = \dfrac{894.4}{2000-400} = \mathbf{0.559}$

**8. (a) Design the narrow bandpass filter for a centre frequency 2 *kHz*, $Q$ = 5 and $A_f$ = 12. (b) Change the centre frequency to 3 *kHz*, keeping $A_f$ and the bandwidth constant.**

**Data:** $f_O = 2\ kHz$, $Q = 5$ and $A_f = 12$. In the next case $f_O = 3\ kHz$

**To find:** Resistances $R_1$, $R_2$ and $R_3$ and capacitances $C_1$ and $C_2$

**Solution:** Choose the values of $C_1$ and $C_2$. Let $C_1 = C_2 = C = 0.01\ \mu F$.

(a) We know
$$R_1 = \frac{Q}{2\pi f_o C A_f} = \frac{5}{2\times3.14\times2000\times10^{-8}\times12} = \mathbf{3.32\ k\Omega}$$

$$R_2 = \frac{Q}{2\pi f_o C(2Q^2 - A_f)} = \frac{5}{2\times3.14\times2000\times10^{-8}(2\times5^2-12)}$$

$$R_2 = \frac{5\times10^5}{12.56(38)} = \mathbf{1.05\ k\Omega}$$

$$R_3 = \frac{Q}{\pi f_o C} = \frac{5}{3.14\times2000\times10^{-8}} = \mathbf{79.62\ k\Omega}$$

Use $R_1 = 3.3\ k\Omega$, $R_2 = 1\ k\Omega$, and $R_3 = 82\ k\Omega$

(b) The centre frequency $f_o$ can be changed to a new frequency $f_o'$ without changing the gain or bandwidth, changing $R_2$ to $R_2'$, we have

$$R_2' = R_2\left(\frac{f_o}{f_o'}\right)^2$$

$$= 1\times10^3\left(\frac{2}{3}\right)^2 = \mathbf{444\ \Omega}$$

Use $R_2' = \mathbf{470\ \Omega}$

**9. Find the phase angle of the all-pass filter if the frequency of the input signal is 1*kHz* Given *R* = 15.9 kΩ, and C = 0.01 μF.**

**Data:** $R = 15.9\ k\Omega$, and $C = 0.01\ \mu F$ and $f = 1\ kHz$

**To find:** Phase angle, $\phi$

**Solution:** The phase angle $\phi$ is given by

$$\phi = -2\tan^{-1}\left(\frac{2\pi fRC}{1}\right)$$

$$= -2\tan^{-1}\left(\frac{2\times3.14\times15.9\times10^3\times10^{-8}}{1}\right)$$

$$= \mathbf{-90^o}$$

**10. Determine the threshold voltages for the comparator circuit if $R_1$ = 1 kΩ, $R_2$ = 8.2 kΩ, and $v_{in}$ = 1$V_{pp}$ sine wave and the op-amp is type 741 with supply voltages = ± 15 V.**

**Data:** $R_1 = 1\ k\Omega$, $R_2 = 8.2\ k\Omega$, $v_{in} = 1\ V$ and $V_{cc} = V_{EE} = \pm 15\ V$

**To find:** Threshold voltages $V_{UTP}$ and $V_{LTP}$

**Solution:** Threshold voltage, $V_{UTP} = \dfrac{R_1}{R_1 + R_2}(V_{out\,max})$

For 741 the maximum output voltage swing is ± 14 $V$, i.e., + $V_{sat}$ = 14 $V$ and − $V_{sat}$ = − 14 $V$.

$$V_{UTP} = \frac{1}{8.2+1} \times 14 = \mathbf{1.52\ mV}$$

$$V_{LTP} = \frac{1}{8.2+1} \times (-14) = \mathbf{-1.52\ mV}$$

**11. Determine the minimum and maximum output voltages of the LM 317 adjustable voltage regulator. Assume $I_{Adj}$ = 50 µ$A$. Given $R_1$ = 220 Ω and $R_2$ = 5 $k$Ω.**

**Data:** $R_1$ = 220 Ω, $R_2$ = 5 $k$Ω, $I_{Adj}$ = 50 µ$A$

**To find:** Minimum and maximum output voltages

**Solution:** $V_{R_1} = V_{ref} = 1.25 V$

When $R_2$ is set at its minimum of 0 Ω,

$$V_{out(min)} = V_{ref}\left(1 + \frac{R_2}{R_1}\right) + I_{Adj}\, R_2$$

$$= 1.25(1) = \mathbf{1.25\ V}$$

When $R_2$ is set at its maximum of 5 $k$Ω,

$$V_{out(max)} = V_{ref}\left(1 + \frac{R_2}{R_1}\right) + I_{Adj}\, R_2$$

$$= 1.25\left(1 + \frac{5k\Omega}{220\Omega}\right) + 50 \times 10^{-6} \times 5 \times 10^{3}$$

= 29.66 $V$ + 0.25 $V$ = **29.9 V**

**12. Design an adjustable voltage regulator to satisfy the following specifications: Output voltage = 5 to 12 $V$, output current = 1.0 $A$ and voltage regulator is LM 317.**

**Data:** $V_o$ = 5 to 12 $V$, $I_o$ = 1 $A$

**To Find:** Value of resistor $R_2$

**Solution:** We know $I_{Adj}$ = 100 µ$A$ (max) for LM 317 comparator.

Choose $R_1$ = 270 Ω. Then for $V_o$ of 5 $V$ the value of $R_2$ can be calculated from

$$V_{out(min)} = V_{ref}\left(1 + \frac{R_2}{R_1}\right) + I_{Adj}\, R_2$$

$$5 = 1.25\left(1 + \frac{R_2}{270}\right) + 10^{-4} R_2$$

or

$$R_2 = \frac{3.75 \times 270}{1.25} = \mathbf{810\ \Omega}$$

Similarly, for $V_o$ = 12 $V$, the value of $R_2$ is

$$V_{out(max)} = V_{ref}\left(1+\frac{R_2}{R_1}\right)+I_{Adj}R_2$$

$$12 = 1.25\left(1+\frac{R_2}{270}\right)+10^{-4}R_2$$

$\therefore$ $R_2 =$ **2.32 kΩ**

Thus, to obtain the output voltage of 5 V to 12 V we need to vary the value of $R_2$ from 0.81 kΩ to 2.32 kΩ. To accomplish this, a 3-kΩ potentiometer can be used.

## QUESTIONS

1. What is an instrumentation amplifier? List three applications of the instrumentation amplifier.
2. Analyse and explain with a diagram the operation of an instrumentation amplifier.
3. Explain how op-amps are connected to form an instrumentation amplifier.
4. Explain briefly the characteristics of a thermistor, photocell and strain gauge.
5. What is the main purpose of an instrumentation amplifier and what are three of its key characteristics?
6. How is the gain determined in a basic instrumentation amplifier?
7. Define a filter. How are filters classified?
8. List the most commonly used filters.
9. Define passband and stopband for a filter.
10. What are the advantages of active filters over passive filters?
11. What is the Butterworth response? Describe the Butterworth characteristic.
12. Explain with a diagram the operation of first-order low-pass Butterworth filter.
13. Explain with a diagram the operation of first-order high-pass Butterworth filter.
14. Explain the critical frequency and bandwidth of a low-pass filter.
15. What determines the bandwidth of a low-pass filter?
16. What limits the bandwidth of an active high-pass filter?
17. How are the Q and the bandwidth of a bandpass filter related? Explain how the selectivity is affected by the Q of a filter?
18. Explain the order of a filter and its effect on the roll-off rate.
19. Explain with a diagram the operation of bandpass filter.
20. Explain with diagrams the operation of narrow bandpass filter.
21. Explain with diagrams the operation of wide bandstop filter.
22. Explain with diagrams the operation of narrow bandstop filter.
23. What is an all-pass filter? Where and why is it needed?
24. Derive expressions for gain magnitude and phase angle for all-pass filter.
25. Name the basic parts of an active filter.
26. Describe a bandpass filter composed of a low-pass filter and a high-pass filter.
27. What determines selectivity in a bandpass filter?
28. What is a comparator?
29. Explain how input noise affects comparator operation.
30. Write the difference between a basic comparator and the Schmitt trigger?
31. Write the important characteristics of the comparator.
32. Describe the operation of zero-crossing detector.
33. Explain with relevant diagrams the operation of a Schmitt trigger.
34. Define hysteresis. Explain how hysteresis reduces noise effects.

35. What is the purpose of hysteresis in a comparator?
36. What is A/D converter?
37. Explain with relevant diagrams D/A converter with binary weighted resistors.
38. Explain with relevant diagrams D/A converter with $R$ and $2R$ resistors.
39. What is D/A converter? What is the difference between A/D and D/A converters? Write one application of each.
40. Explain with a diagram the working of successive-approximation type A/D converter.
41. Define the following terms for D/A converters: (i) Resolution, (ii) Settling time and (iii) Conversion time.
42. What are clippers/clampers?
43. What is the difference between clippers and clampers. Write one application of each.
44. Explain with relevant diagrams the operation of a positive/negative clipper.
45. Explain with relevant diagrams the operation of a small-signal half-wave rectifier.
46. Explain with relevant diagrams the operation of positive/negative clamper using an op-amp.
47. Describe with relevant diagrams the working of an absolute-value output circuit.
48. Describe with relevant diagrams the working of a peak detector that uses an op-amp.
49. What is an absolute value output circuit?
50. What is a sample-and-hold circuit? Why is it needed?
51. Explain with relevant diagrams the operation of sample-and-hold circuit.
52. What is a voltage regulator? Write four different types of voltage regulators.
53. Describe the basic concept of voltage regulation.
54. Explain (a) line regulation and (b) load regulation.
55. Explain with a diagram the operation of positive/negative fixed voltage regulator.
56. Explain with a diagram the operation of positive/negative adjustable voltage regulator.
57. Describe the principles of switching regulators.
58. What are switching regulators? List four major components of the switching regulators.
59. List the advantages of a switching regulator.
60. What is a voltage reference? Why is it needed?
61. What are the advantages of the adjustable voltage regulators over the fixed voltage regulators?
62. Explain with a block diagram the working of a basic switching regulator.
63. Describe the LM 317 adjustable positive regulator.
64. Describe the LM 337 adjustable negative regulator.
65. Describe IC switching regulators.
66. Write the applications of IC switching regulators.

## EXERCISES

1. What value of external gain-setting resistor is required for an instrumentation amplifier with $R_1 = R_2 = 39\ k\Omega$ to produce a gain of 325? **[Ans: 240 Ω]**
2. In the circuit of an instrumentation amplifier, $R_1 = 1\ k\Omega$, $R_2 = 4.7\ k\Omega$, $R_A = R_B = R_C = 100\ k\,\Omega$, $V_{dc} = +5\ V$, and op-amp supply voltages = ± 15 $V$. The transducer is a thermistor with the following specifications: $R_T = 100\ k\Omega$ at a reference temperature of 25°$C$, temperature coefficient of resistance = – 1 $k\Omega$/°$C$ or 1 %/°$C$. Determine the output voltage at 0°$C$ and at 100°$C$. **[Ans: 1.47 V, –4.41 V]**
3. Design a first-order low-pass filter so that it has a cutoff frequency of 2 $kHz$ and a passband gain of 1. Convert the 2 $kHz$ to a cutoff frequency of 3 $kHz$. **[Ans: $R$ = 5.3 kΩ, $C$ = 0.01 μF]**

4. Design a first-order high-pass filter at a cutoff frequency of 1 *kHz* with a passband gain of 2. **[Ans: $R_f$ = 10 kΩ, C = 0.01 μF and R = 15.9 kΩ]**

5. Design a wide bandpass filter with $f_L$ = 400 *Hz*, $f_H$ = 2 *kHz*, and passband gain = 4. Also draw an approximate frequency response plot for the filter.
**[Ans: R' = 7.95 kΩ, C' = 0.01 μF for 2 kHz and R = 39 kΩ, C = 0.01 μF for 400 Hz; For $A_f$ = 4, $R_1$ = $R'_1$ = $R_f$ = $R'_f$ = 10 kΩ]**

6. Design a wide band stop filter using first-order high-pass and low-pass filters having $f_L$ = 2 *kHz* and $f_H$ = 400 *Hz* respectively.
**[Ans: For $f_L$ = 2 kHz; R = 7.95 kΩ, C = 0.01 μF. For $f_H$ = 400 Hz; R' = 39 kΩ, C' = 0.01 μF. For $A_f$ = 2; $R_1$ = $R_f$ = $R'_1$ = $R'_f$ = $R_2$ = $R_3$ = $R_4$ = 10 kΩ.]**

7. (a) Design a wide bandpass filter with $f_L$ = 200 *Hz*, $f_H$ = 1 *kHz* and a passband gain = 4. (b) Draw the frequency response plot of this filter. (c) Calculate the values of *Q* for the filter. **[Ans: (a) C = 0.05 μF, R = 15.9 kΩ, (b) 447.2 Hz]**

8. Determine the phase angle φ between the input and output at *f* = 2 *kHz* for all pass filter. **[Ans: φ = –126.83°]**

9. In the circuit of inverting Schmitt trigger, $R_1$ = 150 Ω, $R_2$ = 68 kΩ, $v_{in}$ = 500 *mV* pp sine wave and the saturation voltages = ± 14 *V*.
(a) Determine the threshold voltages. (b) What is the value of hysteresis voltage?
**[Ans: (a) 30.8 mV, –30.8 mV (b) 61.6 mV]**

10. For the D/A converter using an *R*–2R ladder network;
(a) Determine the size of each step if $R_f$ = 27 kΩ.
(b) Calculate the output voltage when the inputs $b_0$, $b_1$, $b_2$ and $b_3$ are at 5 *V*.
(c) What is the advantage of this type of D/A converter over the one with binary weighted resistors?
**[Ans: (a) 0.84375 V, (b) –12.66 V (c) The advantage of R/2R DAC is that it requires only two sets of resistance values]**

11. Design an adjustable voltage regulator to satisfy the following specifications: $V_o$ = 8 to 20 *V*, $I_o$ = 1.0 *A*. Use $R_1$ = 330 Ω and $I_{Adj}$ = 100 μ*A*.
**[Ans: $R_2$ = 1.78 kΩ and 4.95 kΩ]**

12. Design a step-down switching regulator according to the following specifications: $V_{in}$ = 12 *V*, $V_o$ = 5 *V* at 500 *mA* maximum, $V_{ripple}$ = 50 *mV* or 1% of $V_o$ , μA78540.
**[Ans: $I_m$ = 1A, $R_{SC}$ = 0.33 Ω, 0.5 W, $t_{on}$ = 1.06 $t_{off}$, $C_t$ = 0.0109 μF, L = 151.69 μH, $C_o$ = 125 μF, $R_2$ = 12.45 kΩ, $R_1$ = 36 kΩ, η = 81 %]**

# 10 555 TIMER

## Chapter Outline



## INTRODUCTION

The 555 timer is a versatile integrated circuit used in many applications. Signetic Corporation first introduced this device as the SE/NE 555 in early 1970. An IC 555 timer is used in a number of novel and useful applications which include monostable and astable multivibrators, pulse generators, digital logic probes, waveform generators, analog frequency meters and tachometers, temperature measurement and control, infrared transmitters, burglar and toxic gas alarms, voltage regulators, electric eyes, etc.

### Features of IC 555 Timer

(i) The 555 timer is a monolithic timing circuit which produces accurate and highly stable time delays or oscillation. It basically consists of **two comparators, a flip-flop, a discharge transistor** and **a resistive voltage divider** as shown in Fig. 10.1(a). It is available as an 8-pin metal can, an 8-pin mini DIP or a 14-pin DIP. The pin diagram of IC 555 timer is shown in Fig. 10.1(b).

(ii) The SE 555 timer is designed for the operating temperature ranges from $-55^\circ C$ to $+125^\circ C$, while the NE 555 timer operates over a temperature of $0^\circ$ to $+70^\circ C$.

(iii) It operates on $+5$ to $+18V$ supply voltage. It can operate in both astable and monostable modes.

(iv) It has an adjustable duty cycle, and can produce timing delays from microseconds to hours.

(v) It has a high output current; i.e. it can source or sink $200\ mA$.

(vi) The output can drive TTL and CMOS circuits.

(vii) It has a temperature stability of 50 parts per million (ppm) per degree Celsius change in temperature, or equivalently 0.005%/° C.

In Fig. 10.1(a), the flip-flop is a digital device, i.e. it is a two state device whose output can be at either a high voltage level (set, S) or a low voltage level (reset, *R*). The state of the output can be changed with proper input signals. The resistive voltage divider is used to set various voltage comparator levels. All these resistors are of equal value; therefore, the comparator 1 has reference voltage of (2/3) $V_{CC}$ and the comparator 2 has a reference voltage of (1/3) $V_{CC}$. The comparator's outputs control the state of the flip-flop. When the trigger voltage goes below (1/3) $V_{CC}$, the flip-flop sets and the output jumps to its high level. The threshold input is normally connected to an external *RC* timing circuit. When the external capacitor voltage exceeds (2/3) $V_{CC}$, the comparator 1 resets the flip-flop, which in turn switches the output back to its low level. When the output is low, the discharge transistor $Q_1$ is turned ON and provides a path for rapid discharge of the external

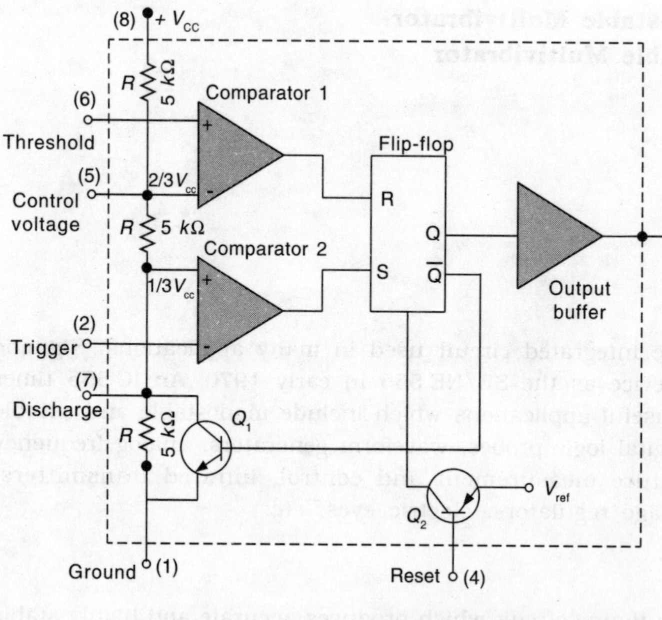


(a) Internal block diagram of a 555 IC timer

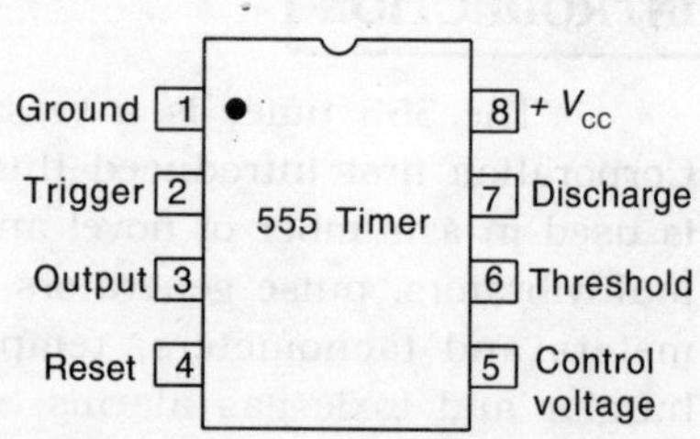


(b) Pin diagram of IC 555

**Fig. 10.1 IC 555 timer**

timing capacitor, *C*. This basic operation allows the timer to be configured with external components as an oscillator, a one-shot or a time-delay element.

## 555 Timer's Pin Functions

**Pin 1: Ground.** All voltages are measured with respect to this terminal.

**Pin 2: Trigger.** The output of the timer depends on the amplitude of the external trigger pulse applied to this pin. If the voltage at pin 2 is greater than (2/3) $V_{CC}$, the output is LOW. When the trigger input is greater than (1/3) $V_{CC}$, the comparator 2 has a LOW output and resets the flip-flop. When trigger input goes below (1/3) $V_{cc}$, the output of comparator 2 is high and the flip-flop is set and the output remains HIGH as long as the trigger terminal is held at LOW.

**Pin 3: Output.** There are two ways of how a load can be connected to the output terminal: (i) Between pin 3 and ground (1) (ii) Between pin 3 and supply voltage $+V_{CC}$ (pin 8). When the output is LOW, the load current flows through the load connected between pin 3 and $+V_{CC}$ into the output terminal and is known as ***sink*** current. But the current through the grounded terminal is zero when the output is LOW. Because of this reason, the load connected between pin 3 and $+V_{CC}$ is known as **normally on load** and that connected between pin 3 and ground is known as **normally off load.** On the other hand, when the output is HIGH, the current flowing through the load connected between pin 3 and $+V_{CC}$ is zero. However, the output terminal supplies current to the normally off load. This current is known as source current. The maximum value of sink or source current is 200 *mA*.

**Pin 4: Reset.** The IC 555 timer can be reset by applying a negative pulse to pin 4. When the reset function is not in use, the reset terminal 4 should be connected to $+V_{CC}$ to avoid any possibility of false triggering.

**Pin 5: Control voltage.** An external voltage must be applied to this terminal which changes the threshold as well as the trigger voltage. Either by applying a voltage to this pin or by connecting a potentiometer between this pin and ground, the pulse width of the output waveform can be varied. When not used, the control pin should be bypassed to ground with a 0.01 $\mu F$ capacitor to prevent noise interference.

**Pin 6: Threshold.** This is the noninverting terminal of the comparator 1, that monitors the voltage across the external capacitor *C*. When the voltage at this pin is $\geq$, threshold voltage (2/3) $V_{CC}$, the output of comparator 1 goes HIGH, which in turn switches the output of the timer LOW.

**Pin 7: Discharge.** This pin is connected internally to the collector of transistor $Q_1$. When the output is HIGH, $Q_1$ is OFF and acts as an open circuit to the external capacitor *C* connected across it. When the output is LOW, $Q_1$ is ON, and acts as a short circuit, shorting out the external capacitor *C* to ground.

**Pin 8: $+V_{CC}$.** The 555 timer can work with supply voltage of + 4.5 to + 18 *V*.

## 555 TIMER AS A MONOSTABLE MULTIVIBRATOR

Figure 10.2 shows the 555 timer connected for monostable (one-shot) operation. It produces a single fixed pulse output each time a trigger pulse is applied to pin 2. The duration of the pulse is determined by the *RC* network connected externally to the 555 timer. Since $\tau (= RC)$ can be changed only by changing resistor or capacitor, the one-shot can also be considered a **pulse stretcher.** In the stable state the output of the circuit is zero or at logic-LOW level. Then the transistor $Q_1$ is ON and the external capacitor *C* is shorted to the ground.

**Operation:** The resistive voltage divider is used to set the voltage comparator levels. Since all the resistors are of equal value, the comparator 1 has a reference of (2/3) $V_{CC}$ and the comparator 2 has a reference of (1/3) $V_{CC}$.

The flip-flop is initially reset, i.e. Q is LOW. The transistor $Q_1$ is ON and capacitor C is shorted to the ground. When a negative trigger pulse less than (1/3) $V_{CC}$ is applied to pin 2, transistor $Q_1$ is turned OFF. This allows the capacitor C to charge up to a value Vcc through RA. When the voltage across C equals (2/3) $V_{CC}$, output of comparator 1 switches from LOW to HIGH, which in turn sets the flip-flop. As soon as Q goes high, it turns ON the transistor $Q_1$ and hence capacitor C quickly discharges through the transistor.

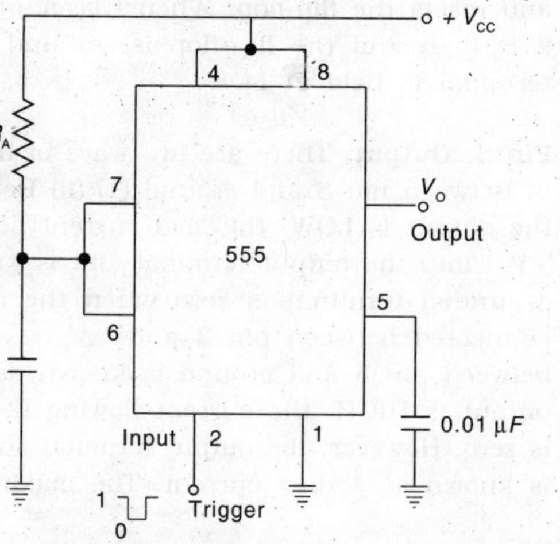


**(a) Monostable 555 timer circuit**

**Fig. 10.2 Monostable multivibrator**

As soon as the output *Q* of FF goes LOW, it turns ON the transistor $Q_1$ and hence capacitor *C* quickly discharges through the transistor. The output of the monostable remains LOW until a trigger pulse is again applied. Then the cycle repeats. The pulse width of the trigger input must be smaller than the expected pulse width of the output waveform.

The capacitor *C* has to charge through $R_A$. The larger the *RC* (= τ) time constant, the longer it takes for the capacitor voltage to reach $\left(\frac{2}{3}\right) V_{CC}$.

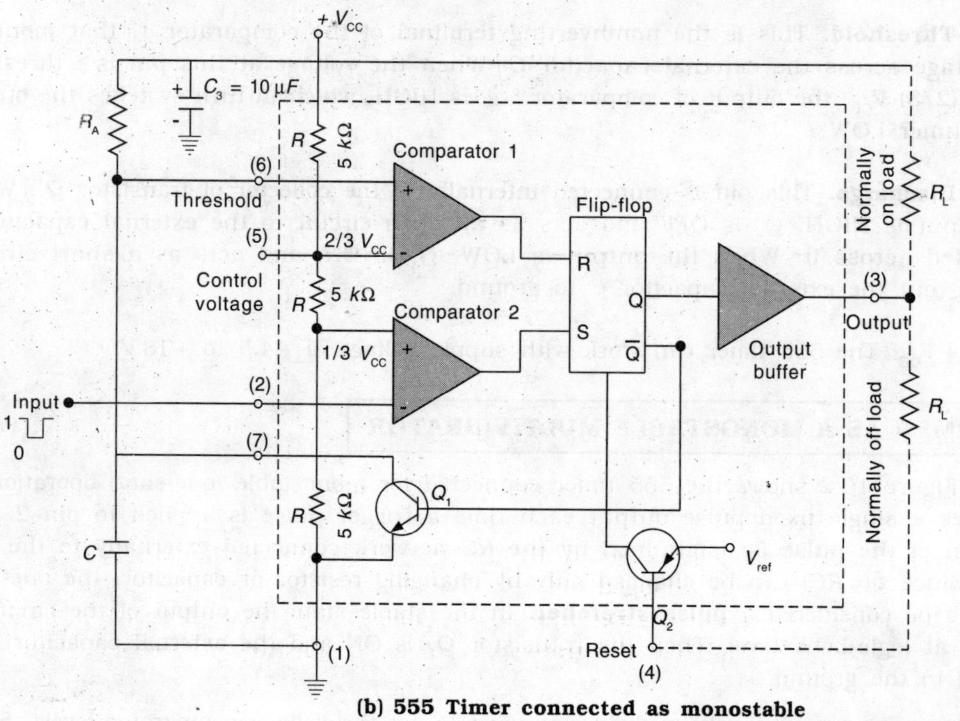


**(b) 555 Timer connected as monostable**

**Fig. 10.2 Monostable multivibrator**

Voltage across capacitor increases exponentially and is given by

$$V_c = V_{cc}\left(1 - e^{-t_p/R_A C}\right)$$

If capacitor voltage, $V_c = \frac{2}{3} V_{cc}$

$$\therefore \quad \frac{2}{3} V_{cc} = V_{cc}\left(1 - e^{-t_p/R_A C}\right) \quad \text{or} \quad \frac{1}{3} = e^{-t_p/R_A C} \quad \text{or} \quad \frac{-t_p}{R_A C} = 0 - 1.0986 \cong -1.1$$

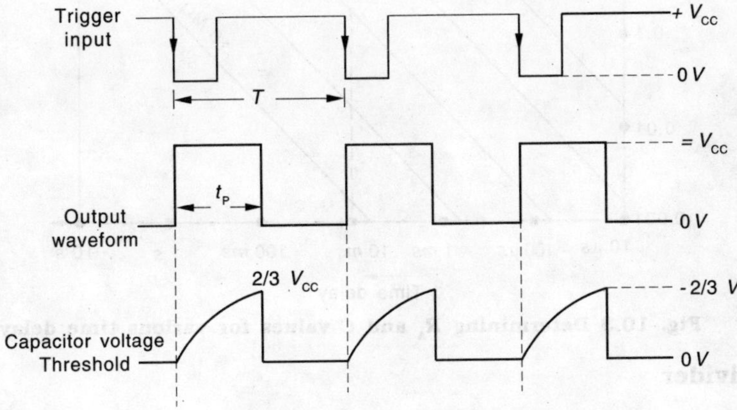


**(c) Input and output waveforms**

**Fig. 10.2 Monostable multivibrator**

The time during which the output remains HIGH is given by

$$t_p = 1.1\, R_A C$$

Figure 10.3 shows a graph of the various combinations of $R_A$ and $C$ necessary to produce desired time delays.

In summary, the monostable 555 timer produces a single pulse whose width is determined by $R_A$ and $C$. The pulse begins with the leading edge of the negative trigger input. Once triggered, the circuit's output remains in the HIGH state until the set time $t_p$ elapses. The output remains in its HIGH state even if an input trigger is applied again during this time interval $t_p$. However, the circuit can be reset during the timing cycle by applying a negative pulse to the reset terminal. The output will then remain in the LOW state until a trigger is again applied.

## Monostable Multivibrator Applications

The monostable operation of IC 555 timer has a number of applications. They are

1. Frequency divider
2. Pulse stretcher
3. Missing pulse detector
4. Linear ramp generator

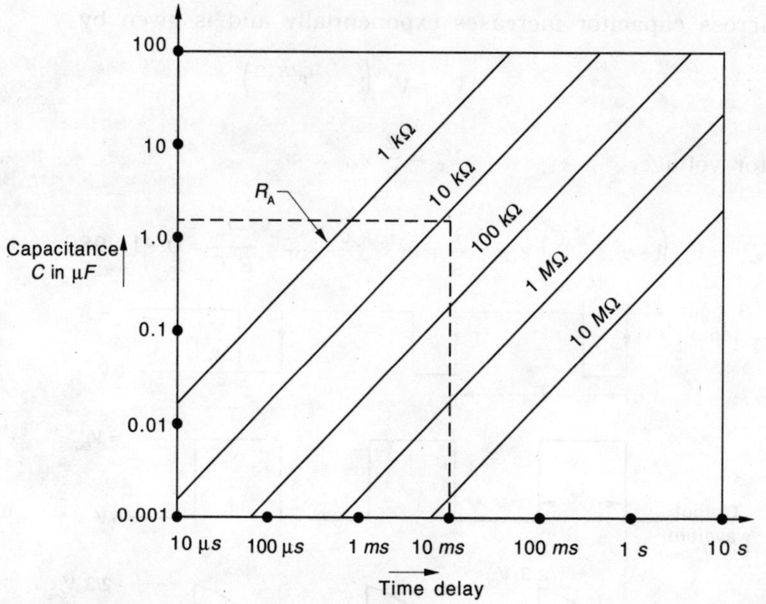


**Fig. 10.3 Determining $R_A$ and C values for various time delays**

## 1. Frequency Divider

The monostable multivibrator can be used as a frequency divider by adjusting the length of the timing cycle $t_p$ with respect to the time period $T$ of the trigger input signal. To use the monostable multivibrator as a divide-by-2 circuit, the timing cycle $t_p$ must be slightly larger than the time period of the trigger input signal, as shown in Fig. 10.4. The monostable multivibrator can also be used as a divide-by-3 circuit provided $t_p$ must be slightly larger than twice the period $T$ of the trigger input signal.

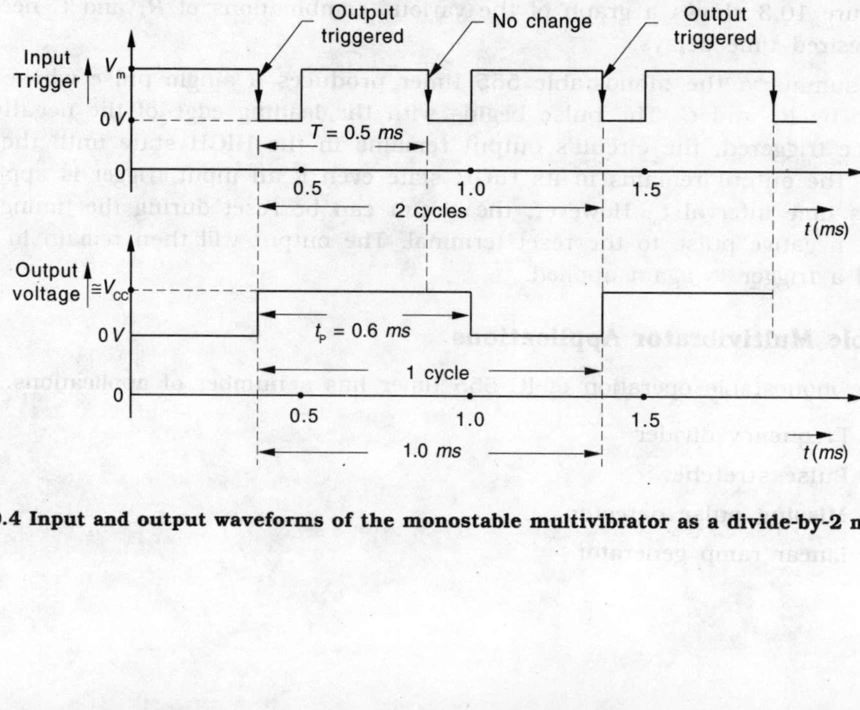


**Fig. 10.4 Input and output waveforms of the monostable multivibrator as a divide-by-2 network.**

## 2. Pulse Stretcher

Since time constant $\tau$ (= $RC$) can be changed only by changing resistors and capacitors, the monostable multivibrator can be considered as a pulse stretcher. As narrow pulse width signals are not suitable to drive LED displays, the 555 pulse stretcher can be used to drive LED.

A basic monostable that is used as a pulse stretcher with an LED indicator at the output is shown in Fig. 10.5. The LED is ON when $t_p = 1.1 R_A C$, which can be varied by changing the value of $R_A$ and/or $C$.

## 3. Ramp Generator

Charging a capacitor through a resistor produces an exponential waveform. If we use a constant source to charge a capacitor, we get a ramp.

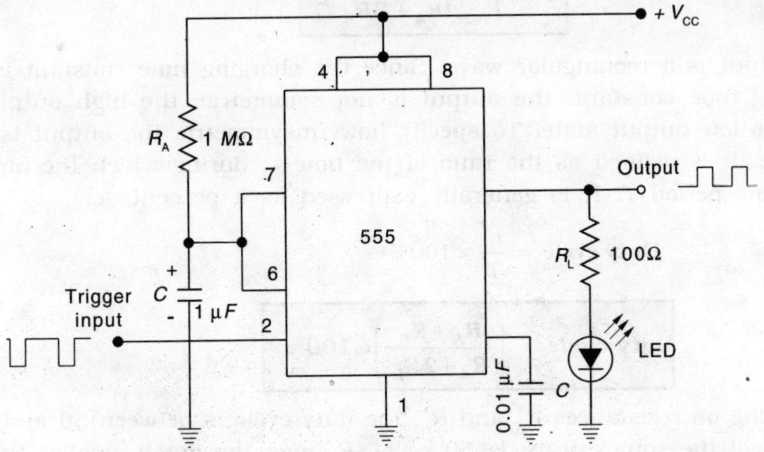


**Fig. 10.5 Monostable multivibrator as a pulse stretcher**

## 555 TIMER AS AN ASTABLE MULTIVIBRATOR

An **astable multivibrator** or **free running** multivibrator is a rectangular wave generator. Unlike the monostable multivibrator, this circuit does not require an external trigger to change the state of the output. However, the time during which the output is either HIGH or LOW is determined by the two resistors $R_A$ and $R_B$ and a capacitor $C$ which are externally connected to the 555 timer.

Figure 10.6(a) shows the 555 timer connected for astable operation. The external components $R_A$, $R_B$ and $C$ form the timing circuit that sets the frequency of oscillation.

Initially, when the output is HIGH, capacitor $C$ starts charging through a total resistance of $R_A + R_B$. Because of this, the charging time constant is $(R_A + R_B) C$. As soon as the voltage across the capacitor $C$ equals (2/3) $V_{CC}$, the comparator 1 triggers the flip-flop and the output goes LOW [Fig. 10.6(c)]. Now the capacitor $C$ discharges through $R_B$ and transistor $Q_1$. Therefore, the discharging time constant is $R_B C$. When the capacitor voltage is slightly < (1/3) $V_{CC}$, output of comparator 2 triggers the flip-flop and the output goes HIGH. The cycle then repeats.

Figure 10.6 (c) illustrates the waveforms. It is seen that the capacitor is periodically charged and discharged between (2/3) $V_{CC}$ and (1/3) $V_{CC}$ respectively. The time that the output is HIGH ($t_H$) is how long it takes capacitor $C$ to charge from (1/3) $V_{CC}$ to (2/3) $V_{CC}$. It is given by

$$t_H = 0.694\,(R_A + R_B)\,C \qquad \longrightarrow (1)$$

The time that the output is LOW ($t_L$) is how long it takes capacitor $C$ to discharge from (2/3) $V_{CC}$ to (1/3) $V_{CC}$. It is given by

$$t_L = 0.694\,R_B C \qquad \longrightarrow (2)$$

The period $T$ of the output waveform is the sum of $t_H$ and $t_L$.

$$T = t_H + t_L = 0.694\,(R_A + 2R_B)C \qquad \longrightarrow (3)$$

Hence, frequency of oscillation is given by

$$\boldsymbol{f = \frac{1}{T} = \frac{1.45}{(R_A + 2R_B)C}} \qquad \longrightarrow (4)$$

The output is a rectangular wave. Since the charging time constant is longer than the discharging time constant, the output is not symmetric; the high output state lasts longer than the low output state. To specify how unsymmetric the output is, we will use the **duty cycle.** It is defined as the ratio of the time $t_H$ during which the output is HIGH to the total time period $T$. It is generally expressed as a percentage.

$$\text{Duty cycle} = \frac{t_H}{T} \times 100\%$$

$$\textbf{Duty cycle} = \left(\frac{R_A + R_B}{R_A + 2R_B}\right) \times 100\% \qquad \longrightarrow (5)$$

Depending on resistances $R_A$ and $R_B$, the duty cycle is between 50 and 100 percent. The duty cycle of the square wave is 50% i.e., $R_A$ must be much smaller than $R_B$.

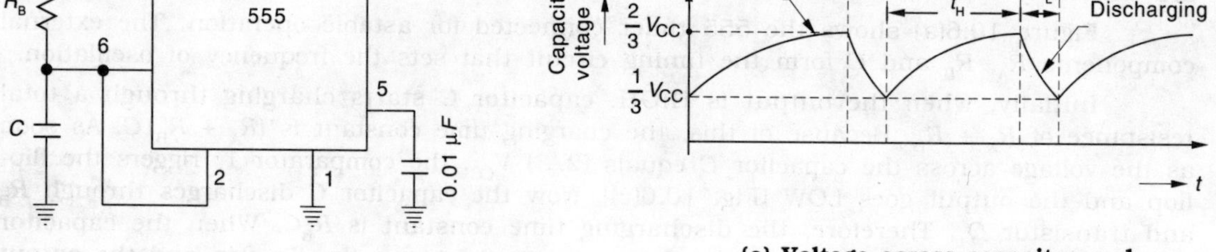


(a) Astable timer circuit

(c) Voltage across capacitor and output voltage waveforms

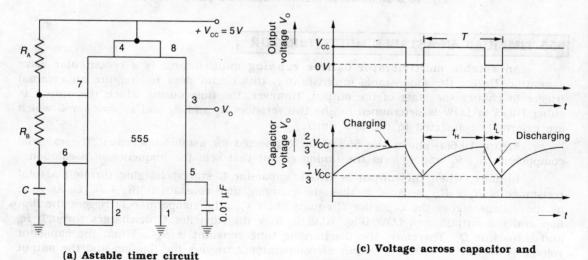

**Fig. 10.6 Astable multivibrator**

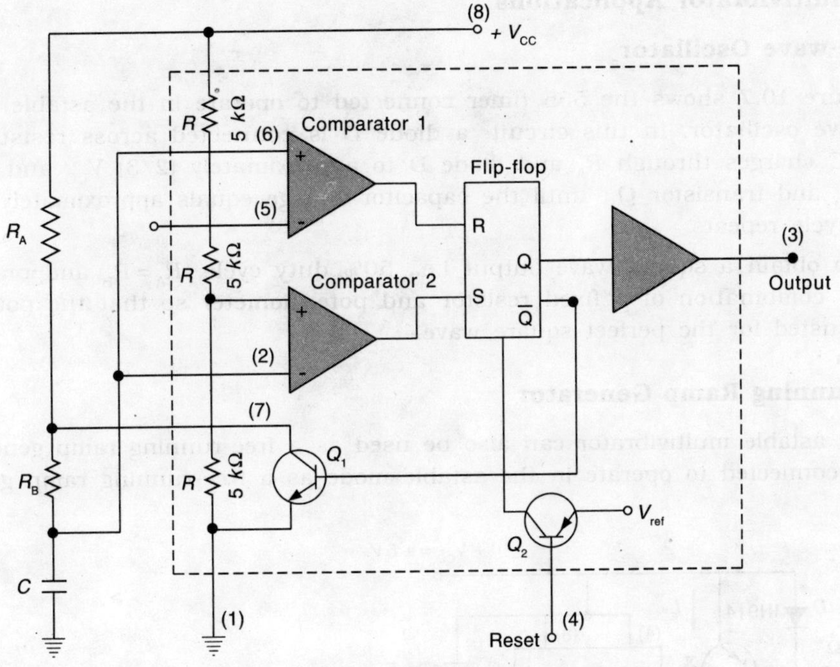


**(b) Circuit of Astable multi-vibrator**

**Fig. 10.6 Astable multivibrator**

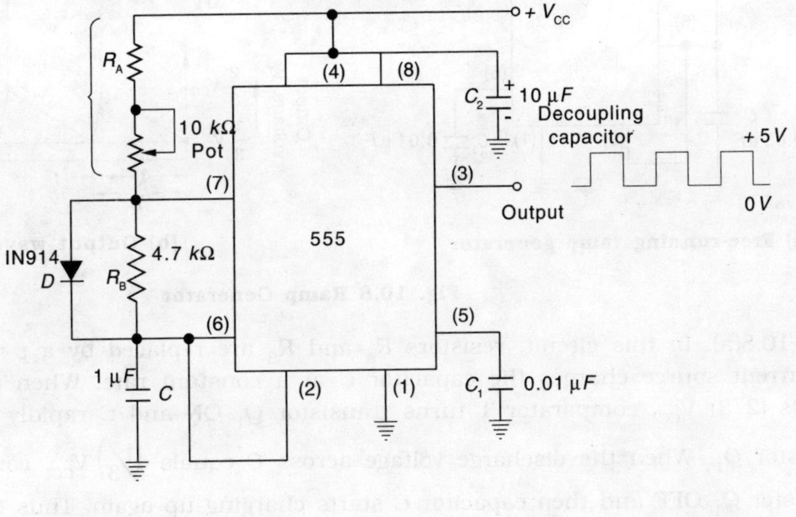


**Fig. 10.7 Astable multivibrator as a square wave oscillator**

## Astable Multivibrator Applications

### 1. Square-wave Oscillator

Figure 10.7 shows the 555 timer connected to operate in the astable mode as a square wave oscillator. In this circuit, a diode D is connected across resistor $R_B$. The capacitor $C$ charges through $R_A$ and diode D to approximately (2/3) $V_{CC}$ and discharges through $R_B$ and transistor $Q_1$, until the capacitor voltage equals approximately (1/3) $V_{CC}$; then the cycle repeats.

In order to obtain a square wave output i.e., 50% duty cycle, $R_A = R_B$ and practically $R_A$ must be a combination of a fixed resistor and potentiometer so that the potentiometer can be adjusted for the perfect square wave.

### (2) Free-running Ramp Generator

The astable multivibrator can also be used as a free-running ramp generator. The 555 timer connected to operate in the astable mode as a free-running ramp generator is

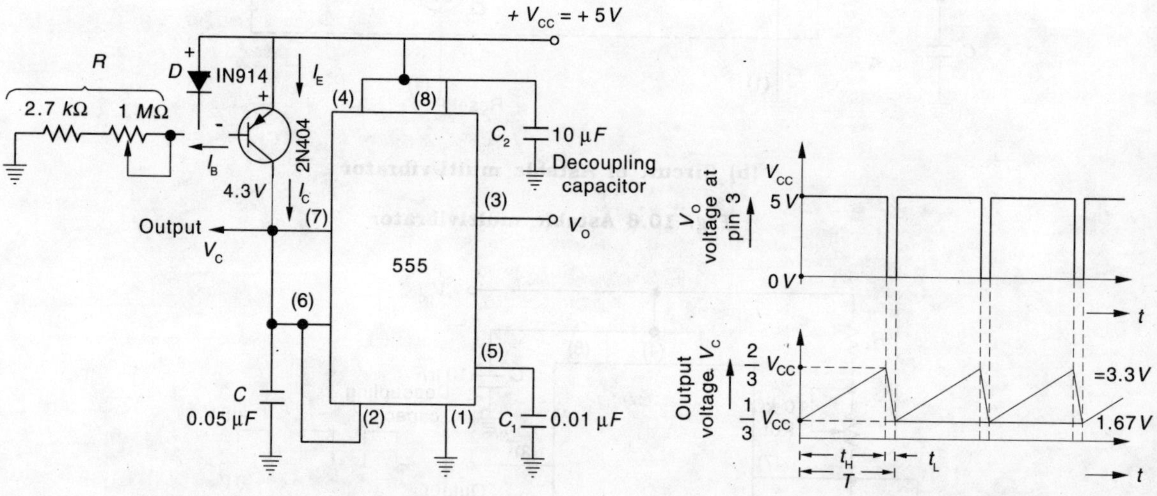


(a) Free-running ramp generator (b) Output waveform

**Fig. 10.8 Ramp Generator**

shown in Fig. 10.8(a). In this circuit, resistors $R_A$ and $R_B$ are replaced by a *p-n-p* current source. The current source charges the capacitor $C$ at a constant rate. When the voltage across $C$ equals (2/3) $V_{CC}$, comparator 1 turns transistor $Q_1$ ON and $C$ rapidly discharges through transistor $Q_1$. When the discharge voltage across $C$ equals $(1/3)$ $V_{CC}$, comparator 2 switches transistor $Q_1$ OFF and then capacitor $C$ starts charging up again. Thus the charge-discharge cycle keeps repeating. Since the discharging time of the capacitor is very small, it can be neglected in comparison with charging time. Hence, for all practical purposes, the time period of the ramp waveform is equal to the charging time.

$$T = \frac{V_{CC}\, C}{3 I_C} \qquad \longrightarrow (1)$$

Charging current, $I_C = \dfrac{V_{CC} - V_{BE}}{R}$

Free-running frequency of the ramp generator is therefore given by

$$f = \frac{3I_C}{V_{CC}\,C} \longrightarrow (2)$$

### 3. Voltage-Controlled Oscillator

Figure 10.9(a) shows a voltage-controlled oscillator, an application of a 555 timer. As shown in Fig. 10.9(a), the control voltage $V_{con}$ changes the threshold values of (1/3) $V_{CC}$ and (2/3) $V_{CC}$ for the internal comparators. Notice that the voltage across the timing capacitor varies between $+V_{con}/2$ and $+V_{con}$. When the control voltage is varied, the output frequency also varies. An increase in $V_{con}$ increases the charging and discharging time of capacitor and causes the frequency to decrease. A decrease in $V_{con}$ decreases the charging and discharging time of the capacitor and causes the frequency to increase.

An interesting application of the VCO is in phase-locked loops, which are used in various types of communication receivers to track variations in the frequency of incoming signals.

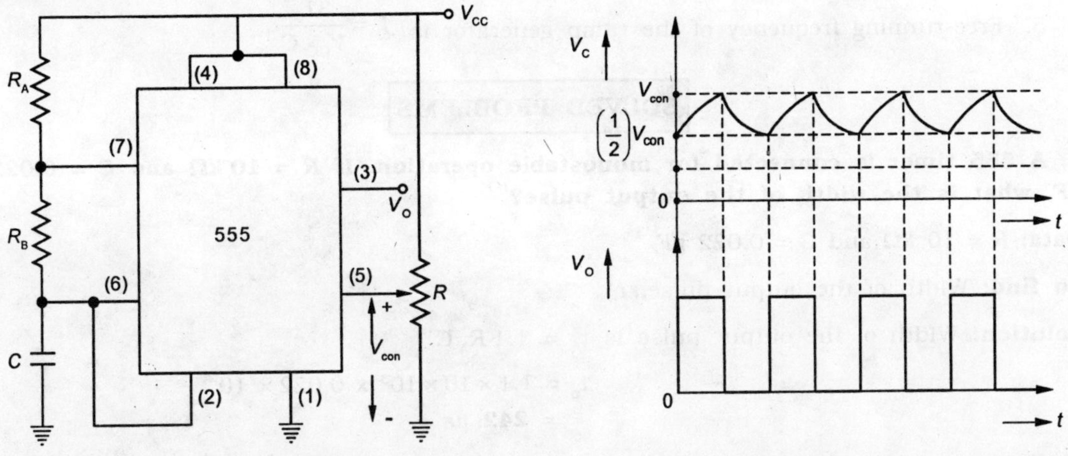


**(a) Voltage-controller oscillator circuit** **(b) Capacitor timing waveform**

**Fig. 10.9 Voltage-controlled oscillator**

## IMPORTANT POINTS TO REMEMBER

1. The 555 timer is an integrated circuit that can be used as an oscillator, in addition to many other applications.
2. The 555 timer combines a relaxation oscillator, two comparators, an R-S flip-flop and a discharge transistor.
3. The 555 timer can be used as either a monostable or an astable multivibrator.

4. Monostable multivibrator produces a single fixed pulse out each time a trigger pulse is applied. It is used as a delay generator or delay line.
5. The monostable multivibrator can be used as a frequency divider and also as a pulse stretcher.
6. An astable multivibrator also called free running multivibrator is a rectangular wave generator.
7. Duty cycle is defined as the ratio of the time during which the output is HIGH to the total time period.
8. An astable multivibrator can be used as a square-wave oscillator, free-running ramp generator and a voltage controlled oscillator.
9. VCO is a type of relaxation oscillator whose frequency can be varied by a *dc* control voltage.
10. The 555 timer can be used in such applications as waveform generators, digital logic probes, infrared transmitters, burglar alarms, toxic gas alarms and electric eyes.

## KEY FORMULAE

1. In monostable multivibrator, width of the output pulse is $t_p = 1.1\, R_A C$.
2. In astable multivibrator, the time period of the output waveform is $\tau = 0.694\,(R_A + 2\,R_B)\,C$
3. Frequency of oscillation, $f = \dfrac{1.45}{(R_A + 2R_B)C}$
4. Duty cycle = $\left(\dfrac{R_A + R_B}{R_A + 2R_B}\right) 100\ \%$
5. Free-running frequency of the ramp generator is $f = \dfrac{3 I_C}{V_{CC}\, C}$

## SOLVED PROBLEMS

**1. A 555 timer is connected for monostable operation. If $R$ = 10 kΩ and $C$ = 0.022 μF, what is the width of the output pulse?**

**Data:** $R$ = 10 kΩ and $C$ = 0.022 μF

**To find:** Width of the output pulse, $t_p$

**Solution:** Width of the output pulse is $t_A = 1.1\, R_A C$

$$\therefore \quad t_p = 1.1 \times 10 \times 10^3 \times 0.022 \times 10^{-6}$$

$$= \mathbf{242\ \mu s}$$

**2. A 555 timer is connected for monostable operation. If $R$ = 10 kΩ, and the output pulse width is 12 *ms*, calculate the value of C.**

**Data:** $R$ = 10 kΩ and $t_p$ = 12 *ms*

**To find:** Capacitance value $C$

**Solution:** Width of the output pulse, $t_p = 1.1\, R_A C$

$$\therefore \quad C = \frac{t_p}{1.1 \times R_A} = \frac{12 \times 10^{-3}}{1.1 \times 10 \times 10^3} = \mathbf{1.09\ \mu F}$$

**3. The monostable multivibrator is used as a divide-by-2 network. The frequency of input trigger pulse is 3.2 *kHz*. If the value of *C* = 0.047 μ*F*, what should be the value of $R_A$?**

**Data:** $f$ = 3.2 *kHz* and $C$ = 0.047 μ*F*

**To find:** Value of $R_A$

**Solution:** For a divide-by-2 network, $t_p$ should be slightly larger than *T*.

Let $t_p = 1.2\,T = \dfrac{1.2}{f}$

$$\therefore \quad t_p = \frac{1.2}{3.2 \times 10^3} = 0.38\ ms$$

We know $\quad t_p = 1.1\, R_A C$

$$0.38 \times 10^{-3} = 1.1 \times R_A \times 0.047 \times 10^{-6}$$

Solving, $\quad R_A =$ **7.35 kΩ**

**4. An astable multivibrator has $R_A$ = 10 kΩ, $R_B$ = 2 kΩ and C = 0.0047 μF. What are the output frequency and duty cycle?**

**Data:** $R_A$ = 10 *k*Ω, $R_B$ = 2 *k*Ω and $C$ = 0.0047 μ*F*

**To find:** Output frequency $f$ and duty cycle

**Solution:** Output frequency, $f = \dfrac{1.45}{(R_A + 2R_B)C}$

$$= \frac{1.45}{(10 + 2 \times 2)\,10^3 \times 0.0047 \times 10^{-6}} = \mathbf{20.71\,kHz}$$

$$\text{Duty cycle} = \left(\frac{R_A + R_B}{R_A + 2R_B}\right) \times 100\%$$

$$= \left(\frac{10 + 2}{10 + 4}\right) \times 100\% = \mathbf{86\%}$$

**5. Design a 555 timer circuit that free-runs at a frequency of 1 *kHz* and a duty cycle of 75 percent.**

**Data:** $f$ = 1 *kHz* and duty cycle = 75%

**To find:** Resistances $R_A$ and $R_B$ and capacitance $C$

**Solution:** Output frequency, $\quad f = \dfrac{1.45}{(R_A + 2R_B)C}$

$$\therefore \quad 1000 = \frac{1.45}{(R_A + 2R_B)C}$$

or $\quad (R_A + 2R_B)\,C = 1.45 \times 10^{-3} \quad \longrightarrow (1)$

$$\text{Duty cycle} = \frac{R_A + R_B}{(R_A + 2R_B)} \times 100\%$$

$$0.75 = \frac{R_A + R_B}{R_A + 2R_B} \longrightarrow (2)$$

From Eq. (2), we get $R_A = 2\,R_B$

Solving Eqs. (1) and (2), we get $R_A$ = **22 kΩ**, $R_B$ = **11 kΩ** and $C$ = **0.033 μF**

**6. Design a monostable for a pulse width of 10 *ms* by using IC 555. [Feb. 2000, V.T.U.]**

**Data:** $t_p$ = 100 *ms*

**To find:** Resistance $R_A$ and capacitance $C$

**Solution:** The pulse width is given by

$$t_p = 1.1\,R_A\,C$$

$$10 \times 10^{-3} = 1.1\,R_A\,C$$

or

$$R_A C = \frac{10 \times 10^{-3}}{1.1} = 9.0909\ ms$$

Choose $C$ = 0.01 μ*F*

Then,

$$R_A = \frac{9.0909 \times 10^{-3}}{0.01 \times 10^{-6}} = 909.09\ k\Omega \cong \mathbf{910\ k\Omega}$$

**7. Design a timer which should turn ON heater immediately after pressing a push button and should hold heater in 'ON' state for 5 seconds.**

**Data:** $t_p$ = 5 *s*

**To find:** Resistance $R_A$ and capacitance $C$

**Solution:** Pulse width is given by

$$t_p = 1.1\,R_A\,C$$

$$5 = 1.1\,R_A\,C$$

Choose $C_1$ = 10 μ*F*, then

$$5 = 1.1\,R_A \times 10 \times 10^{-6}$$

Solving, $R_A$ = **454.54 *k*Ω**

Use a 470 *k*Ω potentiometer.

## QUESTIONS

1. Explain what the 555 timer is.
2. Name the five basic elements in 555 timer IC.
3. List important features of the 555 timer.
4. What are the two basic modes in which the 555 timer operates?
5. Briefly explain the differences between the two operating modes of the 555 timer.
6. Explain with relevant diagrams the monostable operation of the 555 timer.
7. Derive the expression for the pulse width of a monostable multivibrator using 555. **[Feb. 2000, V.T.U.]**

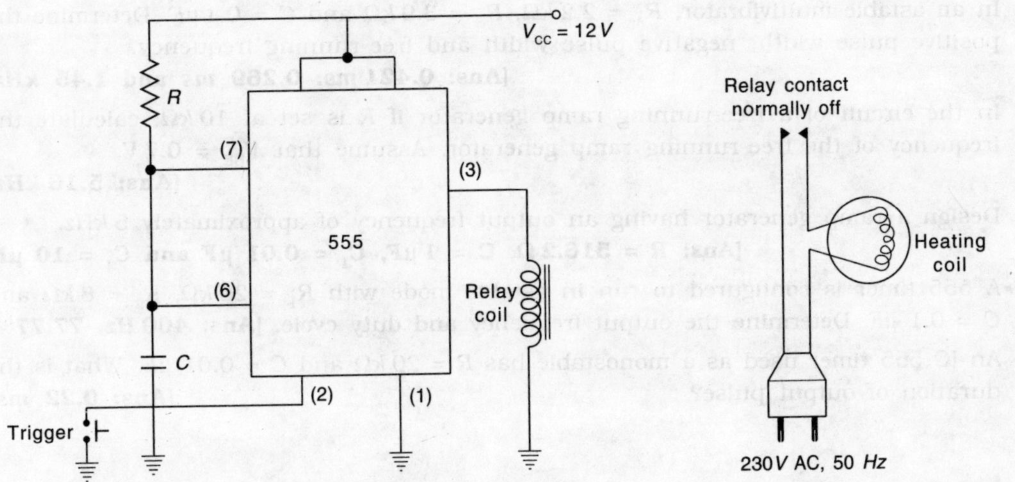


**Fig. 10.10 Monostable multivibrator used to switch 'ON' relay for specific time**

8. Explain briefly with an internal block diagram the features of 555 timer.
9. Explain with a diagram how to use the 555 timer as a frequency divider.
10. Explain with a diagram how to use the 555 timer as a pulse stretcher.
11. What must the relationship between the pulsewidth $t_p$ and the period of the input trigger signal if the 555 is to be used as a divide-by-4 network?
12. Explain with relevant diagrams the astable operation of the 555 timer.
13. Explain with a diagram how to use the 555 timer as a square-wave oscillator. **[Feb. 2000, V.T.U.]**
14. Explain with a diagram how to use the 555 timer as a free-running ramp generator.
15. Explain with a diagram how to use the 555 timer as a VCO.
16. When the 555 timer is configured as an astable multivibrator how is the duty cycle determined?
17. Explain the functions of various pins of IC 555.

## EXERCISES

1. Using the 555 timer, design a monostable multivibrator having an output pulse width of 100 *ms*. Verify the designed values of $R_A$ and $C$ with the graph of Fig. 10.3. **[Ans: $R_A$ = 909.09 kΩ, C = 0.1 μF]**
2. The monostable multivibrator is to be used as a divide-by-3 network. The frequency of the input trigger is 12 *kHz*. If the value of $C$ = 0.05 μF, what should be the value of $R$? **[Ans: 3.33 kΩ]**
3. Design an astable multivibrator having an output frequency of 10 *kHz* with a duty cycle of 25%. **[Ans: $R_A$ = 3.6 kΩ, $R_B$ = 1.8 kΩ, C = 0.02 μF]**
4. In a monostable multivibrator, $R_A$ = 10 kΩ and the output pulse width is 20 *ms*. Determine the value of $C$. **[Ans: 1.818 μF]**
5. The monostable multivibrator is to be used as a divide-by-2 network. The frequency of the input trigger signal is 2 *kHz*. If the value of $C$ = 0.01 μF, what should be the value of $R_A$? **[Ans: 54.5 kΩ]**

6. In an astable multivibrator, $R_A = 2.2\ k\Omega$, $R_B = 3.9\ k\Omega$ and $C = 0.1\ \mu F$. Determine the positive pulse width, negative pulse width and free-running frequency. **[Ans: 0.421 *ms*: 0.269 *ms* and 1.45 *kHz*]**
7. In the circuit of a free-running ramp generator if $R$ is set at $10\ k\Omega$, calculate the frequency of the free-running ramp generator. Assume that $V_{BE} = 0.7\ V$. **[Ans: 5.16 *kHz*]**
8. Design a ramp generator having an output frequency of approximately 5 *kHz*. **[Ans: $R = 516.2\ \Omega$, $C = 1\ \mu F$, $C_1 = 0.01\ \mu F$ and $C_2 = 10\ \mu F$]**
9. A 555 timer is configured to run in astable mode with $R_1 = 20\ k\Omega$, $R_2 = 8\ k\Omega$ and $C = 0.1\ \mu F$. Determine the output frequency and duty cycle. **[Ans: 400 *Hz*, 77.77%]**
10. An IC 555 timer used as a monostable has $R = 20\ k\Omega$ and $C = 0.01\ \mu F$. What is the duration of output pulse? **[Ans: 0.22 *ms*]**

# INDEX